The Digital Ecosystem of Kenya

Carolin Stötzel

The Digital Ecosystem of Kenya

Transforming Governance for a Sustainable Digital Development

Springer VS

Carolin Stötzel
Munich, Germany

Dissertation, University of Passau, Passau (Germany)

ISBN 978-3-658-45552-1 ISBN 978-3-658-45553-8 (eBook)
https://doi.org/10.1007/978-3-658-45553-8

This Springer VS imprint is published by the registered company Springer Fachmedien Wiesbaden GmbH, part of Springer Nature.
The registered company address is: Abraham-Lincoln-Str. 46, 65189 Wiesbaden, Germany

To my family

Acknowledgements

I would like to express my gratitude to my doctoral advisors Prof. em. Dr. Winand Gellner and Prof. Dr. Bernhard Stahl. Furthermore, I would like to thank my family, my friends, my colleagues, and my mentors, for their unwavering support throughout this journey.

Abstract

The impact of the digital transformation on countries' sustainable development is increasingly discussed in both theory and practice. This dissertation falls within the highly dynamic research field of 'Information and Communication Technologies for Development (ICT4D)' or 'digital development', and especially the emerging study of ecosystems. It analyzes the implications of digital technologies and services for developing countries in particular by using the concept of digital ecosystems. This approach offers a combination of analyzing implications of the digital transformation on sustainable development from both a macro (context), meso (ICTs as a variable) and micro level (interactions and activities of stakeholders). It uniquely combines digital development, sustainability and network governance research to answer the research question of "how can digital ecosystems in developing countries be coordinated in such a way that they contribute to a sustainable development?". A theoretical framework, 'Digital Ecosystem Governance', is developed that consists of five determinants for achieving sustainable digital development. For this purpose, theories of ICT4D and network governance, along with sustainability research and development theories were used. Indepth case study data was gathered through qualitative guideline-based expert interviews with representatives of multi-stakeholders in the Kenyan digital ecosystem (e.g., the private tech sector, bilateral and multilateral donor organizations, NGOs and academia) from October 2020—December 2021. This dissertation identifies factors characterizing digital ecosystems and challenges to their sustainable digital development (e.g., a lack of common understandings, power asymmetries and dependencies of multi-stakeholders). For addressing these challenges, this dissertation discusses coordination mechanisms including governance roles of multi-stakeholders as well as governance formats such as multi-stakeholder partnerships (MSPs). Moreover, it proposes two alternative governance forms: a meta organization and a coordinative digital platform. These findings emphasize a deeper discussion around the global emphasis on 'partnerships' and the digital transformation as a 'driving force', if these are to foster and not impede sustainable development.

Contents

Abbreviations

2G	Second-generation
3D	Three-Dimensional
3G	Third-generation
4G	Fourth-generation
5G	Fifth-generation
ADSL	Asymmetric Digital Subscriber Line
ANT	Actor-Network-Theory
AI	Artifical Intellience
AU	African Union
B2C	Business-to-Customer
B2B	Business-to-Business
BIEA	British Institute in Eastern Africa
BMZ	Bundesministerium für wirtschaftliche Zusammenarbeit und Entwicklung / Federal Ministry for Economic Cooperation and Development
BRI	Belt and Road Initiative
CSR	Corporate Social Responsibility
D	Development
DAAD	Deutscher Akademischer Austauschdienst / German Academic Exchange Service
D4D	Digital for Development
DAC	Development Assistance Committee
EU	European Union
FDI	Foreign Direct Investment

FSM	Free Software Movement
GAFA	Google, Apple, Facebook, Amazon
GDP	Gross Domestic Product
GDPR	General Data Protection Regulation
Gbps	Gigabit per second
GIZ	Gesellschaft für Internationale Zusammenarbeit
GPEDC	Global Partnership for Effective Development Cooperation
GSM	Global System for Mobile
HDI	Human Development Index
IATI	International Aid Transparency Initiative
ICT	Information and Communication Technology
ICT4D	Information and Communication Technologies for Development
ICTA	Information, Communications and Technology Authority
IGF	Internet Governance Forum
IMF	International Monetary Fund
IoT	Internet of Things
IT	Information Technology
ITU	International Telecommunication Union
IS	Information Systems
KEPSA	Kenya Private Sector Alliance
KINGS	Kenya, Ivory Coast, Nigeria, Ghana and South-Africa
KPI	Key Performance Indicators
Kshs	Kenya Shillings
MDG	Millennium Development Goals
MSME	Micro, Small and Medium Enterprises
MSP	Multi-Stakeholder Partnership
NACOSTI	National Commission For Science, Technology and Innovation
NAO	Network Administrative Organization
NGO	Non-Governmental Organization
NPM	New Public Management
ODA	Official Development Assistance
OECD	Organisation for Economic Co-Operation and Development
PPD	Public-Private-Dialogue
PPP	Public-Private-Partnership
PPI	Private Participation in Infrastructure
PR	Public Relations
PS4D	Private Sector for Development
R&D	Research & Development
SADA	Smart Africa Digital Academy

SDG	Sustainable Development Goal
SME	Small and Medium-sized Enterprises
SSC	South-South Cooperation
STEM	Science, Technology, Engineering and Mathematics
TVET	Technical and Vocational Education and Training
UN	United Nations
UNCTAD	United Nations Conference on Trade and Development
UNDP	United Nations Development Programme
USAID	United States Agency for International Development
USSD	Unstructured Supplementary Service Data
VTI	Vocational Training Institute
WSIS	World Summit on the Information Society

List of Figures

List of Tables

1 Introduction

In 2020, only 21 percent of the population in Sub-Saharan Africa used safely-managed sanitation services (The World Bank n. d. n), 30 percent used safely managed drinking water services (The World Bank n. d. m), and 48 percent had access to electricity (The World Bank n. d. b). In contrast, 46 percent of the population in Sub-Sahara Africa subscribed to mobile services (GSMA 2021a:3) and 30 percent used the internet (The World Bank n. d. i). These data suggest that more people in the region had access to mobile services and the Internet than to some basic services for living.[1] Moreover, diffusion of internet connections or mobile phones is often associated with increases of the Gross Domestic Product (GDP) (Avgerou 2017:16) and commonly referred to as 'leapfrogging' stages in economic development[2]—similar to leapfrogging technological stages in moving from a few landlines to massive mobile phone use in just a decade (Osiakwan 2017:60). More specifically, the significance of digital technologies and services in Sub-Saharan Africa is further mirrored in the associated economic value, especially since mobile technologies and services generated more than USD 130 billion of economic value added (8 percent of GDP) in Sub-Saharan Africa (GSMA 2021a:3). Consequently, this attracts tremendous business interest and investment. Taking the example of fintech, more than 50 fintech companies in the region raised around USD 821 million in funding in 2020, which is a third more than the previous year (GSMA 2021a:26).

[1] This is just a limited selection of data. It will be elaborated in the following that this needs to be assessed critically, for instance with regard to quality of access, affordability and skills.

[2] 'Development' in this dissertation refers to the development of developing countries from a development theory and ICT4D perspective. It will use this terminology while highlighting the difficulties around the terminology of 'development' as well as of 'developing countries' (see Sect. 2.1.1). Instances where 'development' refers to the development of digital technologies and services (instead of development of countries), are indicated as such by the context.

C. Stötzel, *The Digital Ecosystem of Kenya*,
https://doi.org/10.1007/978-3-658-45553-8_1

This is only one example of the associated narrative of "Africa Tech Rising" (Osiakwan 2017:60) with Kenya, Ivory Coast, Nigeria, Ghana and South-Africa (KINGS) at the forefront. More specifically, the digital economy in Kenya has been a long-standing leader in technology innovation across the Sub-Saharan Africa region (USAID 2020) with Kenya also being labeled as the "Silicon Savannah" (Graham and Mann 2013; Akamanzi, Deutscher and Guerich 2016).

Against this background, digital technologies and services, also referred to as Information and Communication Technology (ICT)s, are attributed with the potential as 'being the force of change'. Change here is understood in the sense of transformation and development. The digital transformation[3] is all-encompassing, influencing economic, political and social contexts worldwide. Given its currently ongoing and also rapidly moving developments, effects are not entirely clear yet. Hence, analyzing the effects of the digital transformation requires a critical inquiry to "move beyond hype and hope" (Graham 2019:2). Therefore implications of the digital transformation will be analyzed in this dissertation in an exploratory design using qualitative expert interviews (Chap. 3) and a multidisciplinary approach. One of such effects for instance is the demonstrated interest in the digital transformation of developing countries[4] from various stakeholders, especially international ones. Expanding on the economic and business interest in Sub-Saharan digital economies, engagements of international tech companies are very prominent. For instance, nearly every company of the so-called Google, Apple, Facebook, Amazon (GAFA) has a presence in Sub-Sahara Africa and is active in the region's digital development.

[3] In academic literature, it is usually distinguished between 'digitization', 'digitalization', and 'digital transformation' (see e.g., multidisciplinary review of Verhoef, Broekhuizen, Bart, Bhattacharya, Qi Dong, Fabian and Haenlein 2021), with the first two needed for the third one. While 'digitization' describes "the action to convert analog information into digital information" (Verhoef et al. 2021:891), 'digitalization' describes how Information Technology (IT) or digital technologies can be used to alter existing processes i.e., changing of simple organizational processes and tasks. 'Digital transformation' on the other hand is more pervasive as it affects the whole organization such as companies or public sector agencies and their ways of operating that goes beyond digitalization. Moreover, in utilizing digital technologies, interactions across borders are enabled (borders beyond organizational level to also national or global level) (ibid.). In this dissertation and more specifically considering the academic and empirical discourse, the terms 'digital transformation' and 'digitalization' are often used interchangeably. Such differences in definitions as well as in understandings and usage of these three concepts are important to highlight (see e.g., Sect. 2.2, Sect. 4.4.4).

[4] Further distinctions in line with the current development discourse are terminologies of 'Global North' and 'Global South'.

Besides the presence of international tech companies in developing countries, in this specific context, stakeholders such as bilateral and multilateral donor organizations additionally join the mix. More specifically with regard to digital development, these actors are increasingly engaged in the ICT sector with UNCTAD (2021b) suggesting that the share of aid for ICT in total aid for trade rose from 1.2 percent in 2017 to 2.7 percent in 2019 (see also UNCTAD 2021a). Hence, interests of private sector actors such as profit maximizing e.g., through access to new markets, trade, access to data and geo-strategical interests (e.g., Madianou 2019; Fuchs and Öhler 2020; Chipidza and Leidner 2019), collide with public implementing of development policy goals such as capacity building or following the approaches of 'do-no-harm' and 'leave no-one behind'. Especially in the digital transformation, concerns around the role of 'the West' here start to grow with narratives and concepts of "Digital Colonialism" (e.g. Kwet 2019; Young 2019; Hill 2014) arising.

Against this background, the universal Sustainable Development Goal (SDG)s, especially SDG 17 ('Global Partnerships'), aim to function both as an umbrella for these various of multi-stakeholders[5] and to counteract these notions of inequality for the overall goal of sustainable development. Moreover, the SDGs specifically incorporate digital technologies and the digital transformation. Ultimately, sustainable development in the line with the SDGs as the globally agreed core goal for all countries, "[...] is about creating balances between the environmental, economic and social dimensions of life, also with a view to future generations" (Meuleman 2019:5). In this dissertation, sustainable digital development is understood as effective measures, inclusion and equity in the digital transformation.[6]

1.1 Problem Statement, Research Goals and Research Questions

What the previous passage illustrates is that challenges for developing countries arise both from their development context as well as the current digital transformation. On the one hand, these challenges that emerge from the development contexts themselves, already pose a threat to sustainable development of these countries. On the other hand, this dissertation will demonstrate that these challenges are fur-

[5] In the reviewed literature, there are various spellings of 'multistakeholders' or 'multi-stakeholders'. For consistency, this dissertation uses the spelling 'multi-stakeholders' except in cases of direct quotations.

[6] Here, the distinction between 'sustainable development' and 'sustainability' is similarly important, as both terms are prone to misconceptions, which will be discussed in more detail in both the theoretical background (Chap. 2) and empirical analysis (Chap. 4) of this dissertation.

ther interlinked and hence influence each other. It therefore argues that both forces require coordination. For this purpose, this research project will use the emerging concept of *Digital Ecosystems* (see e.g, Nguyen and Mahundi 2019; Dhungana, Groher, Schludermann and Biffl 2013; Boley and Chang 2007; Li 2009; Li, Badr and Biennier 2012, as further discussed below). For answering its research question of "how can digital ecosystems in developing countries be coordinated in such a way that they contribute to a sustainable development?", the three areas of sustainable development, digital transformation and governance will be uniquely synthesized in a framework[7] for coordination in digital ecosystems.[8]

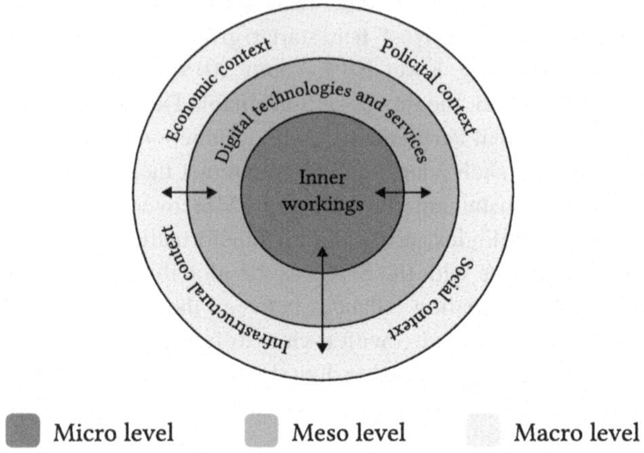

Fig. 1.1 Levels of analyzing digital ecosystems (own illustration)

[7] This dissertation has chosen to use the terminology 'framework' for theoretical and empirical reasons as in both instances 'framework' is the commonly used term. Nevertheless, it is important to note that, from a methodological perspective, the developed framework could also be understood as a model, since both the logic of levels and determinants is followed.

[8] Governance and coordination will be used interchangeably throughout this dissertation both in academic literature (Chap. 2) as well as in the empirical analysis (Chap. 4). This will be discussed in more detail in the respective chapters. Exceptions are instances where 'governance' refers to governance theories. Similarly, innovation is often used synonymously with digital development and ICT4D (Madianou 2019).

This dissertation demonstrates that sustainable digital development is influenced by both external factors such as the development context of a country (macro level), digital technologies and services (meso level) as well as internal actions of multi-stakeholders in the ecosystem (micro level). Moreover, it argues that coordination of and between the levels is crucial for achieving and fostering sustainable digital development. On the one hand, this three-level-logic is derived from the composition of ICT4D research that consists of ICT (here meso) and Development (D) (here macro) dimension. On the other hand, it is derived from a network perspective, specifically with regard to flows of exchange between the three levels as exchange in an ecosystem both occurs *within* the network (e.g., flows of information or resources; micro level) and *between* the network and its environment (e.g., Castells 2004; Monkelbaan 2019, macro and meso level). This is illustrated in Fig. 1.1.

The research goals of this dissertation revolve around both a theoretical and empirical desideratum and will be discussed for each identified problem area and supported by guiding research questions. It will explore the timely concept of digital ecosystems in contributing to its theoretical but also practical understanding, especially in the context of developing countries such as Kenya.

1. **Problem: Development context problems** Firstly, developmental challenges such as limited infrastructure, political instability and corruption, limited GDP per capita and gaps in education attainment pose challenges both to sustainable development and to governing the digital transformation. In development discourse, (good) governance is seen as a precondition for sustainable human development (e.g., UNDP 1995; Fukuyama 2016; TWI2050 2018). More specifically, ineffective governance or limited state functions influence development (The World Bank 2017) since factors such as state authority, state capacity and state legitimacy are pre-conditional for the functioning of governmental institutions (Grävingholt, Ziaja and Kreibaum 2015:1286). For instance, state capacity and legitimacy can be limited by high levels of corruption. Taking the example of Kenya, it ranks 124th of 180 countries in the 2020 Corruption Perception Index (Transparency International n. d.). Similarly, these developmental challenges on a governmental level are intertwined with economic and social problems, as indicated for instance by low rates of GDP per capita that influence the affordability of living costs such as electricity. In Kenya for instance, a large proportion of the population experiences chronic poverty as 36.1 percent of the population live below the national income poverty line (UNDP 2020). This is similarly reflected in the low GDP per capita, purchasing power parity of USD 4,452.2 (2020) in comparison, e.g., to USD 12,095.9 in South Africa and USD 54,263.6

in Germany (The World Bank n. d. g), but rather high in comparison to other developing countries.

In addition to these developmental challenges of countries such as Kenya, many developing countries face the related challenge of severe *donor and aid fragmentation* that likewise constitutes the context in which digital ecosystems operate. Aid fragmentation describes the phenomenon of a multitude and variety of stakeholders engaging in development cooperation activities (both public and private), whose activities are often poorly coordinated. In development research this has long been identified as one major hurdle to sustainable development (e.g., Gehring, Michaelowa, Dreher and Spörri 2017; Fuchs, Nunnenkamp and Öhler 2015; Aldasoro, Nunnenkamp and Thiele 2010; Anderson 2012; Knack and Rahman 2007; Annen and Kosempel 2007; Djankov, Montalvo and Reynal-Querol 2009; Kimura, Mori and Sawada 2012; Klingebiel, Mahn and Negre 2016; Öhler 2017; Davies and Klasen 2019). This dissertation will further explore the impact of the digital transformation on aid fragmentation with increasing sets of stakeholders in form of (multinational) tech companies. With regard to development cooperation actors, so far there is only an "[...] incomplete assessment of the extent to which donors specialize and coordinate" (Nunnenkamp, Sotirova and Thiele 2016:832). This dissertation will analyze these interactions between multi-stakeholders using the concept of digital ecosystems, where actors interact in the ecosystem (micro level), but are likewise influenced by the stated macro factors of the development context (macro level) as well as the digital transformation as a force in-between (meso level).

a) **Research Goal 1** This dissertation will demonstrate the importance of a country's development context for its performance in the digital transformation. For this purpose, contextual factors that influence the digital ecosystem will be analyzed using development theories such as modernization theory, dependence theory and postcolonial theory, but mostly concepts of ICT4D theory such as leapfrogging. Stakeholder theory will help identifying and analyzing the multi-stakeholders present in and shaping the digital ecosystem (from outside the ecosystem and from within). Specific attention will be drawn to the multi-stakeholders active in digital development of a country (here Kenya) such as donor organizations, partner governments, private tech actors or Non-Governmental Organization (NGO)s.
b) **Guiding Question 1** For this purpose it will follow the guiding question: "What characterizes digital ecosystems as a network form? Are these ecosystems restricted and if so how or by what circumstances?".

2. **Problem: Implications of the digital transformation** Secondly, the digital transformation is accompanied by rapid technological advancements that come with specific requirements. On the one hand, seizing the opportunities of the digital transformation for development includes 'hard' factors such as necessary infrastructure in terms of electricity and connectivity as well as funding for innovations and adoption of technologies. On the other hand, 'soft' factors such as digital literacy and skills are required for harnessing potentials and shaping sustainable digital development. These factors, among others, are however shaped by power and knowledge asymmetries. Power imbalances in ICT4D is debated, but not to great extent in ICT4D literature (Chipidza and Leidner 2019:146). However, for achieving sustainable digital development, especially in the digital technology context, the interlinkages for instance between knowledge and power are crucial (Monkelbaan 2019). Hence, the digital transformation in this research project is an independent variable that affects both the context in which the ecosystem operates, but also its inner workings. Moreover, it is debated in the literature whether ICTs are a means to an end (i.e., to development) or an actor group themselves (e.g., Zheng, Hatakka, Sahay and Andersson 2018). From one angle, the digital transformation is dynamic, rapidly progressing and still ongoing. On a different note, it incorporates a technical angle as well as implications for social, economic and political processes. Against this background, there is a "plethora of theories" (Sein, Thapa, Hatakka and Øystein 2019:8) and no coherent narrative with regard to development context.

    a) **Research Goal 2** This dissertation takes a multidisciplinary approach to the impact of digital technologies and services on development. It analyzes the current digital landscape in developing countries including respective digital technologies such as ICTs in general, and more specifically with regard to 'basic' technologies such as Unstructured Supplementary Service Data (USSD) as well as 'innovative' technologies such as Artifical Intellience (AI). The goal here is to identify potential contributions of digital technologies to the development of countries (ICT4D), while also highlighting dangers to sustainable development, especially with regard to the suitability of ICTs. It aims at explaining unique factors of the digital transformation as a form of 'revolution'. Thereby it aims at contributing to the current academic and practical discourse around the digital transformation in exploring the role of ICT, for instance whether ICT are a driving force or also an actor themselves through offering approaches from ICT4D literature. For this purpose, it uses concepts of leapfrogging and of digital ecosystem as well as digital economy research. These will be synthesized with ICT4D research and will embed digital technologies and services in a wider conceptual framework that takes account of social, economic and political dynamics. Additionally, it will explore whether the digital transformation increases (or decreases) dependencies arising from the development context (see e.g., concept 'Digital colonialism').

b) **Guiding Question 2** What factors from both digital transformation and development context enable and hinder sustainable development of digital ecosystem in a developing country?

3. **Problem: Unsustainability due to development context, digital transformation and impeded coordination** Thirdly, combining these two stated problems emerging from the development and digital transformation contexts, both affect and trickle down to multi-stakeholder activities. While the general development discourse, but also the digital development discourse, focuses on partnerships (e.g., SDG 17) such as on Public-Private-Partnership (PPP) and Multi-Stakeholder Partnership (MSP), these are prone to failures that impede sustainable development (e.g., The World Bank 2016a; Choi and Chandler 2020; Beisheim, Ellersiek, Goltermann and Kiamba 2017; Mirza 2013; Ismail, Heeks, Nicholson and Aman 2018; Chipidza and Leidner 2019). For instance, the The World Bank (2016a) states that only 13 percent of ICT projects succeed, whereas the majority of 58 percent partially fail and almost one third of projects (29 percent) even fail completely. Examples of projects that suffer chronic failure (Chipidza and Leidner 2019:145) include a variety of ICT projects, for instance projects on e-government systems (e.g., Choi and Chandler 2020; Anthopoulos, Reddick, Giannakidou and Mavridis 2016) or donor-funded ICT4D projects (e.g., Chipidza and Leidner 2019). This dissertation will deliver further insights into why multi-stakeholders' ICT projects fail and how these failures can be overcome. It will emphasize the relevance of coordination in the form of governance for addressing these challenges. With regard to the digital transformation in the sustainability context, this is in line with the global understanding that "[...] governance approaches are needed to adequately balance the potential benefits and risks posed by digitalization and ensure a sustainable digital economy" (Linkov, Trump, Poinsatte-Jones and Florin 2018:2).

a) **Research Goal 3** The research goal here is to contribute to governance research, especially network governance in reference to the analytical unit of digital ecosystems by offering coordination 'solutions'. More specifically, network governance literature distinguishes between network forms of coordination that "may or may not be brokered" (Provan and Kenis 2008:233).

This theoretical lens is chosen for both offering governance 'solutions' to these problems, but also to explain digital ecosystems as a form of network. Here, a natural affinity between networks and ICT exist: On the one hand, the internet and digital infrastructure itself follow a network structure (Heeks 2020a:7). This is further reflected for instance in 'knowledge networks' (see e.g., Clegg, Josserand, Mehra and Pitsis 2016) or 'knowledge society' and 'innovation networks' (see e.g., Clegg et al. 2016; Desmarchelier, Djellal and Gallouj 2019). Besides this technical association of networks, networks on the other hand are also formed through collaborating sets of stakeholders (e.g., Bevir 2011).

In analyzing digital ecosystems, this dissertation aims at contributing to network governance in expanding the focus on networks as ecosystems where networks are seen as a "valuable framework for enhancing collaborative innovation" (Sørensen 2014:1). For this purpose, it is important to highlight that network governance here is chosen as a coordination mechanism, since the network is understood as a governance approach treating networks as the unit of analysis. Hence, networks also serve as a mechanism of coordination (Provan and Kenis 2008:232). The goal here is to identify possible governance roles that multi-stakeholders in the ecosystem can fulfill—either an already existing or a new organization in the ecosystem. Moreover it aims at exploring possible institutionalized ('brokered') and less institutionalized ('non-brokered') governance formats for coordination of multi-stakeholders' digital development activities in the ecosystem.

b) **Guiding Question 3** Who coordinates in a digital ecosystem, hence which governance roles are to be identified and ascribed to which actor?
c) **Guiding Question 4** How can ecosystems be governed, i.e., through which governance form?
d) **Guiding Question 5** Are new or alternative forms of governance needed in order to achieve an equitable and sustainable development of digital ecosystems in developing countries?

Theoretical and Empirical Desideratum Lastly, the research goals of this dissertation revolve around both a theoretical and empirical desideratum. To begin with the *theoretical desideratum*, it will contribute to ICT4D or digital development research, especially regarding sustainable digital development and digital ecosystem research. The research field of ICT4D, and as only currently framed 'digital development', is relatively new (e.g., Bon and Akkermans 2019), which will be discussed in more detail in the subsection below. Hence, no theory has been developed to-date that encompasses digital transformation, development theory (ICT4D), sustainability and governance research. Thus, this research project follows the theoretical desideratum of contributing to ICT4D and governance research in developing a holistic framework that combines each of these fields, especially with regard to digital ecosystems. More specifically, given the multidisciplinary nature of ICT4D research and its currently still ongoing development, no universal definition for instance of digital ecosystems exist in academic (or practical) literature (e.g., Li et al. 2012). This dissertation argues that the study of digital ecosystems requires the use of interdisciplinary theories. Hence, the theoretical goal is to firstly develop an understanding of digital ecosystems as a concept. In particular, it aims at exploring and explaining a) the context in which digital ecosystems operate and how they are influenced by it (macro level); b) what influence the digital transformation has on both the development of the ecosystem but also on development of the country in which it is embedded; and c) how digital ecosystems work internally i.e., how multi-stakeholders interact. This is why the proposed framework is designed around an analysis of macro, meso and micro levels (Sect. 2.5). Through this approach, these levels of analysis likewise mirror the network perspective on digital ecosystems, where exchange and interaction flows of multi-stakeholder activities illustrate the interlinkages of the three levels: Exchange in an ecosystem both occurs *within* the network (e.g., flows of information or resources; micro level) and *between* the network and its environment (meso and macro levels) as similarly discussed for instance by Castells (2004) or Monkelbaan (2019).

Secondly, this research project aims at delivering insights into how these three levels can be coordinated—coordinated in the sense of actively governed for addressing derived challenges of digital ecosystems in order to achieve a sustainable development. It will contribute to an understanding of the roles of actors in the digital ecosystem and their decision making, especially of governments as elected bodies and how this is influenced both by other multi-stakeholders and the digital transformation. For this purpose, it will use (network) governance theory, especially in order to explain the governance role of developing countries' governments. The ultimate theoretical goal is to develop a framework for coordination in digital ecosystems for sustainable digital development.

Besides a theoretical desideratum, this dissertation likewise pursues an *empirical desideratum* that is twofold: On the one hand, the empirical desideratum refers to the empirical case of the digital transformation as one influential variable. Ecosystems as a network form can be constituted by the sole presence or interaction of actors that together build a network. They can exist in various constellations and contexts across industries and countries. The digital transformation however here is a unique phenomenon as it affects all of these contexts as a cross-cutting development. Hence, the first empirical desideratum refers to contributing to a further understanding of implications of digital technologies on countries' developments, i.e., the digital transformation itself can be seen as an empirical case.

On the other hand, the empirical case refers to the study, not of a transformational variable but of a country. Here this is represented by the case of Kenya, precisely the Kenyan digital ecosystem. The case of Kenya represents one example of a developing country and its dynamics that constitute a context for empirical analysis. This dissertation will analyze the Kenyan digital ecosystem and its macro, meso and micro levels. The goal here is to derive recommendations for development cooperation actors and Kenyan multi-stakeholders in achieving sustainable digital development. Thereby, it indirectly likewise follows the desideratum of delivering further empirical insights into the SDG, especially SDG 17.

Against this background, this dissertation offers a unique framework to analyzing the highly complex interplay of factors and actors in digital ecosystems in not only identifying ICTs contributions to development, but also the opposite effect which tends to be overlooked. The framework can be used by each multi-stakeholder for not only recognizing the importance of contexts, but also their roles in shaping a sustainable digital development. In increased complexity, this framework includes a development theory perspective and an actor focus, but can likewise be transferred to developed countries (just with different contexts).

1.2 Literature Review

From a theoretical perspective, this research project falls within the scope of the timely and highly dynamic research field of ICT4D or, as most recently framed, "the new 'Digital Development' " (Bon and Akkermans 2019:35).[9] ICT4D entails the two components of ICTs and D for explaining their interrelation as well as the impact of the digital transformation on development.

Beginning with a literature review on the research field of ICT4D, Walsham (2017) provides a brief overview of the history of ICT4D research, from early beginnings (mid–1980s to mid–1990s) to present (mid–2000s to present). Similarly, Heeks (2020a) for instance distinguishes between three ICT4D research phases: ICT4D 0.0 (1960s–mid-1990s), ICT4D 1.0 (mid 1990s–mid-/late-2000s) and ICT4D 2.0 (mid-/late-2000s onwards). Heeks (2020a) further debates whether the current phase from 2020 onwards can be classified as a ICT4D 3.0 phase. Against this background, a broad review of the field leads to dominant and frequently cited works for instance of Heeks (2002, 2007); Heeks and Stanforth (2007); Heeks (2008,

[9] This dissertation will use both terms interchangeably.

2010); Heeks, Amalia, Kintu and Shah (2013); Heeks, Foster and Nugroho (2014); Heeks (2020a,b), the annual conference 'IFIP WG 9.4 International Conference on Social Implications of Computers in Developing Countries (ICT4D)' with the conference proceedings of Choudri, Islam, Wahid, Bass and Priyatma (2017), Nielsen and Kimaro (2019a,b) and Bass and Wall (2020), as well as publications of the Oxford Internet Institute (for this dissertation specifically Graham and Mann 2013; Graham 2019; Friederici, Ojanperä and Graham 2017; Friederici 2019).

Within the current ICT4D literature, the study of digital ecosystems is in its infancy. Reminiscent of ecological ecosystems, the ecosystem metaphor has only lately been adopted for understanding and analyzing the ICT field in academic literature (Nguyen and Mahundi 2019) as for instance further discussed in Dhungana et al. (2013); Boley and Chang (2007); Li (2009); Li et al. (2012). Here, species in the biological ecosystem can on the one hand be interpreted as actors in the networks (Boley and Chang 2007). On the other hand, this ecosystem logic can likewise be transferred to the digital transformation context since any digital artifact is embedded in a wider and constantly shifting ecosystem (Kallinikos, Aaltonen and Marton 2013). Synthesizing these two perspectives of actors and ICTs, Boley and Chang (2007) provide a definition of digital (or here rather software) ecosystems from an IT perspective as "[...] open, loosely coupled, domain clustered, demand-driven, self-organizing agent environment, where each agent of each species is proactive and responsive regarding its own benefit/profit but is also responsible to its system" (Boley and Chang 2007:399). This definition further emphasizes an analysis of "self-organizing" dynamics in the (digital) ecosystem, which this dissertation will analyze using a network governance approach and discuss governance formats from different formalization degrees in terms of self-governance to centralized governance (see Sect. 2.5.4, Sect. 4.4.6). Li et al. (2012) provide a further definition highlighting interactions and relationships between actors: "[...] Digital Ecosystems are composed of multiple and independent entities such as individuals, organizations, services, software and applications that share one or several missions and focus on the interactions and inter-relationships among them" (ibid.:117).

More importantly for this dissertation's analysis of sustainable digital development, Stuermer, Abu-Tayeh and Myrach (2017) share the synthesis of digital sustainability and ecosystem perspective in arguing that sustainability of digital artifacts and their ecosystem is needed while including a (good) governance perspective (without using a specific empirical case study). Similarly, Linkov et al. (2018) propose the concept of 'Digital Sustainability' combining governance approaches with digitalization fields for sustainable digital development. This dissertation follows a similar approach, but uses a network governance approach to the ecosystem and Kenya as an empirical case. Sustainability in this dissertation does not (only) refer to environmental sustainability, but to inclusive and effective development as the overlap of social, economic and environmental dimensions (Monkelbaan 2019:11) in relation to the SDGs. Sustainability in the context of the digital transformation is further discussed by Krueger and Levin (2022) analyzing governing sustainable transformations of urban social-ecological-technological systems. The role of stakeholders in the context of sustainability and digital transformation is similarly discussed by Lock and Seele (2017) using stakeholder theory. Meuleman (2019) and Monkelbaan (2019) here offer additional governance perspectives on sustainable development.

This directly leads over to a literature review of governance, particularly of network governance research. Here, Provan and Kenis (2008), Kenis and Provan (2009) and Klijn and Koppenjan (2016) are the most prominent works. A synthesis of this dissertation's components of digital transformation, governance and coordination is similarly offered for instance in Meyerhoff Nielsen and Jordanoski (2020). In contrast to this dissertation, the authors do not focus on sustainability and development and use different case studies (Australia, Denmark and the Republic of Korea). Similarly, Vallet, Locatelli, Barnaud, Makowski, Quispe Conde and Levrel (2020) discuss power asymmetries in social networks of ecosystem services governance, but use social network analysis and the case of Peru. A framework similar to this dissertation's is proposed by Molin and Masella (2016a,b) that identify determinants for coordination from a cross-over between policy networks, network governance and network management resulting in a conceptual framework for network governance. This framework is similar to this dissertation's framework in that it incorporates starting conditions (in this dissertation 'preconditions'), modes of network governance (in this dissertation 'governance forms'), meta tools (which in this dissertation is subsumed under governance forms) and outcome (in this dissertation on theoretical basis is sustainable digital development and outcome in empirical analysis are two governance forms)—each with respective determinants. In contrast to the framework by Molin and Masella (2016a,b), this dissertation will further incorporate 'challenges' for coordination and thereby challenges to sustainable digital development. Moreover, it will identify 'governance roles' that (all) actors take in the digital ecosystem in addition to governance forms. More importantly, in contrast to the one of Molin and Masella (2016a,b), this dissertation's framework is applied to the developing context and thereby includes respective development theory components and is further transferred to the context of the digital transformation and digital ecosystems, here in the case of Kenya.

Further expanding on related empirical work with regard to Kenya, the ICT ecosystem or ICT sector is discussed by Diga and May (2016), Ndemo and Weiss (2017), Song (2019) and Akamanzi et al. (2016) among others (mostly focusing on a economic perspective). Digital development in Kenya is a further subject in the research for instance of Prieto-Martin, Faith, Hernandez and Ramalingam (2017), Osiakwan (2017) and Graham and Mann (2013). More similar to this dissertation, Ominde, Ochieng and Omwenga (2020) and Ominde, Ochieng and Owenga (2021) conduct a stakeholder management in ICT projects in Kenya. While the authors analyze these on a project level, this dissertation takes a more macro approach using network governance perspective that has ICT projects as one determinant of the inner ecosystem workings. A specific governance approach to sustainable development in Kenya is further offered by Beisheim et al. (2017). Nan and Markus (2019) and Onyango and Ondiek (2021) further discuss sustainable digital development in similar reference to this dissertation's focus on the Kenyan case (but instead focusing on a specific organization and public organizations). This dissertation's approach of analyzing the Kenyan digital ecosystem on the nation state level relates to Nguyen and Mahundi (2019)'s study of national ICT ecosystems. In contrast however, the authors use the case of Tanzania and focus on district health information systems. While the authors discuss a dynamic in the ecosystem in the sense of evolution, this dissertation's analysis of dynamics in the ecosystem refers to dynamics in form of interactions between actors in the ecosystem and especially their asymmetries in power and knowledge in the digital transformation.

1.3 Investigation Period and Case Selection

Against the background of the discussed "Silicon Savannah" narrative and Kenya being a lighthouse example for digital development in Sub-Saharan Africa (e.g., Graham and Mann 2013; Akamanzi et al. 2016), this dissertation focuses on Kenya as a 'best-practice' case study for analyzing its digital ecosystem (Chap. 3). The study period of this research project mainly concentrates on the most recent state of ICT4D research spanning from 2016 to 2022, in which the concept of using the ecosystem logic in the digital development context has been introduced (see e.g., Nguyen and Mahundi 2019). With regard to the empirical investigation period, the case of the Kenyan digital ecosystem has been approached through a qualitative data gathering phase using guideline-based expert interviews spanning from October 2020 to December 2021. During this time, 23 interviews were conducted with experts representing organizations of multi-stakeholder groups in the Kenyan digital ecosystem. These include international and Kenyan private tech sector, bilateral and multilateral donor organizations, international and Kenyan NGOs and academia (for further details see Chap. 3). This approach is chosen due to the multidisciplinary and exploratory nature of the research project. A full anonymization of interviewed experts is provided, so that only the respective actor group is stated and interviews are numerated (e.g., I1). Instances where organizations are named, they are referred to or commented on by interviewees without being personally affiliated to the institution.

1.4 Outline

For answering this its research question, this dissertation is structured as follows. Following up on this introduction, Chap. 2 of this dissertation is both a theoretical background as well as a literature review of ICT4D and (network) governance research given the limited state of research and the multi-disciplinary approach of this dissertation. This chapter further mirrors the discussed three-level design of analysis and is structured accordingly: It is firstly divided into the macro level of analysis focusing on the development context (the D in ICT4D) of the digital ecosystem. This is followed by the meso level of analysis (the ICT in ICT4D) operating between the macro and micro level, which is mostly an ICT4D literature review discussing the implications of the digital transformation for development (Sect. 2.3). The micro level (Sect. 2.4) further provides an analysis of the multi-stakeholder interactions and activities in the digital ecosystem using the governance research, especially network governance and sustainability governance. These sections of the theoretical background culminate in the theoretically developed framework with five determinants for coordination of and in digital ecosystems for a sustainable digital development.

In Chap. 3, this dissertation's research design and methodology will be elaborated. The chapter will firstly introduce this dissertation's qualitative methodological positioning in discussing the chosen case study approach and cross-cultural data collection as well as research ethics in the field study of Kenya. Moreover, the fieldwork preparation phase of this research project will be discussed, especially its sampling strategy, preparation of interview guide, interview pretest and field access. The actual collection of verbal data through guideline-based expert interviews will be presented and also the style and settings of interviews. Finally, the analytical process of gathered interview data will be discussed including the transcription process, data management as well as the method of qualitative content analysis and qualitative network analysis will be explained.

The model developed in Chap. 2 then serves as a heuristic framework for the empirical analysis of the Kenyan digital ecosystem in Chap. 4. For this purpose, the structure of the empirical analysis mirrors the theoretical one. Hence, the empirical analysis begins with an analysis of the macro level contexts of the Kenyan digital ecosystem (economic, political, infrastructural and social context as well as development cooperation activities in Kenya). Secondly, the meso level analyzes the implications of the digital transformation on Kenya's development, for instance in terms of leapfrogging before highlighting barriers to sustainable development (such as hype around digital transformation, power and knowledge asymmetries). The main part of the empirical analysis in Chap. 4 is the analysis of the Kenyan digital ecosystem in reference to the theoretically developed framework, here mirroring its components of determinants 1 to determinant 4. In contrast to the theoretical section, determinant 5 has been identified in the empirical analysis as so important that it is analyzed as an individual subsection. It further emphasizes the outcome of this dissertation in proposing two governance 'solutions' or coordination approaches for achieving a sustainable digital development (meta organization and coordinative digital ecosystem platform). Thereby, it simultaneously functions as an interim conclusion (Sect. 4.5). Given that the theoretical background likewise serves as a literature review, the empirical analysis is mainly based on expert interview data. For also demonstrating interlinkages between the two chapters and especially their findings, respective sections are cross-referenced. This dissertation's theoretical and empirical findings will be further discussed and summarized in Chap. 5, specifically regarding characteristics of the (Kenyan) digital ecosystem, factors influencing sustainable digital development and governance mechanisms for coordination.

In Chap. 6, research limitations will be discussed. More specifically, theoretical, methodological and empirical limitations to this research project are elaborated. Lastly, Chap. 7 concludes this dissertation by answering the research question as well as guiding questions, while also providing contributions to the research fields and an outlook on future research.

2 Theoretical Background and Literature Review: Governing Digital Ecosystems for Sustainable Development using Network Governance

After introducing the relevance of this dissertation, this chapter serves as both a theoretical background and literature review to the highly dynamic research field of ICT4D and of network governance. This combination has been chosen due to the close relation between the network logic in governance research and the network logic of digital technologies. ICT4D lies at the intersection between ICT, development and the transformation process by which ICTs may lead to development. For expressing digital (for) development, the acronym Digital for Development (D4D) is also found in the research and practical discourse (UNCTAD 2019). Before analyzing the intersection between these three, it is firstly necessary to provide an understanding of their theoretical premise (Sein et al. 2019). Therefore, both contextual factors of a digital ecosystem will be discussed: the D in ICT4D and the ICT in ICT4D. These factors influence the micro level, namely the inner working of a network, which in this case is specifically represented in the form of digital ecosystems. Digital ecosystem and digital network here are understood as synonymously.

For answering this dissertation's research question, how digital ecosystems can be coordinated so that they contribute to a sustainable digital development, it will demonstrate factors that enable coordination as well as factors impeding coordination with special focus on governance and government action. In line with Provan and Kenis (2008), it argues that networks are mechanisms of coordination, which are at the center of network governance. Hence, this dissertation has chosen a network governance approach for the analysis of digital ecosystems. It will introduce a framework that uniquely combines ICT4D research and governance while highlighting their interlinkages. It argues that sustainable digital development is influenced by both external factors such as the development context of a country (macro level),

C. Stötzel, *The Digital Ecosystem of Kenya*,
https://doi.org/10.1007/978-3-658-45553-8_2

digital technologies and services (meso level) as well as internal actions of multi-stakeholders (micro level) that need to be governed for a sustainable development. Hence the theoretical background follows this logic and is structured around the proposed framework design (Fig. 1.1), which is complemented by the theoretical development of the network governance framework for sustainable development in Sect. 2.5. Thereby it mirrors the network perspective, here with exchange flows as an example, for illustrating the interlinkages of the three levels of macro, meso and micro as exchange in an ecosystem both occurs *within* the network (e.g., flows of information or resources) and *between* the network and its environment (e.g., Castells 2004; Monkelbaan 2019).

Firstly, it will provide theoretical insights into the macro level of analysis, namely development of developing countries and development theory—the D in ICT4D (Sect. 2.1). Secondly, it will analyze the influence of the the current digital transformation (the ICT in ICT4D) on the development of developing countries and the working of digital ecosystems (Sect. 2.2). Thirdly, specific challenges to sustainable digital development emerging from synthesis of the previous two chapters are discussed (Sect. 2.3) before fourthly, solutions for governing these challenges will be proposed (Sect. 2.4). For this purpose, this dissertation has developed a framework for governing digital ecosystems for sustainable digital development, which will be introduced in Sect. 2.5. Lastly, the targeted outcome of a fair and inclusive sustainable digital development will be analyzed in Sect. 2.5.5.

2.1 Development Context (Macro Level)—The D in ICT4D

To begin with, and in order to provide context for the following analysis (both theoretically and empirically), the research field of development theory in the context of ICT4D will be elaborated. More specifically, this section will elaborate why the (macro) development context in which digital ecosystems operate is important for sustainable development and coordination. This is in line for instance with Diga and May (2016)'s argument that ICTs should be viewed as embedded in the global, national, and local socio-economic context in which it is utilized instead of seeing ICTs as isolated technical systems. This section hence analyzes the macro level of digital ecosystems as illustrated in the outer layer of Fig. 1.1 and thus from the D part of ICT4D.

2.1.1 Development Theory in ICT4D

Firstly, the relationship between ICTs and development is constantly evolving (Heeks 2020a). One element is increasingly drawn attention to in ICT4D research: the understanding of development. In line with Sein et al. (2019) it is necessary to understand the context in which ICT is applied. This is why the macro layer of the here proposed ecosystem analysis (Fig. 1.1) is labeled around context. According to Avgerou (2017), context can be defined as "the processes and conditions, other than the constituent causal sociomaterial interactions (or intra-actions) of IS [information systems] phenomena, that affect their formation and are effected by them." (ibid.: 13). In this dissertation, understandings of development, political stability and aid fragmentation among others serve as contextual factors for digital development and ICT4D, together forming conditions and processes.

For understanding ICT4D, Sein et al. (2019) for instance distinguish between three groups of theories: "Group 1. Development theories: What is development (D)? Group 2. Theories conceptualizing ICT: What is ICT in the context of development? Group 3. Theories on transformative processes linking ICT to D: How does ICT make D happen?" (ibid.: 8). This also helps to differentiate ICT4D from technical fields like *Computer Science* and *Human-Computer-Interaction*, including technical research in developing contexts, as they are mainly concerned with the design and development of digital technologies and their practical applications, while largely ignoring broader social processes and outcomes of development (Zheng et al. 2018). Moreover, the ICT4D research field has increasingly "shifted from a technical to a more development oriented issue, with gradually greater emphasis on the '4D' than the 'ICT'" (Uimonen and Hellström 2015:4). To provide an overview of the ICT4D literature evolvement, Heeks (2020a) provides a summary of ICT4D phases.

In that sense, in ICT4D research (also in development research in general), there is a need to first define what is meant by 'development'. However, there is much debate and no consensus so far on how development should be understood and what development really is as for instance discussed by Sein et al. (2019), Zheng et al. (2018), Avgerou (2017), Chipidza and Leidner (2019) and Walsham (2017). Development studies per se is a multi-disciplinary field drawing upon economics, sociology, political science among others (Zheng et al. 2018). Ultimately, as Heffernan, Lin and Thomson (2012) highlight, "no single theory is likely to accommodate this diversity of objectives" (ibid.:903) so that a unifying construct is missing.

Similarly, the question of "[w]hose development do we study?" (Sein et al. 2019:21) is crucial. When dealing with the digital transformation, especially in development cooperation and development research, it is necessary to distinguish between different levels of analysis. Using the typology of Qureshi (2015), level of analysis can be for instance individual, organization, country, region or world, and based on that different indicators and roles played by ICT (Sein et al. 2019). Due to the case study research design of this dissertation, the country level will be analyzed here while highlighting associated implications for the world level and organizations. Moreover, "[e]ven as scholars ascribe different meanings to development in the context of ICT4D, so too do the various stakeholders involved" (Chipidza and Leidner 2019: 145). This underlines the importance of common understanding between multi-stakeholders active in digital development, here with regard to common understanding of 'development'. This dissertation further analyses the common understanding of multi-stakeholders in digital ecosystems both with regard to development but also in the context of digital transformation, which synthesis will be further analyzed in Sect. 2.5.1. On the one hand, in their review of the development understanding in ICT4D literature, Chipidza and Leidner (2019) identified four categories of understanding 'development': development as increased freedom, development as expanded inclusion, development as improved well-being, and development as increased economic productivity (ibid.:150). On the other hand, Sumner and Tribe (2008) for instance propose three definitions of "development" from a rather 'classical' development theory perspective in terms of a long-term process of structural societal transformation, a short- to medium-term outcome of desirable targets and a dominant discourse of Western modernity. These definitions are closely linked to different development schools, here especially *modernization theory*. In this theory, the role of 'the West' is crucial for development due to the underlying assumption that developing countries do not having the same knowledge and resources to advance modes of production, which leads to the hypothesis that they need to become more 'Western' and copy developed countries for their development (e.g., Avgerou 2017; Sein et al. 2019; Pieterse 2010; Zheng et al. 2018). This is one perspective that allows an analysis of digital development of developing countries, which is in contrast for instance to perspectives of leapfrogging (Sect. 2.2.2).

While modernization theory can be seen as one traditional paradigm, Zheng et al. (2018) also see dominance in *neoliberalism* "and its consequences which arguably is the dominant development model in the world today" (ibid.:2). Other schools of thought addressing digital development include dependence theory, (post-)colonialism, capability approach, sustainable development or Actor-Network-Theory (ANT) among others (Zheng et al. 2018; Sein et al. 2019; Walsham 2017; Hatakka, Øystein and Thapa 2019; Chipidza and Leidner 2019). Due to this variety, it is impossible to use all of these theoretical approaches. Hence, and due to the fact that ICT4D lies at the intersection of several disciplines, a multidisciplinary approach is needed (Chipidza and Leidner 2019). This dissertation will contribute to this research claim by synthesizing network research perspectives and sustainable development approaches. Post-colonial aspects will also be highlighted with regard to the role of external actors (see also Sect. 2.3.2) in the digital ecosystem of a developing country and in the concept "Digital Colonialism" (Sect. 2.3.6).

Nonetheless it is important to emphasize that the digital transformation is a worldwide force that poses opportunities and challenges to all countries—both in the Global North and the Global South. In a sense, all societies are still 'developing' here. Zheng et al. (2018) for instance highlight that "[...] all societies, no matter how economically developed and technologically advanced, have to address issues of employment, inequality and social exclusion, and continue to face challenges in education, healthcare, public administration, civil society and so on" (Zheng et al. 2018: 1). However, the different developing contexts here need to be recognized, especially considering the before mentioned importance of understanding development in the digital transformation context and with regard to a different set of multi-stakeholders (e.g., donor organizations) before implications of ICT can be analyzed (Zheng et al. 2018).

Conclusively, three emerging themes in the changing patterns of development priorities are identified: transformation, inclusion and sustainability (Heeks 2020a). Hence, this dissertation revolves around these interwoven themes by analyzing the impact of digital transformation for sustainable and inclusive development. Inclusion and sustainability are the main goals to be achieved in sustainable (digital) development (see further Sect. 2.1.3).

2.1.2 ICT4D and the Role of Governance in Development Research

As mentioned in the previous subsection, development in early development research was mainly synonymous with economic growth. From modernist to dependency theory in the 1970s and 1980s, the development discourse evolved towards neoclassical economic thoughts in the 1990s with markets (and less state) being the decisive force (see e.g., Pomerantz 2011). Since then, the role of governance has evolved in the development discourse with increased attention. Governance, in its simplest sense, explains patterns of rule, regulation and coordination. It is derived from both a economic perspective with the market as the central element of coordination (e.g., Coase 1991; Williamson 1985) and from a political perspective with hierarchy being the decisive force (Benz,Lütz, Schimank and Simonis 2007:11). Hence, governance research is conducted in various disciplines (e.g., economics, political science, geography and development studies). Governance research in political science is characterized by a debate around the role of the state. After research around cybernetic planning and regulation (*Steuerung*) in the 1970s and 1980s (Deutsch 1973; Scharpf 1975; Mayntz 1987) with a mainly institutional perspective, there has been in shift in recent years towards governance as coordination of increasing interdependencies (Benz et al. 2007; Bevir 2011). Interdependencies, not only in governance research but also in the context of the digital transformation are further crucial for this dissertation's research and will be discussed in more detail in Sect. 2.5.1.

More importantly for this dissertation's focus on ICT4D, international development institutions such as The World Bank likewise started to use their own definitions of governance, for instance: "[...] governance is the process through which state and non-state actors interact to design and implement policies within a given set of formal and informal rules that shape and are shaped by power" (The World Bank 2017:41). Here, power and power dynamics are highlighted as one important element. Power in ICT4D is similarly important, so that power will be discussed in more detail for instance in Sect. 2.3.6. As the research question of this dissertation revolves around governance with regard to ecosystems and coordination in the context of digital development, the following definition of governance will be used here: "Governance has been defined as an attempt to improve coordination between relatively dependent actors for the purpose of solving societal problems (Klijn 2008), and as the solution that individuals and organizations devise for addressing issues of coordination (Markus and Bui 2012)" (Medaglia, Hedman and Eaton 2018:2–3).

The importance of governance in development is further emphasized by the emergence of *good governance*. Since the mid-1990s, good governance has been a key concept in development strategies aiming at fostering and enhancing political stability and strengthening of governmental institutions. Here, factors such as government capacity and effectiveness, transparency, citizens' participation, and how governments exercise power are considered, although "there is disagreement over which elements are included, which are most important, what constitutes good governance, and how much it matters for economic growth." (Pomerantz 2011:160). This is likewise reflected in the multitude of statistical indicators addressing good governance such as the *Corruption Perceptions Index* (Transparency International n. d.) or the *Governance Indicators* by the World Bank (voice and accountability, political stability and absence of violence, government effectiveness, regulatory quality, rule of law and control of corruption, see The World Bank n. d. t). Despite these difference in the development discourse around the role of (good) governance, it is seen as a precondition for sustainable human development (e.g., UNDP 1995; Fukuyama 2016; TWI2050 2018).

This dissertation argues that this challenge of good governance in developing countries is one factor that influences the country's digital development (and that is different from the political context in most developed countries). Highlighting the importance of the political (governance) context for (sustainable) digital development, ineffective governance or limited state functions among others influence development (The World Bank 2017). More specifically, in state fragility research, the following three factors are preconditional for the functioning of governmental institutions (Grävingholt et al. 2015:1286):

- State authority (the ability to uphold the state monopoly of violence)
- State capacity (the ability to deliver basic services to the population);
- State legitimacy (the general acceptance by the citizenry of the state institutions' claim to set and implement binding rules)

If none of these factors are fulfilled, then governmental institutions are ineffective and with only limited function. On the other hand, if states turn too strong, they are more likely to concentrate power and limit the freedoms of their people (such as in autocratic states). In that case, state authority and state capacity is given, but the state lacks legitimacy (e.g., TWI2050 2018). Hence, legitimacy is identified as one important precondition that is derived for the theoretical framework (Sect. 2.5.2). With regard to this dissertation's research goal, this dissertation argues that it is first of all necessary to understand this (general) political context before analyzing the influence of the digital transformation. This is further underlined by SDG 16 incorporating strong institutions, digital innovations and ICTs. Moreover, there is currently a shift away from good governance to the notion of 'inclusive governance' in international development debates (TWI2050 2018:108).

Conclusively, (good) governance therefore also plays a vital role in digital development, especially with regard to sustainable digital development as it influences also the performance of ICT4D projects. Especially in the context of governance for a sustainable development (in line with this dissertation's research question), governance and sustainability in the development context are closely linked:

> "Good, effective and equitable governance is critical for implementing the Sustainable Development Goals (SDGs). To achieve the SDGs, it is essential to address the fragmentation of governance along the vault lines of governance levels, sectors, interests and approaches in order to take unified action." (Monkelbaan 2019:v)

This is one reason why this dissertation has chosen a governance approach to the analysis of digital development in developing countries, which will be further discussed in Sect. 2.4.3.

2.1.3 Goal: Fair and Inclusive Sustainable Digital Development in line with SDGs

To further elaborate on the interlinkages between governance, digital transformation and sustainable development, the SDGs serve as a guiding frame. In contrast to the Millennium Development Goals (MDG), the SDGs emphasize global cooperation between both the 'developed' and 'developing countries', where inclusivity is seen as vital for the achievement of the goals. Inclusiveness describes notions of empowerment and non-discrimination (Monkelbaan 2019). It is further at the center of the 2030 Agenda for Sustainable Development, for instance with regard to the paradigm of 'leave no one behind' and the global vision for economic prosperity, social inclusion and environmental sustainability around the globe. Sustainable development therefore further incorporates permanently sustainable structures and avoiding short-term perspectives (Klingebiel 2013:19). Moreover, the SDGs compromise of 17 goals and 169 targets or subgoals (UN n. d.). Here, digitalization, in correspondence to this dissertation, is a key element which can be found in nearly every sub-goal. Hence, this subsection will further elaborate on the interdependence of digitalization, sustainability and development.

Digital technologies are recognized in the SDGs as factors influencing sustainable development. The International Telecommunication Union (ITU) for instance has published 'ICT indicators for the SDG monitoring framework' (ITU 2015) in 2015, where the role of ICTs for every 17 goals is emphasized. Moreover, (digital) infrastructure here is crucial for sustainable digital development, which not only needs to be reliable, sustainable and resilient, but also focuses on affordability and access for all (ITU 2015). This is why infrastructure, affordability and access (especially in developing countries) among others are derived as preconditions for sustainable digital development (Sect. 2.5.1, Sect. 2.1.4, Sect. 2.2.1). Moreover, digital infrastructure is one part of the 'Digital Economy Framework' proposed by the Organisation for Economic Co-Operation and Development (OECD) (OECD 2018a).

In light of the increasing importance of sustainable development, Linkov et al. (2018) for instance propose the concept of 'Digital Sustainability' combining governance approaches with digitalization fields for sustainable digital development. Taking this as an example, calls among the reviewed literature grow louder as "[...] there is an urgent need to bring the sustainability and the digital and technology communities together to align the direction of change with the 2030 Agenda and a sustainable future beyond." (TWI2050 2018:6). Moreover, it further emphasizes the interlinkage between digital transformation and sustainability, with ICT being ascribed different roles in the SDG context:

> "Digital technologies support the sustainability transition and cut across all the other transformations. They are both indicators used for tracking the SDGs in themselves (e.g. Internet or mobile coverage) as well as enabling technologies to harness efficiency gains and inducing life style changes" (TWI2050 2018:75).

Expanding on this role of digital technologies in the sustainability context, this dissertation transfers this perspective to its analysis design of the meso level emphasizing its influence to the adjacent levels (macro and micro) (see Fig. 1.1).

So far the discussions around sustainable (digital) development seem to run in parallel according to Seyle (2018) noting that one discussion revolves around digital technologies and innovation to support the SDGs, while the other revolves around partnerships (ibid.:14). This dissertation synthesizes both perspectives as it delivers further insights into the role of partnerships in digital development (see further Sect. 2.5.4). Hence, this dissertation argues that governance is needed in order to link the transformation towards sustainability and digitalization, which is in line with the current digital development discourse (e.g., TWI2050 2018; WBGU 2018; The World Bank 2016a).

2.1.4 Multi-stakeholder in Digital Development using Stakeholder Theory

After discussing the importance of the political development context and international goals for sustainable digital development of developing countries, it is further important to analyze the role of the involved stakeholders in digital sustainable development. More specifically, this dissertation's research question aims at an analysis how digital networks with their interactions can be coordinated. In order to do so it is firstly necessary to analyze the multi-stakeholders in the digital ecosystem, before assessing how they can be coordinated. As already indicated in the beginning of this section, a key role in the D in ICT4D is played by involved stakeholders. Together, these stakeholders form a network of multilateral and bilateral donor organizations, private sector, governmental and non-governmental actors. These stakeholders are also the ones which will later be classified as members of the digital ecosystem in Kenya (see Sect. 4.4.2). This is especially relevant and of most recent interest, since current development cooperation is characterized by "an increasingly diverse array of governmental, multilateral and non-state actors working in different ways—government-to-government, people-to-people, peer-to-peer and public-to- private" (OECD 2019a:21). Hence, and with specific regard to developing countries, partner country governments and development partners consult with a variety of stakeholders.

In this dissertation, stakeholders in ICT4D will be analyzed according to these mentioned groups (multilateral, bilateral, private sector, academia, civil society organizations and hubs, see Fig. 2.1)[1] in order to illustrate the interconnections, not only of ICTs and development, but also of their actors in the formed network (ecosystem). It is argued that developing countries face additional challenges from the amount and variety of actors. This will be firstly addressed by insights into each stakeholder group. Ultimately, in the context of the digital transformation and sustainability, stakeholders have further differing roles in sustainability, especially with regard to the digital transformation (Lock and Seele 2017:235). Here, this dissertation follows Lock and Seele (2017)'s argument that there is a "need to combine stakeholder theory with sustainability incorporating the new possibilities of big data" (ibid.:237), or digital technologies in general for that matter.

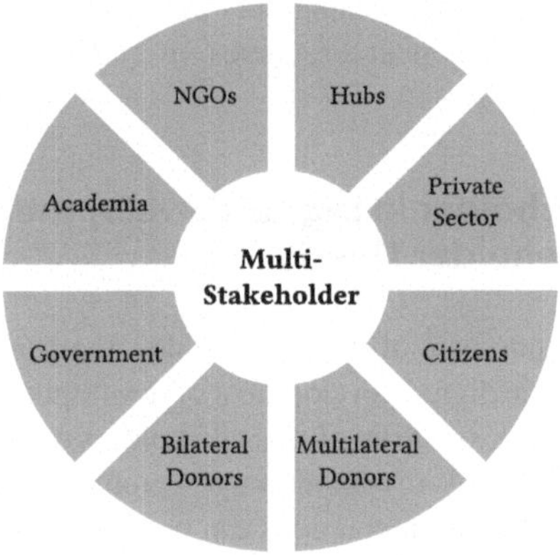

Fig. 2.1 Key multi-stakeholders in digital ecosystems of developing countries (own illustration)

[1] For full disclosure, it is noteworthy that citizens are a further important actor group. This actor group however is not part of the expert sample of this dissertation and will hence only be analyzed indirectly. Academia as an actor group has likewise been identified in the literature review, although with only limited reference. Nonetheless, it is part of this dissertation's sample and empirical analysis.

Firstly, it is necessary to define this dissertation's understanding of 'stakeholder' before elaborating on each stakeholder group. As a standard definition, a stakeholder is defined as "[...] any group or individual that can affect or be affected by the realization of an organization's purpose" (Freeman, Harrison, Wicks, Parmar and De Colle 2010:26). Addressing the complex interplay of societal actors, stakeholder theory emerged in the 1980s and is among the most influential theories for theorizing and managing the complex relations (Lock and Seele 2017). Moreover, stakeholder theory developed "from a simple, but rather static theory focusing on stakeholder groups and their roles to more dynamic, yet more complex, relationship-focused models that incorporate interdependencies, conflicts, and intergroup perceptions" (Lock and Seele 2017:237). Against this background, the stakeholder theory perspective has been chosen for this dissertation's analysis of multi-stakeholder digital development activities in digital ecosystems.

In the context of ICT4D stakeholders, they are here grouped in four stakeholder groups: multilateral, bilateral, private sector and civil society. While Lock and Seele (2017) for instance identify 10 stakeholder groups in digital sustainability (governments, intergovernmental organizations, companies, media, non-governmental organizations, academia, charitable foundations, grassroots organizations, private citizens, and future generations), due to the design and goal of this research project with an ICT4D focus, a limited number and slightly different composition of stakeholder groups is chosen (see Fig. 2.1).

In the following, the role and interests of each group in digital ecosystems will be discussed, as not only their presence forms a network but also their interests pose challenges for coordination of the ecosystem, which will be discussed in more detail in Sect. 2.3.3. This subsection serves the purpose of a stakeholder introduction as illustrated in Fig. 2.1, while their interactions and dynamics will be further analyzed throughout this dissertation.

Bilateral and Multilateral Donor Organizations

As a first stakeholder group, both bilateral and multilateral donor organizations are important stakeholder groups in ICT4D. They can be distinguished as follows. Bilateral development cooperation is carried out by a single state in cooperation with two governments, while in multilateral development cooperation, cooperation takes place indirectly via an international institution (Klingebiel 2013). Multilateral organizations are characterized for instance as less politicized by a pooling of resources that allows projects at a larger scale and different tasks, such as standard-setting functions of the United Nations (UN) or a coordinative function of members such as the European Union (EU) (Klingebiel 2013:24). More than one-third of all Official Development Assistance (ODA) is channeled through the multilateral development system (OECD 2020).

Moreover, governments of developing countries further engage with bilateral donors engaged or present in their country. Here, it is important to distinguish between 'traditional donors' and 'new' donors, the so-called *South-South Cooperation South-South Cooperation (SSC)*. To begin with, on a bilateral level, the main development cooperation actors involved are 'traditional donors'. These are mostly members, participants or associates of the Development Assistance Committee (DAC) of the OECD. The DAC has 30 members. In 2018, ODA by DAC members totaled USD 153.0 billion in the form of grants, official aid loans or contributions to multilateral institutions. As of 2018, the United States continues to be the largest donor (USD 34.3 billion), followed by Germany (USD 25.0 billion), the United Kingdom (USD 19.4 billion), Japan (USD 14.2 billion) and France (USD 12.2 billion) OECD (2019b). Most of official ODA (around 70 percent) is conducted through bilateral aid OECD (2019b). Bilateral development cooperation aims at "implementing development policy goals with appropriate strategies and using specific funding instruments either in direct cooperation with selected partner countries or through international organizations" (Stockmann et al. 2016:9). Since development policy is an integral part and a special form of foreign policy, it is thereby often characterized as linked with strategic, ideological and foreign economic motives, goals and instruments of foreign policy (Stockmann et al. 2016). Development research has long recognized that aid serves as an instrument of state-craft used to advance developmental and non-developmental goals (Dietrich 2016). Motives that states can pursue in development cooperation span from motives of (inter-)national security, ecological, economic and moral respectively altruistic motives (Degnbol-Martinussen and Engberg-Pedersen 2003; Davies and Klasen 2019). The interests of donor organization are of special importance for this dissertation's research as these are one factor challenging coordination and hence sustainable digital development, which will be discussed in more detail in Sect. 2.3.3. By working with the recipient state, donors can continue to exert policy influence, which is often critiqued as 'intervention' in development literature (Bon and Akkermans 2019). Allocation patterns of official aid are determined by recipient countries' needs and performance but also by the political and economic interests of the government of the donor country, among others (Fuchs and Öhler 2020).

Besides the traditional DAC donors, development assistance (especially in digital development) is increasingly provided by new emerging partners from the Global South under the paradigm "From Aid to Development—New partners for development". 'New' donors are emerging, such as China, India, Brazil among others. Especially with regard to the African continent, China is an important actor in development cooperation. In the literature, SSC is seen as "bringing more fresh resources to the table at a time when the volume of traditional aid has been eroded by acute financial and budgetary distress in the North." (Bracho and Grimm 2016:121). Additionally, supply opportunities hereby arise along with new practices, mechanisms and ideas. Moreover, the before mentioned attributed "intervention thinking" of traditional donors is claimed as running counter to working in a Southern-led collaborative partnership way (Bon and Akkermans 2019). Notwithstanding the advantages of SSC, nonetheless more actors emerge, adding yet another actor layer in developing countries and in their digital ecosystems. Not only is the number of bilateral donors increasing with emerging SSC providers, but this also accounts for new SSC formats and SSC multilateral actors (Bracho and Grimm 2016). As a consequence, and not surprisingly, "[...] recipient governments are facing the challenge of retaining an overview of old and new opportunities and their respective potential pitfalls" (Bracho and Grimm 2016:129). Hence, "a common complaint is that the emerging powers are adding to the problem of fragmentation of the aid system" (Bracho and Grimm 2016:122). Aid fragmentation is one central problem of international development aid, including digital development and ICT4D. Therefore it will be discussed in more detail in Sect. 2.1.5.

Private Sector Stakeholders

After establishing the role of multilateral and bilateral donor organizations, the private sector stakeholder group (non-governmental) will be elaborated in this section with a special focus on digitally active private sector companies. Development cooperation is not only achieved in bilateral or multilateral terms, but also varies between governmental and non-governmental donors. These non-state actors include local and international NGOs, and private sector actors (Dietrich 2016). Non-governmental donors are increasingly engaged in international development cooperation and therefore increase in importance. Non-governmental donors are also called private donors, including private sector companies, NGOs, PPPs, charitable individuals, foundations and global philanthropy (Madianou 2019). For-profit actors such as companies frequently provide aid, for instance in form of cash donations or access to critical infrastructure, especially in cases of emergency (Fuchs and Öhler 2020). Hence, non-governmental development assistance is also called 'private governance', 'private aid' or 'corporate', which varies in definition, but mostly encompasses both for-profit private sector and non-profit engagements.

In this dissertation's analysis of sustainable digital development in developing countries, relations between corporate and governmental actors will be focused on in more detail due to their position in the digital transformation. This includes both multinational tech companies as powerful actors, while also acknowledging the important role of (tech) startups. In agritech for instance, "[p]rivate business start-ups account for the largest share of solely-owned digital platforms" (Ezeomah and Duncombe 2019:240). Moreover, productivity growth (and innovation) is likewise driven by start-ups. Early-stage instruments for fostering innovation and supporting (tech) startups include incubators, accelerators, or hubs (Cirera and Maloney 2017:172) that are pursued by both private and public actors (see implications on digital development in Sect. 2.3.1). In this dissertation incubators and accelerators are summarized as part of ICT4D projects in the sense of an instrument, while hubs here serve as an individual actor group that is set between private and public actor groups (see below).

The involvement of the private sector in public service provision has been promoted in development discourses and practices since the early 1980s (Bayliss and Van Waeyenberge 2018) and is most recently highlighted in the Agenda 2030 and SDGs. It is commonly agreed that states alone are not able to address the complexity of sustainable development challenge and achieving the SDGs (Wehrmann 2018), so that over the last 30 years "the rate of outsourcing public responsibilities to nonstate entities has increased" (Dietrich 2016:68). Due to growing challenges such as increasing humanitarian crises and natural disasters, but also challenges emerging from the digital transformation, governments increasingly tend to address theses challenges together with/in reliance on non-governmental actors. More specifically, private sector engagement and their active participation in development results is also called 'Private Sector for Development (PS4D)' (OECD 2018b). As an practical example, corporate representatives of tech companies often participate in panels discussing the future of ICT developments such as at the 13th meeting of the Internet Governance Forum (IGF) in Paris. Here private sector representatives comprised 20 percent of the participants (onsite) as compared with 45 percent for civil society, 16 percent for government, 7 percent for intergovernmental, 1 per cent for press/media and 11 percent for the technical communities (IGF 2019). Nonetheless, the engagement of the private sector in development cooperation and partnerships with private-sector actors is highly debated for instance in Ellersiek and Beisheim (2017) and Pattberg and Widerberg (2016) (see further Sect. 2.5.4 and Sect. 2.5.4) as well as in terms of private sector domination in ICT4D partnerships in many developing countries (Ismail et al. 2018).

Private Sector in ICT4D Advantages of private aid in contrast to official aid, identified by Fuchs and Öhler (2020), lie in the fact that official donors are often unable to provide the required funds as discussed for instance by Riddell and Robinson (1995) and Fuchs and Öhler (2020). This criticism mainly refers to the common critique that official bilateral aid is perceived as bureaucratic, slow, and politically-driven, while private donors, in particular NGOs, have often been perceived as more need-oriented than official donors (Fuchs and Öhler 2020; Easterly 2002).

More importantly for this dissertation, the IT sector is not only gaining importance globally, but also in development cooperation. Due to the research design of this dissertation, the relations of various multi-stakeholders in digital ecosystems are of interest, including relations between public and private actor groups. Fuchs and Öhler (2020) for instance demonstrate that huge differences exist between donor countries to what extent the government engages with the business sector. For instance, the public development actors in the United States are perceived as more open to business engagement than their more skeptical counterparts in Germany and the United Kingdom as argued by Fuchs and Öhler (2020) based on Rieth (2009). This dissertation will transfer this question specifically to IT corporations and public sector actors for instance from the US and Germany, in order to both identify these relations and possible underlying motives of the private sector in providing aid to developing countries. Moreover, interests or motivations for private (tech) actors in developing countries such as Kenya includes boosting the image of the company and distribution of branded products (business interest) among others (Fuchs and Öhler 2020), which will be discussed in more detail in Sect. 2.3.3.

With special reference to the engagement of private IT companies in international digital development and ICT4D, it is not of much surprise that "[a]lmost all major tech companies have developed partnerships with United Nations and other agencies: Accenture, Amazon, Facebook, Google, IBM, and Microsoft to name a few" (Madianou 2019:5). The increasing importance of ICT and related infrastructure provision, is likewise emphasized by the private sector engagement. According to Bayliss and Van Waeyenberge (2018), ICT and electricity have accounted for over 70 percent of total private infrastructure investment between 1990 and 2015.[2] The increasing reliance of governments on the private sector, is likewise emphasized by the increasing emergence of IT PPPs, which format will be discussed in further detail in Sect. 2.5.4.

Addressing the digital transformation foreign tech companies are attributed the role of colonizing ('digital colonialism', see Sect. 2.3.6) as only a handful of mostly US multinationals (specifically GAFA) dominate in the digital sector, for instance in terms of (Kwet 2019:6):

[2] Other infrastructure sectors in 'The World Bank Private Participation in Infrastructure (PPI)' database as cited in Bayliss and Van Waeyenberge (2018) include railways, airports, water and sewerage, ports, natural gas, roads.

- Search engines (Google)
- Web browsers (Google Chrome)
- Smartphone and tablet operating systems (Google Android, Apple iOS)
- Desktop and laptop operating systems (Microsoft Windows)
- Office software (Microsoft Office)
- Cloud infrastructure and services (Amazon, Microsoft, Google, IBM)
- Social networking platforms (Facebook, Twitter, Instagram)
- Transportation (Uber)
- Business networking (Microsoft LinkedIn)
- Streaming video (Google Youtube, Netflix, hulu)
- Online advertising (Google, Facebook)

It is asserted that technology companies in the digital sector can have significant influence in the countries they invest in (Wakunuma 2019).

The role of the private sector as an 'intermediary' can hence also have deregulatory effects of stamping out competition in the (local) tech market (Kwet 2019). Another factor for determining notions of digital colonialism in the private sector is seen in their interests (e.g., emerging markets, profit maximization). In more detail, the author lists implications of foreign tech multinationals' dominance in various political, social, economic and military functions mediated by their technologies (Kwet 2019:8):

- Accumulation of profits from revenues derived from rent (in the form of intellectual property or access to infrastructure)
- Possible surveillance (in the form of Big Data, see also Zuboff 2019)
- Exercising control over the flow of information (such as the distribution of news and streaming services)
- Social activities (like social networking and cultural exchange)

This stated US dominance of digital architectures is argued to endanger other countries' sovereignty (Kwet 2019) and hindering competition as well as a possible inclusiveness. Driven by private sector interest, many digital technologies (software, hardware and even access and affordability) can hinder inclusiveness and participation. Here Kwet (2019) sees a fostering of the digital divide as poorer people can obtain these technologies less and it reduces communication between rich and poor. This will be further discussed in the following paragraph and in Sect. 2.3.6.

Private Sector Activities and the SDGs Moreover, with regard to the sustainability component of this dissertation, the importance of private actors allocating aid is also emphasized by SDG 17, which aims at global partnerships. According to a latest publication of the OECD and United Nations Development Programme (UNDP), the current topics relevant to the SDGs most frequently covered by public-private-dialogue Public-Private-Dialogue (PPD) in partner countries are (OECD/UNDP 2019:73):

- Regulations for doing business
- Infrastructure development
- IT infrastructure (e.g., mobile coverage and Internet)

Accordingly, IT infrastructure builds the third most important topic in SDG related PPD. Accordingly, PPD is attributed the potential to accelerate achievement of the SDGs by leveraging the potential of the private sector's contribution to sustainable development. The Agenda 2030 recognizes the important role of a diverse private sector in achieving sustainable development and calls "on all businesses to apply their creativity and innovation to solving sustainable development challenges" (OECD 2018b:1). More specifically,

> "[t]he private sector can contribute both financial and non-financial resources for sustainable development. Maximizing these contributions requires a conducive operating environment for business, however, and this in turn requires effective engagement of the public and private sectors based on open and transparent dialogue" (OECD/UNDP 2019:69).

However, partnerships with corporate actors are often criticized as facilitating neoliberalization in developing countries, for instance in Wehrmann (2018) and Pattberg and Widerberg (2016), which will be discussed in more detail in Sect. 2.3.6. With specific regard to the digital transformation, concerns are raised for instance with regard to the collection of datasets by private sector actors in terms of data ownership and accountability (Madianou 2019).

Civil Society Organizations: NGOs and Hubs

Fourthly, the group of (international) civil society organizations will be elaborated in this section. Here mainly NGOs and hubs are of interest here. Following the common definition of NGO as "self-governing, private, not- for-profit organizations that are geared to improving the quality of life for disadvantaged people" (Vakil 1997:2060), various forms of NGOs are included here. The for instance range from small local groups operating on a largely voluntary and informal basis to large-scale formal development agencies with multi-million-dollar budgets and high number of professional paid staff (Lewis and Madon 2004:119–120).

On the one hand, NGOs as civil society organizations serve the purpose of providing specific services, improving democratic processes and widen citizen participation in civic life (Fuchs and Öhler 2020; Lewis and Madon 2004) as well as broaden the information base and bring in types of professional knowledge states often lack, for instance on local circumstances, applicable technical standards (Glasbergen 2011:196). Thereby they often serve as a counterweight to public and private actors, especially in the context of the digital transformation. Moreover, in contrast to the before mentioned donor organizations, NGOs are perceived as being more altruistic since their aid is not shaped by commercial or political interests of donor countries (Nancy and Yontcheva 2006; Fuchs and Öhler 2020). On the other hand, there is debate in the academic literature around NGOs for instance with regard to impact and efficiency to the extent that NGOs even are not included in (policy) processes or partnerships (Glasbergen 2011). In sum, the actor group of NGO is understood as a subgroup of civil society organizations that is here treated as an individual actor group. This dissertation here will deliver empirical insights into the role of NGOs in the Kenyan digital ecosystem (especially their impact on sustainable digital development, see Chap. 4).

As special actor within the non-governmental actor group in this dissertation further includes hubs that are here understood as a hybrid between public and private actors and hence summarized under NGOs. International Trade Centre (2019) defines a tech hub as "a space where technology enthusiasts, innovators and entrepreneurs gather to share ideas and build" (ibid.: viii) with the purpose of building businesses and creating a community. While hubs as an organizational format are discussed in the literature such as Friederici (2019), this dissertation follows International Trade Centre (2019)'s understanding of hubs as a hybrid organization that builds networks and is an intermediary actor. More importantly for this dissertation, hubs "operate in ecosystems, which are typically defined by nine key pillars: markets, human capital, funding, support systems, infrastructure, culture, government, policy and regulatory framework, and education" (International Trade Centre 2019:viii), which is in line with this dissertation's ecosystem framework (Fig. 1.1).

In academic literature, innovation and entrepreneurship are crucial for poverty alleviation and long-term economic growth and shapes a "[...] narrative of driving development through innovation and entrepreneurship", especially in Africa (Jiménez and Zheng 2017:96). This narrative is for instance reflected in Africa experiencing a boom in technology entrepreneurship with a rise of innovation hubs (Friederici 2019), as well for both development organizations and private sector actors supporting hubs (Jiménez and Zheng 2017:97). Hubs as a timely organizational format for fostering collaboration is likewise represented in the emergence of *SDG Hubs* (Monkelbaan 2019).

To conclude this section, digital ecosystems in developing countries are characterized by a variety (and high amount) of multi-stakeholders with different interests and responsibilities. The impact of this setting, especially for sustainable digital development as proclaimed by this dissertation, will be discussed in the next subsection.

2.1.5 Aid Fragmentation

Synthesizing the previous section and the various stakeholder groups, it has been shown that development cooperation settings have become more complex and interactive, and are characterized by a multitude of actors in constant and changing interaction with each other (Keijzer, Klingebiel, Örnemark and Scholtes 2018). International discussions reflect an increased recognition of the relevance of development contributions from diverse actors and a rising heterogeneity among development actors (Lundsgaarde and Keijzer 2019). This amounts to a highly fragmented aid landscape, where *aid fragmentation* creates challenges for developing countries. In order to measure aid fragmentation, various statistical approaches and indices exist. Although the *Herfindahl Index* is the most common index (e.g., Gehring et al. 2017), measuring aid fragmentation is often criticized as volatile since "in some sectors, five donors may appear to be many, in others, 15 donors may be a number dealt with easily" (Gehring et al. 2017:322). Besides this quantitative classification, Gehring et al. (2017) propose further determinants for aid fragmentation, for instance different aid outcomes, different recipient country contexts (which is in line with this dissertation's emphasis on D in ICT4D), different levels of donor alignment, different time periods, and different channels through which fragmentation could affect development in recipient countries (ibid.:321).

In the literature, aid fragmentation is widely recognized as a significant challenge for many aid-recipient countries and effective development cooperation, since it is often accompanied by the lack of donor coordination and alignment to recipients' policies and processes. The most prominent and agreed upon challenges according to the reviewed literature associated with aid fragmentation are for instance (Gehring et al. 2017; Fuchs et al. 2015; Aldasoro et al. 2010; Anderson 2012; Knack and Rahman 2007; Annen and Kosempel 2007; Djankov et al. 2009; Kimura et al. 2012; Klingebiel, Mahn and Negre 2016; Öhler 2017; Davies and Klasen 2019):

- Increased number of actors in changing interactions
- Differing donor interests
- Rising transaction costs
- Transparency issues
- Administrative burden for recipient countries
- Less effective impact of aid on economic development
- Aid volatility (aid darlings vs. aid orphans)
- Diffused responsibilities
- Overlaps and duplication

Firstly, the number of donors has dramatically increased in past decades, more and more actors are involved in developmental relationships with partner countries (Klingebiel, Mahn and Negre 2016; Davies and Klasen 2019; Lundsgaarde and Keijzer 2019; Keijzer et al. 2018). In addition to the traditional OECD-DAC members and the so-called 'new donors' (Sect. 2.1.4), multilateral organizations, global funds and private philanthropic platforms have added to the complexity of the funding landscape (Lundsgaarde and Keijzer 2019). Not only does the increased number of actors manifest aid fragmentation, but also additional levels of interaction between actors in development landscape (national and international level) and sectors (governmental and non-governmental/private) emerge (Klingebiel, Mahn and Negre 2016). This development also affects 'classical' bilateral development cooperation, since today's development cooperation increasingly moves away from direct 'government to government' relations towards more intermediated relationships involving a multitude of stakeholders on various levels of action (Keijzer et al. 2018). Moreover, this context becomes even more complex when factoring in the digital transformation.

The ICT4D field itself consists of many stakeholders, including bilateral and international donors as well as businesses, foundations, and philanthropists. Moreover, it is rather fragmented, with technological advancements and complex financial arrangements that have likewise led to a diverse field of actors (Uimonen and Hellström 2015). However, it has to be noted that a vast number of actors also offers advantages (especially with regard to inclusivity and sustainability): "At the same time, a greater number of donors active in a country may bring up more innovative ideas, allow the recipient government to benefit from a variety of experiences, and put the recipient government in a more powerful position to take its own decisions, thereby enhancing ownership" (Gehring et al. 2017:321). Taking both sides into account, these arguments indicate reasons why a network governance approach including multi-stakeholder perspectives is proposed as one solution form for governing the digital development sustainably (Sect. 2.4.3).

Secondly, and closely linked to the increasing number of stakeholders, is the increased selectivity of donors in terms of serving only a subset of eligible countries (*aid darlings*), leaving other countries as *aid orphans*. This is likewise reflected in focusing on specific purposes of aid (*aid sectors*), for instance on education, infrastructure, social structures or digitalization (Aldasoro et al. 2010). These foci of donor organizations on specific topics, especially digital transformation topics, will be discussed further in Sect. 2.2.3 as one challenge to sustainable digital development of ecosystems. Another factor contributing to the phenomenon of *aid darlings* and *aid orphans* is proposed by Davies and Klasen (2013) demonstrating that bilateral donors respond to aid flows from other donor countries by increasing their own aid funds. Moreover, and one disadvantage of public-private cooperation, a close relation ('clumping together') of official and private donors and their aid activities can lead to a recipient countries being grouped as donor darlings and orphans (Davies and Klasen 2019; Klingebiel, Mahn and Negre 2016).

Thirdly, the inclusion of additional actors from various sectors (such as the IT sector) as well as the increased number of donors raises issues of policy consistency of donors. Current development cooperation faces a decline in joined-up approaches and diverging approaches, which often lack consistency (Klingebiel, Mahn and Negre 2016; Keijzer et al. 2018)—not only between 'traditional' DAC donors, but also with regard to SSC and the increasing emergence of non-governmental actors such as the private sector:

> "Moreover, when additional sector agencies of donor countries enter the arena, they not may follow established cooperation frameworks and procedures, and may instead operate in parallel. This means new sector agencies not only bring in additional topics, approaches and they do so without a administrative or other requirements, but structure that would support the partner country in keeping control over the overall cooperation portfolio." (Keijzer et al. 2018:38)

Arguably, the digital transformation has brought additional and relatively new actors: (multinational) IT companies. They serve as a new (donor) community with the (digital) private sector becoming more and more engaged in development cooperation. Hence, the missing policy consistency and few joint-up and coordinated approaches of donors, add a layer of complexity and challenges for developing countries, further fostering aid fragmentation.

Fourthly, aid fragmentation is associated with higher transaction costs, which can "result from the presence of multiple actors pursuing different priorities and cooperation approaches" (Lundsgaarde and Keijzer 2019:217 based on Lithman 2014). Hence, aid fragmentation is costly for both recipients and donors, since it "causes incomplete information which, in turn, may lead donors to concentrate too much on priority sectors and to underfund other, still important sectors" (Aldasoro et al. 2010:922). Moreover, "multiple donor missions, different sets of policy conditions and inconsistent reporting requirements absorb scarce administrative resources in recipient countries, especially the poorest [...]" (Aldasoro et al. 2010:921), which overburden the capacities of partner governments. Hence, this has a further impact on the general capacity of partner governments, which will be discussed in more detail in Sect. 2.5.2.

Fifthly, all these factors pose challenges on the administrative capacity of developing countries. In terms of the increased number of actors, the number of potential 'owners' on the side of recipients increases, which for instance poses managerial challenges to aid-dependent countries (Keijzer et al. 2018). Can administrations of developing countries deal with the vast number of (mostly international) actors? And if so how, and what are differences? And is the 'burden' of external international IT organizations any different to actors from the development cooperation landscape? This will be further explored in Sect. 2.5.2 and the empirical analysis of Kenya here will deliver further insights (see Sect. 4.4.4).

In sum, aid fragmentation in this dissertation is understood as another variable in order to understand the D in ICT4D, i.e., the development context of the digital transformation. Here, it is argued that the digital transformation is a force that has the potential to both increase aid fragmentation due to 'new' actor groups from the tech scene, but also to decrease aid fragmentation through implementing and using digital technologies (see further Sect. 2.2, Sect. 2.5.3, Sect. 2.5.4). More specifically with regard to the research question, aid fragmentation is not only a factor hindering sustainable digital development, but also coordination of multi-stakeholder activities, especially with regard to donor organization, which will be further discussed in Sect. 2.1.6.

These challenges associated with aid fragmentation have hence been derived from the theoretical literature review and are incorporated into this dissertation's developed digital ecosystem framework (see Sect. 2.5).

2.1.6 Need for Aid and Donor Coordination

As previously highlighted, aid fragmentation and especially a lack of coordination in donor activities, emphasizes the (global) need for aid and donor coordination. In order to analyze aid fragmentation, aid effectiveness and aid coordination (and their linkages), it is firstly necessary to establish what is understood by *coordination*. Coordination is an essential part of this dissertation's research question and detrimental for the analysis of sustainable digital development. While coordination in this dissertation encompasses all multi-stakeholder activities in the digital ecosystem, coordination can be further distinguished from related forms of *cooperation* and *collaboration* (Lundsgaarde and Keijzer 2019).

Firstly, collaboration can be distinguished from coordination and cooperation, all of which are closely related and hence often used synonymously. All of these three will be addressed throughout this research project. This dissertation follows the distinction of Keast et al. (2007) who have established a spectrum of collaborative relationships from 'fully fragmented' to 'fully integrated' as illustrated in Fig. 2.2. This logic is also incorporated into this dissertation's framework as it indicates different degrees of closeness and intensity of the relations between multi-stakeholders (Sect. 2.5.1). Here, cooperation demonstrates the least integration, followed by coordination in the middle and collaboration as the most integration (Keast et al. 2007). With reference to the context of digital innovations, collaboration can further involve forms such as co-investigation, co-design and co-creation (Bon and Akkermans 2019). In the context of development cooperation, collaboration is seen as opposite to intervention (Bon and Akkermans 2019). Cooperation is understood as a (institutional) means or instrument to achieve coordination for sustainable digital development. Following the logic of Keast et al. (2007), collaboration would be the ultimate goal where all involved actors collaborate, hence work together in an inclusive way, for instance in the sense of the Global Partnership for Effective Development Cooperation (GPEDC) (Lundsgaarde and Keijzer 2019).

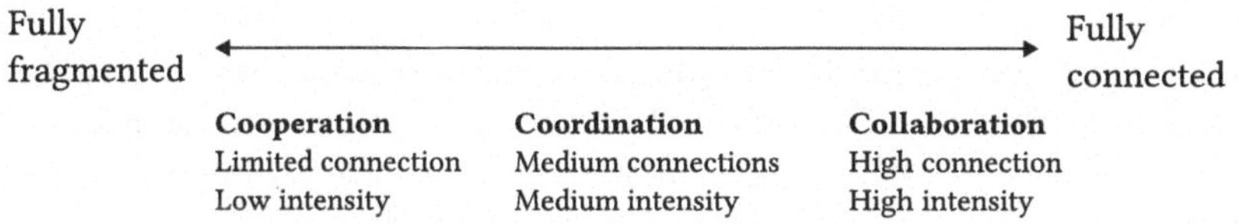

Fig. 2.2 Horizontal integration continuum (own illustration based on Keast et al. 2007:12)

Taking a development perspective on coordination, coordination is a form of cooperation among multi-stakeholders with the goal of promoting mutually beneficial outcomes, while cooperation "relates to collaborative efforts involving agreement on common goals and joint planning to achieve them" (Lundsgaarde and Keijzer 2019:219). Hence coordination can be understood as "a technical exercise at the level of implementation to avoid the negative consequences of overlap with the activities of other actors" (Lundsgaarde and Keijzer 2019:219). Examples of cooperation and also forms of coordination in this dissertation can for instance be found in formalized relationships between multi-stakeholder such as in the case of PPPs, which will be further discussed as a governance format in Sect. 2.5.4. Coordination in this dissertation is understood in terms of coordination of involved stakeholders, especially with regard to donors (and their activities) as well as tech companies and further multi-stakeholders introduced in Sect. 2.1.4. More importantly for this dissertation's analysis of digital ecosystems from a network governance perceptive, Frey, Lohmeier, Lee and Tollefson (2006) further add networks to the mix as they distinguish between networking, cooperation, coordination, coalition and collaboration as levels of relationship characteristics.

Secondly, and similar to aid fragmentation, the number of stakeholders in development cooperation is one factor that underlines the need for coordination (see further Sect. 2.5.1). Fragmentation of aid and donors as well as of programs and projects, contributes to a transparency problem: "[...] it becomes difficult to trace sources of aid and promote accountability around particular issues" (Lundsgaarde and Keijzer 2019:221). Hence, aid fragmentation increases with an increasing number and variety of actors likewise posing a challenge to coordination and governance. Moreover, responsibility is rather diffused when there are many donors involved (Aldasoro et al. 2010). Contrary to this perspective, the number of donors might not be a hindering factor, if they cooperate and speak with one voice:

> "When donors cooperate well with one another, so that they essentially speak with one voice, the number of donors involved should no more be of any concern for the recipient country. From this perspective, donor coordination could be a substitute for the concentration of aid flows" (Gehring et al. 2017:322).

Hence, this is derived for this dissertation's framework, especially in Sect. 2.5.1 (determinants 1 and 2). Similarly, unilateral action of donors "would help overcome coordination failure—but only if donors concentrated on different recipient countries and aid sectors, rather than all engaging with the same 'aid darlings' and crowding in the same high-publicity sectors." (Aldasoro et al. 2010:921). This however is argued to be hindered, not only by individual donors interests, but also due to competition (e.g., for export markets and political support) that prevents donor countries from coordinating their aid activities more closely (Fuchs et al. 2015). Moreover, this dissertation argues that another hindering factor can be seen in the digital transformation, especially in terms of the associated ICT hype along with actor proliferation (Sect. 2.2.3).

Thirdly, the development landscape is still described as an "inter-related but uncoordinated environment" (Davies and Klasen 2019:244). So far there is only an "[...] incomplete assessment of the extent to which donors specialize and coordinate" (Nunnenkamp, Sotirova and Thiele 2016:832). The question of how to organize and manage development cooperation is still debated (Lundsgaarde and Keijzer 2019). This dissertation argues that governance of (digital) multi-stakeholder ecosystems can be a solution when additionally factoring in the digital transformation, which will be explained in the following (Sect. 2.5.4). This is especially relevant since academic research has similarly revealed limited global progress in improving coordination (Aldasoro et al. 2010; Nunnenkamp, Rank and Thiele 2016; Nunnenkamp, Sotirova and Thiele 2016; Davies and Klasen 2019; Lundsgaarde and Keijzer 2019). Against this background, calls for a new agenda for aid effectiveness and associated donor rankings as well as for coordination grow louder (Davies and Klasen 2019:246). While past approaches for coordination aimed at a top-down, centralized decision-making process, Lundsgaarde and Keijzer (2019) highlight a shift towards more flexible platforms for promoting collaboration among diverse actors, reflecting the network character of the initiative (ibid.:230). This dissertation aims to contribute to these network formats of coordination for more effective (donor) collaboration (Sect. 2.4.3) while incorporating digital platforms (Sect. 2.5.4). Hence, aid and donor coordination is closely linked to the overall goal of aid effectiveness that contributes to a sustainable (digital) development, which will be analyzed in more detail in the following subsection.

2.1.7 Coordination for Decreased Aid Fragmentation and Aid Effectiveness

To further elaborate on the role of development cooperation in the context of ICT4D/digital development, it is important to highlight the role of aid effectiveness for sustainable (digital) development and the role of coordination, which is essential for this research project. Firstly, "calls to improve the coordination of development cooperation have a long history" (Lundsgaarde and Keijzer 2019:216), highlighting coordination as a core guiding concept to increase development or aid effectiveness. Aid effectiveness is and has been a crucial element of development cooperation, which has been fostered by global initiatives such as the *Paris Declaration on Aid Effectiveness (2005)* and the *Accra Plan of Action (2008)*. These aimed at "donors to coordinate their aid better, to make aid flows more predictable, to pool aid flows in country-led programs, and to use country systems for aid management" (Davies and Klasen 2019:246). In development cooperation, coordination "takes place at multiple levels and in different actor constellations, encompassing both donor self-coordination and partner-driven coordination activities" (Lundsgaarde and Keijzer 2019:220). Hence, this underlines this dissertation's argument that coordination is crucial for a sustainable development, especially with regard to digital development.

Coordination will therefore be discussed throughout this dissertation. Another prominent global initiative worth mentioning in this context includes the SDGs since effectiveness and sustainability here are closely linked. Although the concept of sustainable development is not an original development policy topic, its universal dimension also plays a central role in development policy (Stockmann et al. 2016:427, see also Sect. 2.1.3).

The importance of aid effectiveness is further emphasized in the GPEDC principles (2015). According to theses principles, aid effectiveness is compromised of ownership, alignment, harmonization, results and accountability, actions both required by partner countries and donors (Palagashvili 2019). Of special importance for the research design of this dissertation is private sector engagement in the digital development landscape. The importance of the private sector is similarly highlighted by the GPEDC, which has published the *Kampala Principles on Effective Private Sector Engagement in Development Co-Operation* (Global Partnership for Effective Development Co-operation 2019:5):

- Principle 1: Inclusive Country Ownership
- Principle 2: Results and Targeted Impact
- Principle 3: Inclusive Partnership
- Principle 4: Transparency and Accountability
- Principle 5: Leave No One Behind

In sum, collaboration between actors builds a core principle of effective development cooperation and the targeted outcome of sustainable development. These principles serve the purpose of guiding development interventions and projects and hence are important determinants for this dissertation's research goal as to how sustainable digital development can be achieved. Their role in digital development will be further analyzed and expanded by empirical findings in Chap. 4. While the here mentioned principles likewise apply to private sector actors, there are challenges to achieving effective development cooperation. Still, "[...] governments must do more to ensure meaningful and inclusive participation in planning and tracking of development efforts" (OECD/UNDP 2019:59).

In line with academic research, the question arises of how coordination should be achieved and by whom. This will be further discussed in Sect. 2.5.3 and Sect. 2.5.4. By its research design, this dissertation aims at contributing to and supporting a multi-stakeholder approach to addressing aid fragmentation and coordination in the digital development landscape. After establishing the rather general context of the D in ICT4D, the focus will now shift to the second component of ICT4D, namely ICT in ICT4D. Hence, the digital transformation as a driving force and factor affecting the context D as well as the outcomes ICT4D will be explored in the following (as the meso level of analysis).

2.2 Driving Force Digital Technologies (Meso Level)—The ICT in ICT4D

Following the logic of the previous section (D in ICT4D), the potential driving force of the digital transformation and ICTs are to be further established. ICTs as a driving force for development here are understood as a factor to enables, encourages or reinforces digital development and here represents the ICT in ICT4D.

Firstly, in order to assess the possible governance of the digital transformation through coordination, understandings of ICTs as digital tools need to be elaborated. Providing an understanding of ICTs beyond a mere definition is difficult similarly to the definition of development (D in ICT4D, see section above) since "few studies have conceptualized what ICT means" (Sein et al. 2019:12). Generally, and as this section will demonstrate, the role of ICT in fostering development of developing countries and a theoretical framing of ICT4D research is still being highly debated (Avgerou 2017; Walsham 2017; Sein et al. 2019).

Such a multi-faceted field as ICT4D and its complexity, increasingly calls for a more holistic view of how ICTs can foster development (ICT4D) (Avgerou 2017; Walsham 2017; Sein et al. 2019). For instance, it is criticized in the literature that there is a focus on the IT artifact rather than its implications beyond their adoption and diffusion in the ICT4D landscape (Zheng et al. 2018). This dissertation aims at contributing to this holistic view by synthesizing sustainability, governance and ICT. Moreover, innovation is often synonymous with digital development and ICT4D (Madianou 2019). In this research project, innovation is understood in more general terms as the ability of countries to innovate and Research & Development (R&D) advances, with a special focus on digital technologies such as ICTs. While innovation research has been closely related to the organizational learning literature (Choi and Chandler 2020), this dissertation approaches innovation from a macro perspective focussing on its effects on countries' developments.

More specifically, technologies and innovations can be powerful drivers of development, simultaneously supporting and threatening development as well as the ability to achieve sustainability in terms of the SDGs. In the development cooperation landscape, there is an urgent need to bring sustainability and digitalization with their respective communities together to align with the 2030 Agenda. Until now, there is a lack of structures, especially governance structures that allow the mitigation of potential trade-offs of a digital revolution, particularly relating to its impact on governmental institutions and their respective relationships with private sector and civil society (TWI2050 2018). More importantly for this dissertation, collaboration and cooperation become increasingly important with regard to digital technologies: "ICT's role in development includes its ability to create effects, generating a network of knowledge that can be shared and is open to a larger stakeholder group." (Bass, Nicholson and Subrahmanian 2013:22). However, current trends in technological innovation "[...] have stimulated a trend of 'disintermediation' as recipients become available to directly source expertise from the global market. It has been suggested that development cooperation may in the future become disintermediated too" (Keijzer et al. 2018:26). Therefore, this dissertation argues that digital development activities of multi-stakeholders in digital ecosystems need to be actively governed in order to achieve sustainable development.

2.2.1 ICT4D Discourse around ICTs as a Driving Force for Development

In the current ICT4D discourse (combining the previous two sections), a focus innovation initiatives can be explained by the underlying expectation that technologies can provide solutions to complex challenges confronting developing countries, such as "grossly inefficient government administration, inadequate provision of health care and education and inability to compete in a global economy" (Avgerou 2017:10), or to enhance the efficiency and transparency of humanitarian operations (Madianou 2019). Generally, research on ICT in developing countries emerged in the 1980s (Avgerou 2017; Walsham 2017). Walsham (2017) for instance provides a brief overview of the history of ICT4D research, from early beginnings (mid–1980s to mid–1990s) to present (mid–2000s to present).

Discourse is very important for this dissertation arguing that common language and understanding are essential factors in governing the digital transformation. The importance of a common language and understanding between multi-stakeholders in the ecosystem is for instance emphasized as factor for achieving coordination and collaboration in social, environmental and economic topics (Monkelbaan 2019:67). Consequently, a lack of common understanding can be a barrier to sustainable (digital) development (see e.g., Monkelbaan 2019:132), which is hence further derived for this dissertation's framework for determining the potential of coordination in ecosystems (see further Sect. 2.5.1). With regard to specific topics in ICT4D, the ICT4D literature covers a wide range of topics (Walsham 2017; Zheng et al. 2018):

- Economic well-being and growth
- Systemic poverty
- Global health
- Environment and climate change
- Humanitarian crises
- Wars and terrorism
- Digital divide
- ICT readiness and access

Currently, there is now broad consensus in ICT4D academic literature "that ICTs are no silver bullets providing simple and linear solutions to complex and wicked social problems" (Zheng et al. 2018:2). Moreover, ICT4D should not be understood as "a static destination to be reached, but [as] multifaceted and dynamic social processes which are often full of tensions, contradictions and always contested and situated in time and place" (Zheng et al. 2018:1–2) based on (Avgerou 2010). More detailed criticism of ICT4D will be provided for instance in the next subsection (Sect. 2.2.2).

Ultimately, there is a "plethora of theories" (Sein et al. 2019:8) and no coherent narrative in ICT4D. Moreover, there is a knowledge gap in the link between ICT intervention and development in the context of developing countries (Sein et al. 2019).

2.2.2 Leapfrogging: ICT4D and (Economic) Growth through Innovation

Digital technologies such as ICTs, software and hardware are diffusing to all parts of the world, also described as the 'digital revolution' or 'digital transformation'. Both in academic research as well as (development) practice, the digital diffusion process has been ascribed as the potential for all countries to *leapfrog*. This usually refers to both 1) leapfrogging or overtaking developmental stages, in the sense of economic catch-up as well as 2) in terms of innovation stages such as leapfrogging generations of earlier technology. Although the concept of leapfrogging is not new (literature dates back to 1985), it is becoming increasingly important due to the exponential rise of mainly internet based digital technologies around the 2000s that allowed "the small and new to compete on equal terms with the large and the well-established, and permit leapfrogging to an 'information economy'" (Heeks 2002:1). To begin with, leapfrogging in terms of skipping developmental stages will be discussed, before analyzing leapfrogging of generations of earlier technology.

Leapfrogging Economic Stages

1) Firstly, leapfrogging economic stages can be explained by the five stages of development originally proposed by Rostow (1960) (traditional, transitional, take-off, drive to maturity, and high mass consumption). With regard to this dissertation's focus on development through ICTs, ICTs can potentially contribute to economic growth (Sect. 2.2.1). Potential mechanisms through which ICTs can be economically beneficial are for instance (1) improving access to and use of information; (2) improving productive efficiency, allowing businesses to better manage their supply chains; (3) creating new jobs to address the demand for ICT-related services; (4) facilitating communication within social networks in response to shocks; and (5) facilitating the delivery of financial, business, health, and educational services (Aker and Mbiti 2010).

However, similarly to the development theory focus (Sect. 2.1), achieving leapfrogging is often accompanied by a certain focus on 'the West', which is discussed as a role-model for developing countries in copying methods and processes to leapfrog developmental stages (see e.g., Sein et al. 2019). More concretely with regard to ICT4D, "[w]herever feasible, countries were encouraged to adopt policies to enable them to 'leapfrog' the industrial countries, that is, to invest in the most innovative, leading-edge technologies to boost their economies" (Mansell 2011:114). As innovation is commonly seen as the result of highly educated labour in R&D-intensive countries with strong ties to the scientific community, it has hence been largely perceived as a 'first world' activity (Cirera and Maloney 2017). However, "[w]ith technological integration in sight, many countries of the Global South are rushing to construct policies for twenty-first-century life" (Kwet 2019:5). Conclusively, digital technologies pose the potential for countries to leapfrog:

> "This exchange of ideas will no longer flow only from the West to the rest, since rapid urbanization enables cities in the developing world to leapfrog their global peers by incorporating smart from the start in ways that are cheaper and easier than retrofitting legacy infrastructure." (The World Bank 2016a:242)

Hence, also in terms of leapfrogging developmental stages through digital technologies, economic growth is "underpinned by the assumption of neo-classical economics that technology increases productivity in organizations and, in an aggregate form, the economy at large" (Avgerou 2017:16). Moreover, technologies being associated with contributions to productivity gains and new organizational forms, is a further explanation for investments in ICTs (Mansell 2011). Productivity gains likewise have an effect on overall GDP so that ICTs here have an essential potential: "The economic growth approach to development associates measures of the diffusion of computers, internet connections or mobile phones with increases of Gross Domestic Product GDP and employment." (Avgerou 2017:16). While this section has mainly concentrated on potential benefits of digital technologies and innovations contributing to leapfrogging economic stages, possible dangers are further discussed in more detail further below in this section and in Sect. 2.3.

Leapfrogging Technological Generations

2) In terms of leapfrogging technological stages or generations through technological innovations and ICTs, it is assumed possible for less developed countries to 'leapfrog' generations of technology. As a prominent example, it is commonly referred to skipping fixed landline connections (both phone and internet) and instead directly jumping to wireless networks (mobile phone and data networks). Thereby countries directly use the most up-to-date, cost-efficient and more advanced technologies (Mansell and Tremblay 2013; Ojanperä, Graham and Zook 2016; Konrad Adenauer Stiftung 2019). Similarly, the theory of 'Schumpeterian catch-up' refers to "technology adoption, imitation, and adaptation of existing non-frontier technologies" (Cirera and Maloney 2017:16). Besides a possible benefit for economic development, digital technologies are also seen as tools for peace and sustainable development (Mansell 2015a; Lee 2019).

However, with regard to technological innovations especially mobile and broadband services, there is a significant difference between the countries from the Global North and the Global South. For instance, mobile and broadband are assessed as having more impact in developing economies than in developed economies (The World Bank 2009). Moreover, "[w]hile developed countries have been transformed by information, communications, and technology, the transformation in the developing world is often even deeper, benefiting not only the wealthy, but also middle- and lower-income citizens" (Osiakwan 2017:58). Hence, it can be argued that developing countries could benefit from ICTs in terms of leapfrogging to a more inclusive development. This argument is supported by Nan and Markus (2019) stating that "[d]igital innovations hold the potential to be inclusive by affording users the ability to leapfrog traditional constraints and deploy previously inaccessible socioeconomic resources" (ibid.:468). This potential for inclusiveness is especially ascribed to mobile money technologies.

Criticism towards Leapfrogging

In conclusion, the concept of leapfrogging demonstrates developmental advantages for less developed countries, both in regard to 1) skipping economic developmental stages, and 2) innovation and technology stages, but also demonstrates conceptual weaknesses.

Despite the popularity of the concept and the promising argumentation, the evidence for leapfrogging however is scarce. To date, there is only limited research "on the nature of innovation activities in developing countries, on how countries develop the capacity to innovate, how it evolves over time, and what the potential barriers to innovation are" (Cirera and Maloney 2017:15). Most research has been done through the advanced country lens and at the cross-country level (Cirera and Maloney 2017; Friederici et al. 2017). Most common determinants for growth through innovations are summarized by Friederici et al. (2017), including usage and penetration rate of ICTs, investment in ICTs and telecommunication or broadband infrastructure and impact of ICT investments on growth across country income levels.

Nonetheless, the discourse around ICT interventions for instance in form of technology investments "invariably is reminiscent of the dominant [exogenous] model" (Mansell 2011:109). According to the author, investments in ICTs "are seen as 'exogenous' interventions that can close technology gaps between the rich and the poor" (ibid.:110). This dissertation will deliver insights into this interlinkage between ICT and asymmetries in more detail in Sect. 2.3.6 and in Sect. 2.5.2. Moreover, a tendency towards believing "too much in leapfrogging" (Wennersten and Qie 2018:325) and its potential to accelerate development is asserted. According to UNEP (2011), leapfrogging has to be analyzed within specific development capacities and institutional contexts and

> "will depend entirely on whether the capacity for innovation exists within a particular developing country and whether, in turn, an appropriate set of institutional arrangements are in place to provide incentives and harness innovations that demonstrate economically viable 'leapfrog' technologies" (ibid.:35).

This will be further discussed in Sect. 2.3.5. Moreover, with regard to the difference between technology growth in the Global North and the Global South, the assumption that ICT has higher growth effects in developed than in developing countries would demonstrate one factor against the existence of leapfrogging (at least on a grand-scale) (Konrad Adenauer Stiftung 2019).

The potential of digital technologies to contribute to economic development of developing countries will be further analyzed with regard to the digital economy and digital ecosystems in more depth in the next subsection.

2.2.3 (Technology) Hype in Digital Development

After discussing the general impact and potential of innovations and technologies, this section's focus will shift specifically to digital technologies (also referred to as ICTs) and hence to a more technical perspective. The goal of this subsection is not to analyze technical perspectives such as functionings of digital technologies, but to analyze the handling of such technologies in the wider ICT4D discourse. More importantly for this dissertation, associated with the discussed notions of 'hype' or 'buzz words' is often an overestimation of digital technologies and their impact. An emphasis on promise and hype is for instance referred to as "reeking 'technology boosterism and cyber utopianism'" (Heeks 2010:629).

To begin with, in its 'Digital Dividends' report of 2016, the World Bank identified seven digital technologies for enhancing digital development based on the Gartner (2015) 'Hype Cycle' (The World Bank 2016a:326-327):

- Fifth-generation Fifth-generation (5G) mobile phones
- Artificial Intelligence AI
- Robotics
- Autonomous vehicles
- Internet of Things
- Consumer Three-Dimensional (3D) printing
- Business 3D printing

Taking these technology 'hype' themes together, Heeks (2020a) assesses the potential impact of some of these 'hyped' technologies on development. According to the author, "we already have enough data to know that mobile, broadband and mobile broadband will be a key foundation" (Heeks 2020a:4). This likewise applies to social media, cloud technologies, digital platforms, Internet of Things (IoT) and sets of technologies (ibid.). The impact however is contested in the literature. While the impact of these mentioned technologies on development were seen as rather certain, there are fewer certainties for instance around 3D-printing, blockchain and robotics, AI and machine learning applications. Their attributed impact is assessed as rather medium-term for digital development instead of immediately (Heeks 2020a; Nagano 2018; Schwab n. d.). A 'hype' around IoT is however further highlighted for instance by Cunningham and Whalley (2020) and big data is likewise associated with a further 'hype' (Gandomi and Haider 2015). These 'hyped' technologies however shifted in 2021, with the *Gartner's Hype Cycle* 2021 emphasizing a more complex setting of emerging technologies with a dominance of AI related technologies (Gartner 2021).

Underlying all these technologies is the internet. The internet is constructed decentrally and is a networked technology (Heeks 2020a). Hence, the internet and emerging digital technologies have led to an association of the relationship between digital and development with flat, peer-to-peer network structures (Heeks 2020a). These likewise are reminiscent of network structures as proposed in this dissertation from a governance point of view (see Sect. 2.4.3). The link between network logic from a digital perspective and a governance perspective for sustainable digital development will be further discussed in Sect. 2.4 with special reference to governance of digital ecosystems.

2.2.4 Open Source for Sustainable Digital Development

In line with the potential of open innovations, calls for open source solutions increasingly emerge, especially in the context of addressing inequalities and dependencies between the Global North and the Global South (Sect. 2.3.6). In reference to the previously discussed leapfrogging (Sect. 2.2.2), Larson and Munger (2017) for instance argues that "[f]or the less-developed world to develop further, it will be necessary to leapfrog over the gradual development of physical facilities and infrastructure (such as reliable electrification and Internet connections) and skip ahead to a more decentralized environment where almost everything is open source" (ibid.:151). Hence, in order to leapfrog and develop further, open source is seen as an important instrument for promoting sustainable and inclusive digital development.

This is also emphasized in the Free Software Movement Free Software Movement (FSM)[3] with notions of "freedom to control technology" (Kwet 2019:18). Open source innovations are the essence of collaboration, so that sharing (e.g., of data or code) is discussed as a potential solution for data and information resource inadequacies (Matias, Kidokoro, Menoni, Mejri and Aminoltaheri 2017:222). Decreasing dependencies can hence serve as a tool for inclusive innovation, as everyone can potentially participate. Moreover, open source innovations offer the opportunity for partner governments to control software, also in terms of licenses.[4] Software licenses are here related to costs and possible lock-ins, which are derived as one challenge to coordination for sustainable digital development in this dissertation (see further Sect. 2.3.6). However, challenges of open source innovations themselves for digital development are for instance possible free-riding or more broadly free use of know-

[3] cf. Free Software Foundation, https://www.fsf.org/about/

[4] Kwet (2019) adds control over hardware and control over network connectivity as further means for decreasing digital dependencies of partner countries.

ledge (Mansell 2017; Baldwin and Clark 2006) and interoperability ensuring sharing of data or information (Heeks 2016; Aulkemeier, Iacob and van Hillegersberg 2019, see further Sect. 2.3.6). The role of open source as possible technological means to decrease dependencies and asymmetries between multi-stakeholders in the digital transformation will be further discussed with regard to Kenya in Sect. 4.2.2.

2.2.5 Digital Economy and Digital Ecosystem

Following the economic perspectives on the digital transformation (for development), this is further conceptualized by 'digital economy' (see e.g., Graham 2019; Bukht and Heeks 2017), 'gig economy' (see e.g., Heeks 2017; Zheng et al. 2018), 'platform economy' (see e.g., Ansell and Miura 2020), 'digital capitalism' (see e.g., Zuboff 2019; Pieterse 2010), or 'information capitalism' (see e.g., Cohen 2019). As elaborated in Sect. 2.2.3, (private sector) digital platforms build a current 'hype' (Ansell and Miura 2020). For both 'digital economy' and also 'digital ecosystem' there is no agreed-upon definitions, which will be discussed in more detail in the following.

Digital (Platform) Economy

Beginning with digital economy, there are no agreed definitions of the digital economy or the 'digital sector' as for instance argued by (IMF 2018). A straightforward definition of digital economy for instance refers to digitization affecting businesses through adoption of digital technologies and services including digital platforms (Eurofound 2018). Bukht and Heeks (2017) provide a further overview of definitions of 'digital economy'. Taking the development perspective of digital economy (ICT4D), the 'Digital Economy Framework' (OECD 2018a) for instance is a guideline for providing an enabling digital economy, which consists of the following building blocks: digital infrastructure, digital enablers, market environment (i.e., openness), regulation ensuring effective use of ICT and involvement of stakeholders in policy-making. According to this framework, digital economy is not only associated with an economic perspective, but likewise with its surrounding factors. This underlines this dissertation's argument of the importance of contextual factors for digital development (as discussed in Sect. 2.1), but likewise demonstrates a close relation to digital ecosystems (see passage on digital ecosystem below). However, this dissertation argues that these definitory elements of the OECD (2018a)' framework further describe a digital ecosystem where all of these factors are interrelated and characterized by interaction (not only from economic perspective), which will be discussed in the respective section below.

Moreover, 'digital economy' is often further associated with digital platform economies since platforms are central in business models of the world's largest companies such as Microsoft, Apple, Amazon, Alphabet, Facebook, Tencent and Alibaba (Bonina, Koskinen, Eaton and Gawer 2021; Cusumano, Gawer and Yoffie 2019; Koskinen, Bonina and Eaton 2019). Generally, platforms can be described in terms of their architecture (Ansell and Miura 2020; Cammaerts and Mansell 2020; Dijck, de Waal and Poell 2018), while noting that the definition of digital platforms depends "on the field under which they have been studied" (Koskinen et al. 2019:320). This dissertation follows the definition of Dijck et al. (2018) seeing a platform as "a programmable digital architecture designed to organize interactions between users" (ibid.:4).

Furthermore, different types of platforms can be distinguished, for instance transaction platforms, innovation platforms and integration platforms (Koskinen et al. 2019).[5]

In line with the logic of leapfrogging, digital platforms are attributed a potential of solving societal and developmental challenges such as insufficient information, weak institutions and poor infrastructure (see e.g., Koskinen et al. 2019). Moreover, "[...] digital platforms have the potential to generate social and economic value in the Global South, yet their developmental impacts are not entirely understood" (Koskinen et al. 2019:320), which is discussed similarly in Bonina et al. (2021). Similar to ICT4D in general (as discussed in Sect. 2.2.1), there is a lack of clarity on what digital platforms are and how they should be conceptualized, while their decentral and intertwined nature further challenges its research (Koskinen et al. 2019:320). Additionally, there is only scarce research on "the potential for software platforms to facilitate inclusive innovation at a large scale" (Sanner and Nielsen 2019:220).

Platforms are further characterized by interactions between various actors such as app developers, content producers, consumers, advertisers or buyers and vendors (Sanner and Nielsen 2019; The World Bank 2016a). Considering the associated economic value of platforms, the "[p]rofits of leading digital platforms have experienced a rising trend since 2017" (UNCTAD 2021a:23) with the net income of the leading digital platforms in the United States reaching USD 192.4 billion in 2020, an increase of 21.1 percent compared with the year (ibid.). As a consequence, and not surprisingly, this revenue potential has not been left unnoticed and lead to yet another 'hype' (Sect. 2.2.3).

[5] Other types for instance are software platforms such as Android and Apple iOS. For further information see e.g., Sanner and Nielsen (2019).

Furthermore, following this research's development perspective ICT4D, digital platforms not only affect the Global North but also the Global South, which is of research interest for this dissertation. Notwithstanding the fact that many of these platforms have their origins in the Global North, they are increasingly important in the Global South, especially in terms of employment or social networking (Koskinen et al. 2019). According to UNCTAD (2021a), 45 of the top 100 global digital platforms are based in Asia and Pacific and even two in Africa (ibid.: 22). The rapidly evolving platform business model as emphasized i.e., by GAFA, however bares its challenges affecting the development landscape. For instance, it "is often seen as the inevitable result of technological innovation and the uncontrollable 'winner-takes-all' effects of network dynamics" (Cammaerts and Mansell 2020). Hence, the underlying private sector business model is criticized due to the collection and processing of personal and other data (also referred to as 'datafication') "with little consideration of how, and to what end, data are gathered and used" (Cammaerts and Mansell 2020:136). As a consequence, digital platform economies are discussed in the context of 'platform capitalism' or 'surveillance capitalism' (Srnicek 2017; Zuboff 2019) or even triggering a 'surveillance society' (Lyon 2018). Conclusively, the power of digital platforms and their providers raises concerns about their "digital dominance" (e.g., Srnicek 2017; Dijck et al. 2018; Ansell and Miura 2020). This has lead to discussion around possible measures for regulation providers abilities and accountability (e.g., Mansell 2015c; Moore and Tambini 2018). The empirical analysis of the Kenyan digital ecosystem will provide further insights into such dependencies and asymmetries as well as its consequences for sustainable digital development (Sect. 4.3.4).

The mentioned hype around digital platforms is also called 'platformisation' (Heeks 2020a). In the same vein, the impact of digital platforms in the development context, is also referred to as 'platformisation' of development (Heeks 2020a) as digital platforms are "a major digital technology impacting development" (Heeks 2020a:4). Platformisation of development further follows "the key trend of the emerging digital-and-development relationship since it subsumes within it many of the other technologies [...]" (Heeks 2020a:4). Conclusively, and in reference to their network potential, digital platforms will increasingly act as both central nodes in digital-and-development networks and as the infrastructures that house such networks (Koskinen et al. 2019; Heeks 2020a). What is of importance for the research aim of this dissertation is the question of ownership, governance, management, etc. of platforms, which "will therefore increasingly shape the structure of the relationship between digital and development" (Heeks 2020a:8). Since this dissertation however follows a macro lens on the impact of the digital transformation on governance structures in developing countries, digital platforms are here seen as only one important factor in sustainable digital development. More concretely, in this dissertation digital platforms are not analyzed in terms of their economic gain, but they will serve as a technical tool for coordination (see Sect. 2.5.4).

Digital Ecosystem

Besides this economic angle on the implications of ICT on development, this dissertation more importantly follows the network perspective by using the concept of digital ecosystems in order to understand how these networks of interactions and technologies can be governed.

Firstly, (digital) ecosystem is a concept used inconsistently as highlighted by World Economic Forum (2019). Similarly, in academic literature different perspectives and understandings are to be identified with related concepts here ranging for instance from *entrepreneurial ecosystems* (e.g., Cohen 2006; Osiakwan 2017; Song 2019, *knowledge ecosystems* (e.g., Owen-Smith and Powell 2004), *innovation ecosystems* (e.g., Oh, Phillips, Park and Lee 2016; Jacobides, Cennamo and Gawer 2018), *platform ecosystems* (e.g., Mukhopadhyay and Bouwman 2019; Hein, Schreieck, Riasanow, Setzke, Wiesche, Böhm and Krcmar 2020; Jacobides et al. 2018) or *ICT ecosystems* (e.g., Fransman 2010; Nguyen and Mahundi 2019; Diga and May 2016) to *digital ecosystems* (e.g., Boley and Chang 2007; Heeks 2016; Stuermer et al. 2017; Sein et al. 2019). This dissertation uses the terminology 'digital ecosystem' synonymously with ICT ecosystem as it understands digital ecosystem as more encompassing than a focus only on ICTs similarly to the shifting discourse from ICT4D to digital development. Both concepts are new and emerging (e.g., Nguyen and Mahundi 2019) emphasizing the timely and innovative relevance fo this dissertation.

Against this background, there is no universal definition of the here relevant digital ecosystems as mentioned in the introduction of this dissertation (Chap. 1). While a first common association with the term *ecosystem* may allude to biological or ecological ecosystems (e.g., Dhungana et al. 2013; Boley and Chang 2007; Li 2009; Li et al. 2012), the ecosystem logic is likewise transferred to the similar construct of a network. Similarities here for instance are seen with regard to actors in the network being the 'species' (Boley and Chang 2007), so that the authors define digital (or here rather software) ecosystem from an IT perspective as "open, loosely coupled, domain clustered, demand-driven, self-organizing agent environment, where each agent of each species is proactive and responsive regarding its own benefit/profit but is also responsible to its system" (Boley and Chang 2007:399). Since this dissertation uses network governance for analyzing digital development activities in a digital ecosystem while focusing on actors' interactions, this dissertation follows the definition of digital ecosystems proposed by World Economic Forum (2019) that characterizes digital ecosystems as being comprised of "any set of interacting producers, suppliers, innovators, customers and regulators that shape a collective outcome [...]" (ibid.:14).

Thereby this definition highlights the actor component of digital ecosystem as consisting of "[...] interacting organizations that are digitally connected and enabled by modularity, and are not managed by hierarchical authority (like in a supply chain)" (World Economic Forum 2019:14). In sum, all digitally involved stakeholders engage with each other in this ecosystem and build stakeholder networks as illustrated for instance in Fig. 2.1. What especially constitutes a digital ecosystem is the fact that "[...] it includes human beings interacting with technology to make this a socio-technical system" (Heeks 2016:40). Similarly the concept digital ecosystem is characterized by the "[...] relationship between the digital artifacts and their social ecosystem as acts of creation and use of digital artifacts" (Stuermer et al. 2017:249) with digital artifacts representing technical information while the surrounding ecosystem of multi-stakeholders interacts and holds know-how in creating and using ICTs (ibid.). This dissertation argues that the digital ecosystem concept offers the opportunity to analyze the context in which ICTs are utilized as they are embedded in global, national, and local socio-economic contexts (e.g., Diga and May 2016) as emphasized in Sect. 2.1. This dissertation uses the proposed definition of national digital ecosystems by Nguyen, Nielsen and Sæbø (2017) as they are encompassing "the people, policies, strategies, processes, information and other ICTs that together make up the socio-technical environment surrounding an ICT embedded within a country" (Nguyen et al. 2017:94). Any digital artifact is here understood as embedded in a wider and constantly shifting ecosystem in line with for instance Stuermer et al. (2017).

Moreover, the definition proposed by World Economic Forum (2019) further highlights possible geographic boundaries for instance with regard to the *Silicon Valley entrepreneurial ecosystem*. This dissertation sees such geographic constraints on a national country level (similarly to Nguyen et al. 2017) while also highlighting possible inner centralization tendencies. Due to their diverse nature, digital ecosystems vary, for instance in terms of size, type of actors and, whose determinants are highlighted in reference to their implications for ecosystems, which is further discussed in Sect. 2.5.1.

2.3 Challenge: Barriers to ICT4D and Development of Digital Ecosystems (All levels)

While the relations between ICT4D and governance have already been addressed as a contextual factor in Sect. 2.1.2, this section will elaborate critical arguments towards ICT4D that are here argued to be addressed by means of governance. Hence these challenges to sustainable digital development are both to be understood as stemming from ICT4D in general as well as from (flawed) governance in the digital development setting so that these challenges apply to *all* of the three levels (macro, meso, micro) underlining their interrelatedness. In contrast to the driving potential of ICTs discussed in Sect. 2.2, this section will discuss criticism and possible barriers of ICT4D. In the reviewed literature, criticism of ICT4D is often referred to as the 'dark side' and 'the unbearable lightness of ICT4D' (Pieterse 2010; Zheng et al. 2018). Moreover it is claimed that more attention should be drawn to such 'dark sides' of ICT4D, for instance with regard to the use of ICT for surveillance and control enabled through an increasing use of big data (Zuboff 2019; Zheng et al. 2018). From the literature review, challenges emerging from the ICT4D context are identified as illustrated in Fig. 2.3.

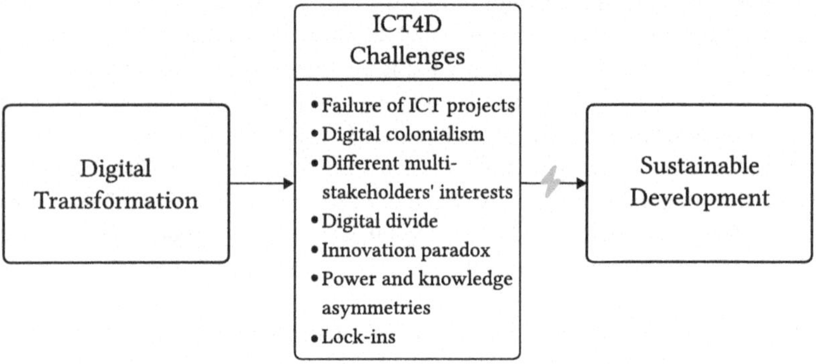

Fig. 2.3 ICT4D challenges for sustainable development (own illustration)

Additionally, these challenges in ICT4D likewise emphasize the relevance of coordination and point of reference for (network) governance, where coordination through network governance will be discussed as a 'solution' to these challenges in Sect. 2.4.3. These will be synthesized with network governance challenges in the proposed framework (determinant 2, see Sect. 2.5.2 and Fig. 2.6).

2.3.1 Failure of ICT4D Projects

As a first barrier to ICT4D and sustainable digital development, critical voices in ICT4D literature refer to the failure of ICT4D projects (possible positive effects of ICTs for development have been discussed in Sect. 2.2). More specifically, a mixed but generally low performance of e-government systems is asserted for instance by Choi and Chandler (2020) and Anthopoulos et al. (2016). A similarly mixed success rate of ICT projects is asserted in the The World Bank (2016a) report. Here it is highlighted that "[m]any public sector digital technology projects fail" (The World Bank 2016a:165). According to the report, and as illustrated in figure Fig. 2.4, only 13 percent of ICT projects succeeded, whereas the majority of 58 percent partially failed and almost one third of projects (29 percent) even completely failed (The World Bank 2016a).

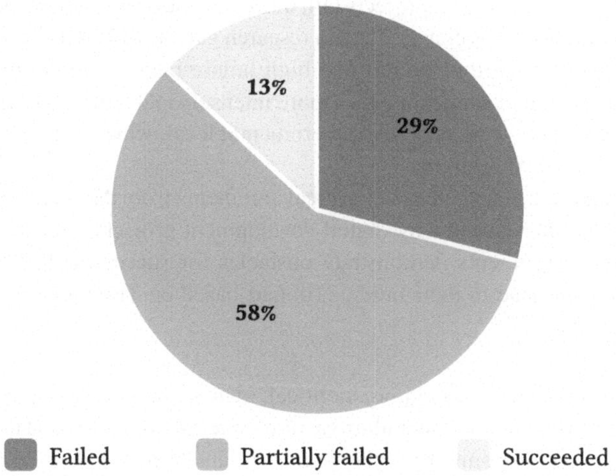

Fig. 2.4 Success rate of large (above USD 6 million) public sector ICT projects (own illustration based on The World Bank 2016a:165 with data based on Standish Group 2014; The World Bank 2015)

Referencing the exact number of ICT4D projects is difficult due to the discussed number of actors (Sect. 2.1.5, Sect. 2.5.1), diverse type of actors with different transparency commitments, and the fact that digital transformation is a cross-sectoral component since "digital components are increasingly included in projects across different sectors such as transport, education, health, agriculture and public sector management" (UNCTAD 2019:23).

While projects in general are not immune to failure, it is to be established if this is an increased problem in ICT4D projects. According to the ICT4D literature, ICT4D projects suffer chronic failure, especially donor-funded ones (Chipidza and Leidner 2019:145). Evidently, "[...] innovation is not determined to succeed or fail a priori when certain factors are found; instead, it depends on how public managers, politicians, and stakeholders effectively respond to the prior conditions and challenges emerging through the implementation process" (Choi and Chandler 2020:1–2). More specifically, ICT4D projects "are often inefficient or ineffective because they follow one specific implementation methodology or follow the interests of a single stakeholder group" (Hussain and Brown 2018:3), which underlines this dissertations identification of the crucial role of multi-stakeholders' activities influencing their interests (Sect. 2.5.1, Sect. 2.3.3) as well as possible lock-ins (here on a project level, see Sect. 2.3.6). Hence, if ICT projects fail, ultimately negative dynamics need to be analyzed and addressed, which is where a research gap prevails (Choi and Chandler 2020). This dissertation thus argues that high failure rates could be diminished if actors would coordinate and cooperate (more intensely and more sustainably), both on the project level as well as on the governmental level, which will be analyzed in more detail in the next sections.

Secondly, this section synthesizes critical arguments from the reviewed literature and offers potential reasons why digital development projects fail. The following selected critical arguments demonstrate obstacles for successful ICT4D projects and can ultimately lead to their failures (in part based on The World Bank 2016a: 165–166):

- Large gap between the regulatory, political, management, process, and skill realities in government and the ambitions of e-government projects (Heeks 2008)
- 'Dangerous enthusiasms' of technological infatuation and in large IT projects (Gauld and Goldfinch 2006, Sect. 2.2.3)
- Rigidities in government procurement rules, IT vendors' lack of understanding of government processes, and a failure to understand the country context (Dunleavy and Carrera 2013, Sect. 2.3.6, Sect. 2.5.1)
- Lack of proper (IT) infrastructure (King'ori Ndegwa et al. 2016; Chipidza and Leidner 2019, Sect. 2.3)
- Lack of digital literacy and skills of staff (King'ori Ndegwa et al. 2016; Chipidza and Leidner 2019, Sect. 2.5.2)
- Bureaucracy (Sect. 2.5.2)

- Design-reality gaps, missing focus, skill issues, unrealistic schedule and planning, politics, education, unclear vision, and system development process (Anthopoulos et al. 2016; Choi and Chandler 2020)

In addition to these factors that are detrimental to the failure or success of ICT4D projects, the literature review of this dissertation has identified further factors that contribute to explaining barriers to sustainable digital development:

Knowledge Vacuum As a first reason, Choi and Chandler (2020) argue that the failure of e-government projects is due to *knowledge vacuum* (which arguably also applies to the digital development projects). A knowledge vacuum

> "[...] refers to an organizational condition in which the pushing force of innovation and the pulling force of organizational inertia both at the structural and behavioral levels such as political pressure, pro-innovation bias, and employee frustration, interact to create a negative spiral that hampers organizational and policy learning." (Choi and Chandler 2020:2)

Hence, a knowledge vacuum contributes to understanding "e-government innovations and their failure by focusing on the learning dynamic of the innovations" (Choi and Chandler 2020:9). Hence knowledge plays a crucial role in the performance and impact of digital development, which is why it has been selected as an individual section for analyzing ICT barriers in more detail (Sect. 2.3.6). Knowledge vacuums may also be due to institutional barriers (e.g., Effah and Nuhu 2017). Knowledge vacuums will be further discussed in more detail in Sect. 2.3.6, Sect. 2.5.1 and Sect. 2.5.2.

Unclear Causality and Impact Secondly, the literature criticizes that a possible correlation of ICTs and development, as a *cause-effect relation*, is still vague. Avgerou (2017) asserts that "[i]t is reasonable to assume that ICT plays an enabling role in socio-economic improvements. But a simple cause and effect relationship between ICT diffusion and development outcomes is misleading because it hides other conditions at play, and the challenge for ICT4D research is to understand and explain the complex processes of ICT-mediated socio-economic change." (Avgerou 2017:11)

Focus on IT as an Artifact As a third reason and critique, rather than focussing on the implications of ICTs for development, a focus on *IT as an artifact* is highlighted in ICT4D discourse (Zheng et al. 2018:1). As a consequence, in ICT4D and in digital development policy, ICT is foremost conceptualized "in terms of infrastructure and devices (access to internet, Web, social media, Internet of Things, etc.)" (Bon and Akkermans 2019:33) which is asserted as "the wrong way to go" (Bon and Akkermans 2019:33). The question here arises whether digital technologies are a tool (means to an end) or whether they are an actor themselves, which will be discussed in more detail in Sect. 2.5.3.

Inequality through ICTs Fourthly, and in opposition to the hopes of inclusive development through the internet (Sect. 2.2), the current research questions the impact of digital technologies on equality as ICTs may even enhance rather than decrease *inequality* due to the necessary high investments of resources such as financial, human and time (Heeks 2010; Zheng et al. 2018). Additionally, project settings where "(high) inequality between those who benefit and those who do not benefit from the presence of ICT4D projects[...] could lead to project failure" (Maail, Kurnia and Chang 2017:81). Another example for failed addressing of inequality through ICTs is the *Global Digital Divide*. The essence of digital inequality is described as the *Digital Divide*, which will be analyzed in the following as one major ICT barrier (Sect. 2.3.4).

Failed Sustainability As a final reason, and with the digital divide prevailing, the impact of ICTs on *sustainable* digital development is questioned in the literature (*failed sustainability*). According to Zheng et al. (2018) "[...] decades of ICT4D research indicates that information technology (IT) intervention has not produced substantial and sustainable impact [...], unless they are embedded in long-term development processes" (Zheng et al. 2018:2). Another factor contributing to failed sustainability is seen in the *increasing number of actors* in ICT4D, as illustrated in Sect. 2.1.4, Sect. 2.1.5 and Sect. 2.5.1. It is agreed upon in the literature that fields of digitalization are accompanied by "increasing marketization, professionalization, pressure for humanitarian accountability, and, crucially, the dynamic entry of the private sector in the humanitarian field." (Madianou 2019:2). These potential barriers are further reflected in the role of 'the West', which will be discussed in more detail in the following.

2.3.2 Role of 'the West' / Digital Colonialism

As elaborated in Sect. 2.1.4, donor organizations play a vital but also challenging role for digital development in developing countries. The 'intervention' of donors in this context is highly debated in the ICT4D literature (e.g., Bon and Akkermans 2019;

Ramadani, Kurnia and Breidbach 2018; Renken 2019), also with regard to their interaction. According to Bon and Akkermans (2019), "intervention thinking presents several current policy obstacles that are in the way of collaborativeness" (Bon and Akkermans 2019:34). More specifically, the authors highlight three challenges of intervention thinking in ICT4D and digital development: 1) the ways programs are spent, 2) imposed operational framework for project and program management (e.g., linear waterfall model), and 3) the associated hegemonic discourse around 'natural' exogenous intervention (Bon and Akkermans 2019:34). As an empirical example, the authors refer to the *Principles for Digital Development*[6] and their value which they argue mirrors this imbalanced approach. Madianou (2019) adds that ICTs become the means for agencies enabling them to secure funding and justify their presence (not only donor organizations) (Madianou 2019:5).

Hence, and, closely linked to this critique of donor intervention in ICT4D is the *dominant role of 'the West'*, which is similarly discussed within the concept of *digital colonialism* (see also Sect. 2.3.6). More specifically, it is highlighted that "[w]hat is striking is that ICT4D and the new 'Digital Development' is foremost conceptualized content-wise as the rolling out of current Western ICT infrastructure of connectivity to the Global South" (Bon and Akkermans 2019:35). Framed in this way, it is asserted that ICT4D is not the hoped force of change but rather the continuation of previous conventional development policy (Bon and Akkermans 2019:35, as further discussed in Sect. 2.1). Similarly, a correlation between "the transformative potential of technologies" and human development aspirations as "symptomatic of a Western-centric and universalist model of economic growth and development" is attested (Mansell 2011:109). This is likewise reflected in the before mentioned hope of technologies as a means for leapfrogging (Sect. 2.2.2) and respective economic emphasis. This is especially relevant in the digital context due to market-led interests of the technology sector in the Global North "whose principal ambition is profits from the sale of digital technologies and content" (Mansell 2011:110). Hence the focus on 'the West' in the digital transformation and digital development is highly debated in the literature, as further emphasized with regard to the West as a role-model for digital development as discussed in Sect. 2.1.1.

Further criticism in the reviewed literature is directed towards a *mislead focus* in digital development activities. Chipidza and Leidner (2019) for instance highlights that a lack of focus on digital infrastructure is one factor for project failure, in line with this dissertation's argument as further discussed in the beginning of this section. More specifically, "[...] one might expect donors to focus on investing funds towards infrastructure building and institutions of higher learning in developing countries"

[6] see *Principles for Digital Development* (n. d.)

(Chipidza and Leidner 2019:145)—however, "[s]uch is not the case" (ibid.). The authors claim that while such investments, especially in digital infrastructure are provided by China, donor seems to continue to invest in "telecentres, healthcare, e-learning and agriculture software application", while infrastructure projects are rare (Delponte, Grigolini, Moroni, Vignetti, Giguashvili and Claps 2015; Chipidza and Leidner 2019). Ultimately, the question of sustainability in the sense of aid effectiveness here arises (Sect. 2.1.7). Providing further practical examples of donor digital development projects, United States Agency for International Development (USAID) for instance launched a strategy paper entitled 'Digital Download' in 2017 with "special emphasis on digital development in the health, food and finance sectors, on education for digital literacy and geospatial technologies, and on support to women and girls" (UNCTAD 2019:11). In this instance, references to digital education are found, but not to infrastructure. Similarly, the German Bundesministerium für wirtschaftliche Zusammenarbeit und Entwicklung / Federal Ministry for Economic Cooperation and Development (BMZ) aims to achieve the following five objectives with its 'Digital Agenda': 1) harnessing digital innovation, 2) reinforcing democratic processes, 3) helping forcibly displaced persons, 4) creating future-proof jobs and 5) safeguarding human rights and ensuring participation (UNCTAD 2019:15). A focus on infrastructure could similarly not be derived from this source. Whether these examples only apply to 'the West' or also to other actors e.g., from 'the East' will be discussed in more detail in Sect. 2.3.6 and in the following section. Moreover, the importance of (donor) ICT projects focussing on digital infrastructure is emphasized with regard to Kenya in Sect. 4.1.3.

2.3.3 Different Interests of Multi-stakeholders

Expanding on the discussed interests of multi-stakeholders in digital (development) ecosystems, not only of donor organizations, this section goes beyond an introduction of each stakeholder group as offered in Sect. 2.1.4. Here the focus lies on the specific interests primarily of donor organizations and private sector organizations. These are important for understanding the here discussed barriers to digital development and to coordination, especially donor coordination (Sect. 2.1.6). Moreover it serves as a basis for understanding the section on power imbalances (Sect. 2.3.6).

Interests of Donors Impeding Coordination

Firstly, donor interests (both similar and diverging interests) are another factor challenging both sustainable digital development and coordination (see further Sect. 2.3.3). One specific factor here is the interaction of donors in their aid allocation, and their dependence on each others allocation decisions. Davies and Klasen (2019) note that "the way donors interact in their aid allocation decisions will critically affect aid levels, the nature and motives behind aid fragmentation across recipients, as well as the interest donors take in donor coordination, alignment, and specialization" (Davies and Klasen 2019:246). Moreover, priorities of individual donor countries and their interpretations of intervention, present challenges to coordination practice on the ground (e.g., Lundsgaarde and Keijzer 2019). For instance, selectivity in terms of recipient countries is one factor determining the before mentioned phenomenon of aid darlings and aid orphans (Sect. 2.1.5). It is further argued that coordination and distributed specialization and selectivity of donors by engaging in specific sectors and regions, would counter coordination failure and aid fragmentation (Nunnenkamp, Sotirova and Thiele 2016; Lundsgaarde and Keijzer 2019; Davies and Klasen 2019). Other shortcomings in coordination originate from the fragmented character of individual donor systems that creates dependencies and lock-ins (Sect. 2.3.6), for instance due to different organizational approaches (Lundsgaarde and Keijzer 2019). Combining this argument with the role of bureaucracy as a factor in project failure (Sect. 2.3.1), it is synthesized in this dissertation's framework (Sect. 2.5.2).

Secondly, and closely linked, similar interests and ideologies of donors are an additional reason for explaining limited coordination among donors are their similar interests—especially with regard to export markets. Fuchs et al. (2015) find that "[...] coordinated aid activities are less likely in recipient countries constituting similarly important export markets of donors and a similar sectoral export structure. This finding holds at both stages of deciding on (i) the presence in a recipient country and (ii) the share of aid granted to this country" (Fuchs et al. 2015:282). Hence, the *presence* in a recipient country and the share of aid granted are important factors in terms of aid coordination and will be discussed further in Sect. 2.5.1. On the other hand, similarities between donors can also be beneficial for coordination, for instance with regard to ideologically closeness of donors that could improve their collaboration (Gehring et al. 2017) and hence coordination of their actions.

Another underlying motive of bilateral (and multilateral) donors, inherently leading to aid fragmentation, is thirdly seen in the global proliferation of bilateral and multilateral agencies, initiatives, and instruments (Lundsgaarde and Keijzer 2019, similarly to Sect. 2.2.3). According to Bigsten (2006) donors tend to weigh global presence more heavily than aid efficiency (see also Sect. 2.5.1). Similarly, Aldasoro et al. (2010) see incentive problems at different levels of the aid relationship as a reason for this global proliferation with poor coordination as a result (see also Lundsgaarde and Keijzer 2019). This applies also to the country level of recipients as "incentives to coordinate have been poor to absent among both African governments and their international partners" (Lundsgaarde and Keijzer 2019:224 based on Whitfield and Fraser 2010). Moreover, a further reason given in the literature for the apparent prevalence of few aid and donor coordination is seen "in the absence of a clear and shared goal for doing so" (Lundsgaarde and Keijzer 2019:222). The importance of goal consensus for sustainable digital development and (network) governance will be discussed in more detail in Sect. 2.5.1.

As a fourth reason, past colonial ties, strategic alliances, and trade relationships affect aid flows (Davies and Klasen 2019). This is further emphasized by: "For former colonial powers like France and Great Britain, development cooperation was (and is in parts) an instrument to maintain special relationships with the colonies, which became independent in the 1960s in particular." (Klingebiel 2013:14). Notions of colonialism with regard to digital development are further discussed in Sect. 2.3.6. Similar motives are followed likewise by the new or emerging donors (SSC) as mentioned in Sect. 2.1.4. Since the early 2000s, numerous providers of SSC increasingly emerged (Bracho and Grimm 2016). In SSC, similar interests to the traditional donors are to be found in the literature, but also new motives emerged. Assumingly, the most prominent SSC actor is China, since it is not only the third largest official donor in Africa, a former developing now middle-income country, but also a global economic power (Lau 2020). In terms of historic ties for instance, China's Belt and Road Initiative Belt and Road Initiative (BRI) depicts Africa as "part of the historical and natural extension of the Belt and Road" (Ministry of Foreign Affairs of the People's Republic of China 2018). Similar to motives of traditional development cooperation as illustrated above, China's interests are for instance expressed in initiatives as part of the BRI with regard to Africa such as industrial promotion, infrastructure connectivity and trade facilitation among others (Lau 2020). Here differences in focal points with regard to infrastructure in contrast to 'traditional donors' become obvious as discussed in Sect. 2.3.2.

Recipient Countries' Interests (in Aid Coordination)

Closely linked to interests of donor organizations, recipient countries' interests are also often not contributing to aid coordination. On the one hand, "[...] one might expect recipients to desire as few constraints as possible in their dealings with donors" (Davies and Klasen 2019:249), so that recipients might prefer coordination among donors. On the other hand, however, "developing-country governments may perceive coordination among donors as restricting their possibilities to exercise leadership and room for maneuver" (Lundsgaarde and Keijzer 2019:221 based on Klingebiel, Negre and Morazan 2016), underlining the linkages between national ownership and coordination (Lundsgaarde and Keijzer 2019). The interlinkage of donor interests and recipient interests is further fostered by existing dependencies, not only between recipient and donor, but also between donors (Sect. 2.5.1). Additionally, with regard to the digital transformation, another layer of dependency might be added by IT stakeholders (Sect. 2.1.4). With regard to dependencies and coordination, information and knowledge play a crucial role Davies and Klasen (2019). This is why information will be discussed in the following as one factor determining the achievement of aid and donor coordination (Sect. 2.3.6).

Private Sector Interests
Lastly, in addition to findings of Sect. 2.1.4, the following interests of the private (tech) sector in ICT4D are identified as follows in the reviewed literature:

- Profit maximizing, for instance through access to new markets, trade, access to data, trade etc. (e.g., Madianou 2019; Fuchs and Öhler 2020; Chipidza and Leidner 2019
- Aid provided by private sector stakeholders is often linked to their home country's interests (Fuchs and Öhler 2020)
- Involvement in humanitarian projects represents branding and Public Relations (PR) opportunities or Corporate Social Responsibility (CSR) (Madianou 2019), especially through corporate philanthropy (Uimonen and Hellström 2015) that can contribute to a competitive advantage (i.e., strategic interest) (Chipidza and Leidner 2019)

Synthesizing the differing interests of multi-stakeholders and underlining this dissertation's argumentation, "[e]ffective coordination would constrain the possibilities of donors to pursue commercial and political self-interest" (Aldasoro et al. 2010:922).

2.3.4 Digital Divide

An additional barrier for (sustainable) digital development and expanding on discussed inequalities, the issue of digital divide emerged simultaneously to increasing digital advances. Using Scheule (2005)'s classification, the following divides can be distinguished (represented in simplified form using binaries):

- Income Divide
- Geographic Divide
- Ethnic Divide
- Education Divide

- Gender Divide
- Age Divide
- Technological Divide
- Global Divide

Quantitative or statistical data to the digital divide, especially with regard to developing countries, can for instance be found in ITU (2019b) and ITU (2020). Taking a theoretical stance on the digital divide in the context ofICT4D, what the concept of digital divide expresses is the tendency that countries with a dynamic telecommunications sector, economic growth, democratic regimes and an educated population, have a higher ICT diffusion than countries with lower computing and telecommunications diffusion and lower human (as well as economic and political) development indicators and failed states, have less ICT diffusion (e.g., Avgerou 2017, as discussed in Sect. 2.1.1). The digital transformation contributes to these preexisting differences and inequalities for instance due to increased electricity generation demand, cybersecurity vulnerabilities, and social discrepancies posed by the widening gap in access to ICT (Linkov et al. 2018). Moreover, inequalities or asymmetries in the digital transformation likewise prevail in the context of culture or language as the use of most digital services and applications requires a knowledge of the English language. De', Pal, Sethi, Reddy and Chitre (2018) for instance questions the role "[...] whether issues of subjectivity—beliefs and values—originate in local, subaltern issues or are driven by colonial values" (De' et al. 2018:75). Widening the digital inequalities is a further instance of exclusiveness instead of inclusiveness as part of sustainable digital development (Sect. 2.1.3). Such barriers can likewise be amplified due to digital technologies and services, especially advanced technologies such as AI (Sect. 2.2). In these cases where advanced digital innovations require complex knowledge digital skills (Sect. 2.5.2), they might further marginalize excluded groups and hence widen existing divisions rather than narrow them through their provision (Nan and Markus 2019:468).

This dissertation argues that contextual factors are crucial for sustainable digital development and here discusses the digital divide mostly with regard to access (see Sect. 2.5.1) due to this project's focus on government action and governance structures. Access likewise includes affordability of digital technologies and services, both on a individual but also governmental level (Sect. 2.5.2). While the digital divide is arguably one of the most pressing barriers for sustainable digital development, this dissertation will further focus on the inherent power and knowledge asymmetries that likewise contribute to the digital divide and hence will analyze underlying factors. More concretely with regard to this dissertation, these imbalances likewise are reflected in power and knowledge asymmetries, which will be discussed in more detail in Sect. 2.3.6. One solution to overcome power imbalances in ICT4D and the educational digital divide is named in the literature for instance through "investing more in infrastructure and education (i.e. building ICT skills)" (Chipidza and Leidner 2019:160). This will be further discussed in this dissertation's sections such as Sect. 2.5.2 and Sect. 4.1.3.

2.3.5 Innovation Paradox

Closely linked to the digital divide and to further discuss on digital development barriers with regard to capability and capacity (Sect. 2.5.2) especially of governments, serves the the so-called 'Innovation Paradox' as proposed by Cirera and Maloney (2017). This includes perspectives on the difference in terms of innovation and innovativeness between countries of the Global North and Global South (similar to regional digital divide). To begin with, after innovation has been previously defined in Sect. 2.2, innovation performance can be determined by "(1) the critical complements to innovation investment needed to realize the high potential returns; (2) the range of firm capabilities required to undertake innovation and take it to market; and (3) the required government capabilities for implementing effective innovation policies." (Cirera and Maloney 2017:xix). The Innovation Paradox here arises from "the coexistence of great potential gains from Schumpeterian catch-up in developing countries with low innovation investment by firms, and the surprising lack of effectiveness by governments in increasing these investments by several orders of magnitude" (Cirera and Maloney 2017:60–61). This has been discussed in this dissertation in Sect. 2.2.2.

From the 'Innovation Paradox' perspective, a difference is seen between 'advanced countries' with financial markets function, friendly business climates, and skilled human capital and 'developing countries' that lack many or all of the above. This can for instance be determined by increases in R&D spending: In advanced countries just a marginal increase can have a high rate of return, while in the expected rate of return to knowledge capital investment in developing countries is often below Schumpeterian models (Cirera and Maloney 2017:60–61). The levels at which this paradox is observed, span from the level of firm capabilities to governmental capabilities. Firm capabilities in innovation range from basic managerial skills to more advanced production systems i.e., more R&D-intensive products or patenting, to technical literacy and to high-quality science and technology institutions. More importantly for this dissertation's research, the government however is the entity overseeing this evolution as it is also tasked with the resolution of market and systemic failures and coordination among various actors (Cirera and Maloney

2017:60). The role of the government in digital ecosystems will be further discussed in Sect. 2.5.3.

However, the innovation paradox describes that these capabilities, both on the firm and government level, are lacking in developing countries (Cirera and Maloney 2017). As this dissertation focuses on governmental institutions, in developing countries they often lack lower institutional quality in terms of rule of law, regulatory quality, or property and patent right protection as discussed in Sect. 2.1. This context "lowers the return to innovation for a large sample of firms in developing countries" as political stability plays an important role for innovation (Cirera and Maloney 2017:60), which is likewise reflected in the *Ease of Doing Business Index*. Moreover, in many developing countries there is "little alignment between the stated goals of policies and actual budgets and impact" (Cirera and Maloney 2017:8). Similarly to the digital divide concept (Sect. 2.3.4), weak firm capabilities, weak financial markets and weak education systems in terms of human capital that might not provide adequately trained workers to run new technologies, demonstrate that the actual returns to innovation are far less than promised by the catch-up effect. Hence, the innovation paradox offers a counterargument to the above mentioned catch-up effect and leapfrogging (Sect. 2.2.2) and emphasizes the importance of the government's capabilities and capacities for a sustainable digital development. More importantly, this highlights not only the importance of the government, but also of the local context since "efforts to build capabilities should aim to solve problems in the local context as nominated and prioritized by local actors." (Cirera and Maloney 2017:117). Factoring in the activities in digital development of external actors such as donors backs the question of an additional challenges of donor instruments, policies and programs, which will be further addressed in Sect. 2.5.2.

A second dimension offered by the Innovation Paradox is the role of intergovernmental coordination for innovation and hence for digital development. According to the authors, innovation policy in developing countries is often formally delegated to a ministry of science and technology, often uncoordinated with other ministries such as the ministry of industry's policies (which will be further discussed in Sect. 2.5.2). Moreover, these policies thereby become fragmented across industries, with different agencies responsible for implementing innovation policies and instruments, which leads to confusion within the private sector regarding responsibility (Cirera and Maloney 2017:128). This likewise includes a technological (communication and coordination) component in terms of interoperability of digital infrastructure (e.g., Heeks 2016). Navarrete, Gil-Garcia, Mellouli, Pardo and Scholl (2010) for instance emphasize the role of interoperability in (multinational) e-government collaboration and information sharing (for e-government see Sect. 2.4.1).

Conclusively, what is crucial for governments in developing countries in terms of shaping technological advances, is effective coordination of actors involved in innovation and policy coherence across governments and within ministries (Cirera and Maloney 2017:127). Here again a difference between 'advanced' and 'developing' countries becomes obvious: Collaboration between actors (e.g., scientific research institutions and private sector) is much higher in the more technologically advanced countries (Cirera and Maloney 2017:136). However, partnering with external agencies (e.g., academia, multilateral organization) for sharing capacity "can help kick-start the evaluation agenda" and can be an initial means (Cirera and Maloney 2017:120). However, collaboration with donor organizations in innovation is an additional challenge, which will be elaborated in the following section.

2.3.6 Power and Knowledge Imbalances between Multi-stakeholders

Closely connected to the other obstacles are ICT power imbalances and dependencies. As mentioned in Sect. 2.1.4, various and diverse sets of multi-stakeholders are involved in the digital transformation in manifold interactions. ICT4D entails intergroup interactions, however in settings of underlying power imbalances (Chipidza and Leidner 2019:146). This section further elaborates power imbalances between the ICT stakeholders, with special reference to the role of donors, specifically from the angle of post-colonial and dependence theory in ICT4D in order to offer one explanation for ICT4D donor related power imbalances. First, power relations will be analyzed followed by a specific focus on the resource of power in terms of knowledge. Then, the role of dependencies contributing to power asymmetries will be addressed from dependence theory perspective and postcolonial theory.

Power imbalances in ICT4D is debated, but not to great extent, in the reviewed ICT4D literature (Chipidza and Leidner 2019:146), which will be elaborated in more detail in the following. Power imbalances between stakeholders become an obstacle to ICT4D impact and success when "some stakeholder groups or individuals take on multiple roles simultaneously, while others do not" (Lock and Seele 2017:238) so that "power is relative to and, therefore, depends on the other actors in the environment" (Lock and Seele 2017:238). Hence power and knowledge are crucial for governing the digital transformation. They are likewise important for the second research interest of this dissertation, namely sustainable development. Monkelbaan (2019) has elaborated similar determinants of power and knowledge in the context of sustainability governance. Most importantly, in order to achieve sustainable (digital) development, the linkages between a) knowledge and norms,

b) power and norms and c) knowledge and power are crucial (Monkelbaan 2019). In this dissertation, these linkages of determinants is transferred to ICT4D. Moreover, power and knowledge asymmetries in ICT4D and digital ecosystems create settings of dependence, which is likewise associated with aid dependence as discussed in Sect. 2.1 (e.g., Knack 2001; Desai and Kharas 2018; Brazys 2016; Mimbi and Kyobe 2017). More importantly, this dissertation derives the following dependencies in digital development in the reviewed literature: dependence on actors (e.g., Wakunuma 2019), dependence on technology (e.g., Heeks 2020a; Pistor 2020), dependence on resources (e.g., Chipidza and Leidner 2019; Bailur 2007; Ismail et al. 2018) and associated lock-ins (see Sect. 2.3.6). This will be further elaborated in the following while incorporating governance of power and of knowledge as a means to achieve and secure sustainable digital development in digital ecosystems.

Power Imbalances in ICT4D using Post-Colonial Theory and Power Dependence Theory

Beginning with power imbalances in ICT4D between recipients and donors pose an obstacle for ICT leading to development (D). The most common theory used to explain dependencies of developing countries in ICT4D with regard to donors, are dependence theory and post-colonial theory (Sect. 2.1). With reference to this dissertation's focus on power asymmetries in digital ecosystems in developing countries, power asymmetries likewise influence governance of services in the ecosystem. In reference to ecosystem literature Vallet et al. (2020) (while having an environmental focus) for instance demonstrate that power asymmetries shape governance due to their influence on the access, use, benefit and participation in the management of ecosystem services (here digital services).[7] These impacts of power asymmetries on these factors are discussed in this dissertation's framework and will be further discussed in Sect. 2.5.

First of all, power needs to be defined. A definition of power is not straightforward and in fact, definitions are diverse and manifold explaining that "[p]ower is one of the most disputed concepts in social and political theory" (Monkelbaan 2019). Power in development cooperation and IT and ICT context is for instance addressed by Mansell (2010, 2011, 2016, 2017); Monkelbaan (2019); Wakunuma (2019); Weber (2014); Medaglia et al. (2018); Adler (2001); Franklin (2013); Moore and Tambini (2018); Büthe and Cheng (2013); Chipidza and Leidner (2019); Cohen

[7] Based on Chaudhary, McGregor, Houston and Chettri (2018); Daw, Brown, Rosendo and Pomeroy (2011); Berbés-Blázquez, M., González, J.A., Pascual, U. (2016); Bodin, Alexander, Baggio, Barnes, Berardo, Cumming, Dee, Fischer, Fischer, Garcia, Guerrero, Hileman, Ingold, Matous, Morrison, Nohrstedt, Pittman, Robins and Sayles (2019); Felipe-Lucia, Martín-López, Lavorel, Berraquero-Díaz, Escalera-Reyes and Comín (2015).

(2019). Generally, "[...] power has to do with relationships between two or more actors in which the behaviour of one is affected by the behavior of the other" (Hall 1999:111). Especially in the context of ICT4D, power needs to be further distinguished from 'influence' and 'impact' as power "always refers to a capacity to do things (…) whilst 'influence' sometimes (and typically) does not" (Morriss 1987:282, see also Monkelbaan 2019). Moreover, power in the context of development cooperation needs to be differentiated from 'empowerment' which is one important goal of development cooperation (Sect. 2.1). More importantly for this dissertation's analysis of digital ecosystems (networks), power also lies in the nature of networks, with Castells (2011) offering four different forms of power in the context of networks (ibid.:773): networking power (the power of the core actors and organizations included in the networks), network power (the power resulting from the standards required to coordinate social interaction in the networks), networked power (the power of social actors over other social actors in the network) and network-making power (the power to program networks according to the interests and values of the programmers). These further mirror the technical or digital network logic (Sect. 2.2).

One important factor in defining power, and especially in the context of ICT4D, are resources. Avelino and Rotmans (2009) for instance define power as the ability to mobilize resources to achieve a specific goal. This dissertation argues that power imbalances here are closely linked to diverging interests of actors and the role of goal consensus for sustainable digital development (Sect. 2.5.1). More importantly for this dissertation, Medaglia et al. (2018) adds that "[i]n all relationships, competitive or collaborative, there is a power struggle" (ibid.:4). Power exerted in IT endeavors with different and often conflicting interests, offers conflict potential and involves negotiation processes based on control of resources and information (Medaglia et al. 2018). This dissertation will analyze these power struggles between multi-stakeholders in digital ecosystem in the context of sustainable digital development. As relationships between multi-stakeholders and their governance in the digital context is the research goal of this research project, power will here by understood in terms of power relations as well as resources of the stakeholders.

The stakeholders in the digital transformation, as illustrated in Sect. 2.1.4, have different degrees of power and resources such as knowledge, especially with regard to digital technologies. Power in the digital context is concentrated "[...] in the hands of a few who come with their own terms and conditions, and their own interests and agendas" (Wakunuma 2019:31), meaning a (almost monopolistic) dominance by major tech corporations (see also Sect. 2.2.5). The argument stated here is that tech companies possess greatest influence in terms of power and a dependency, especially over their users (including public sector actors) (Moore 2016; Wakunuma 2019). Hence, it is asserted that there is "[...] a danger that a culture of dependency on the benefactors and investors of these digital technologies will arise and increase rather than decrease inequalities" (Wakunuma 2019:30). Expanding possible impacts of power imbalances between multi-stakeholders in digital ecosystems, post-colonial theory and power dependence theory will deliver further insights while highlighting challenges for sustainable digital development and coordination of their activities (on an equal and inclusive basis).

Post-Colonial Theory in ICT4D Postcolonial theory will here foremost serve as a basis for analyzing power imbalances between donors (Global North) and recipient countries (Global South). For analyzing power imbalances, especially in reference to ICT4D and digital development, this theory is increasingly used (Chipidza and Leidner 2019; De' et al. 2018), but remains "largely absent from ICT4D research" (De' et al. 2018:74). Post-colonial theory, represented for instance by Escobar (1995) and Spivak (1988), addresses the "post-colonial nature of most developing countries, where former colonial states continue to live with laws, prejudices and over-riding priorities of the developed West" (De' et al. 2018:74). Post-colonial theory also offers insights into issues of power differentials across cultures and nations (De' et al. 2018). This dissertation argues that this (historical) context is a further component for explaining barriers to sustainable digital development and will deliver empirical insights with regard to its impact on Kenya (e.g., Sect. 4.1, Sect. 4.4.4). With reference to ICT4D, post-colonial theory is used "to understand the interactions among stakeholders, technology, and institutions in affecting development outcomes through ICT4D" (Chipidza and Leidner 2019:146). Here, the determinants of "devoicing of subalterns and resource dependency" (Chipidza and Leidner 2019:146), are crucial factors which can be transferred to ICT4D and will serve as a theoretical basis for the so-called *Digital Colonialism* as further explained in Sect. 2.3.6.

(Power) Dependence Theory in ICT4D Dependence theory has been illustrated in the development theory context (Sect. 2.1.1), and is especially crucial with regard to resources such as financial resources (Sect. 2.5.2) and know-how in the digital transformation (Sect. 2.3.6). Power dependence theory has been widely used in management and information science research (Medaglia et al. 2018). The majority of Information Systems (IS) studies uses Emerson (1962)'s concept of dependence focusing on relationships at the organizational level (Medaglia et al. 2018). These power dependency relationships are also highlighted within IT outsourcing (Kern and Willcocks 2000; Tan and Sia 2006; Medaglia et al. 2018), here especially with regard to outsourcing of digital technologies and service provision (mostly to tech companies) by partner governments. Outsourcing in this dissertation is not only analyzed from a dependence perspective for explaining challenges for sustainable digital development, but also for delivering insights into coordination as it determines a degree of intensity in the relationship between multi-stakeholders, especially between public and private actors (Sect. 2.5.1). Existing research on decision-making power in the governance of public-private collaboration in IT-related projects has for instance focused on the extent to which governments retain decision-making power (Ojo and Mellouli 2016). More concretely, regarding the role of ICT for sustainable digital development, new technologies can lead to new dependencies (Chawurura, Manhibi, van Dijk and van Stam 2019). Especially the alignment with people, processes and systems on the local level is often cause for dependencies (Chawurura et al. 2019) and "it is not only resources that one has to look at to explain the distribution of power between actors, but also the contextual factors." (Medaglia et al. 2018:8) Both arguments underline the relevance of this dissertation's digital ecosystems approach. Heeks (2020a) further argues that if technologies, data and users increase, the complexity and thereby dependency on digital technology increases (in contrast to the capabilities of analogue systems). The distribution of power between the two stakeholder groups is hence shaped by contextual factors of market, technology, regulation, and social norms (Medaglia et al. 2018). This also entails the crucial relationship between public and private actors.

However, these dimensions of power and dependence relationships between public and private actors need further research (Medaglia et al. 2018). Hence this dissertation follows Medaglia et al. (2018)'s approach of dependence theory being also used in order to explain ICT power relations. Dependency in power dependence theory can be explained in a straightforward way as "a stakeholder A is more powerful than stakeholder B if A has more control over resources and information than B" (Chipidza and Leidner 2019:153). In ICT4D projects, the relation between agencies undertaking ICT4D projects and intended beneficiaries is largely unequal, "because donor agencies provide the expensive resources required, while beneficiaries are treated as passive recipients of donor largesse" (Chipidza and Leidner 2019:153). However, many ICT4D projects are dependent on bilateral and multilateral donors and private sector organizations for financial resources and technology expertise (Delponte et al. 2015; Chipidza and Leidner 2019). More concretely, dependencies are determinants for sustainable digital development taking the example of ICT4D projects' performances, for instance in cases where "[...] 'the mainstream' withdraws its resources, the ICT4D project might likely fail" (Chipidza and Leidner 2019:153), as further discussed in Sect. 2.3.1) Moreover, in this dependency setting, notions of different levels of activity (i.e., passive or active) are hence likewise underlying, which are further identified in combination with Sect. 2.5.1 and Sect. 2.3.3 as a determinants for this dissertation's research framework using network governance.

In sum, these power imbalances pose challenges to sustainable digital development but also to coordination. This dissertation further argues that if these dependence relationships are not coordinated and governed, further imbalances also to the extent of digital colonialism can be the result (Sect. 2.3.6). Hence, this dissertation follows the literature's proposal of governance as a solution to improve coordination between these dependent actors for solving societal problems (here unsustainable and unequal digital development) and as a solution for addressed issues of coordination (Klijn 2008; Markus and Bui 2012; Medaglia et al. 2018), which will be analyzed in Sect. 2.4.3 and incorporated in Sect. 2.5. Further elaborating on the interlinkage between power and governance specifically in ICT4D, the next section will provide further insights on the dimension of knowledge.

Role of Knowledge Asymmetries in ICT4D

After discussing the impact of power imbalances on sustainable digital development, this section will further discuss the related angle of knowledge asymmetries. Similar to the attempt to define power, knowledge is equally difficult to define. Moreover, in the current 'knowledge society' (see e.g., Mansell and Tremblay 2013; Mansell 2015a) and 'Information Age', knowledge is not only crucial for any innovation, but also for development of countries (in the sense also of leapfrogging). What these terms indicate (similar to 'information' being defining element in ICT), is that information is often used synonymously with knowledge. However, knowledge is more than information or data. Information is defined as data (signs and syntax) put into a context of a problem while knowledge is more the goal-oriented integration of several informational elements (Jansen 2017:16). Knowledge can be further expressed in different forms such as cognitive, economic, social and technical (here especially IT) dimensions, with the latter often being associated with private types often in the context of property rights (Jansen 2017). Similarly to the previously discussed power asymmetries, these likewise apply to knowledge asymmetries between multi-stakeholders in digital ecosystems, which is also described by the term "knowledge vacuum" (Choi and Chandler 2020) as discussed in Sect. 2.3.1. Arguably, knowledge and information are key resources in the digital transformation (see e.g., Alsop and Heinsohn 2005; Kleine 2010) with access to information as a further first step to knowledge acquisition and participation (Kleine 2010:681). Given the increasing importance of (digital) knowledge with the digital transformation, knowledge is highly debated in the academic literature, also in the ICT4D literature (e.g., Mansell 2011; Mansell and Tremblay 2013; Mansell 2015a; Monkelbaan 2019; Choi and Chandler 2020; Adler 2001; Keijzer et al. 2018; Janus, Klingebiel and Paulo 2014; Stuermer et al. 2017; Young 2019).

According to Bass et al. (2013), "ICT's role in development includes its ability to create network effects, generating a network of knowledge that can be shared and is open to a larger stakeholder group" (Bass et al. 2013:22). Therefore, digital communication technologies may enhance entertaining a network of knowledge and therefore foster distribution of knowledge, which is especially relevant in the context of capacity building in developing countries. Stuermer et al. (2017) here agree with regard to the potential of sharing tacit knowledge to reducing dependencies (e.g., on individuals or corporations), but for empowering individuals and organizations to contribute to the development of digital artifacts. Hence, it is argued that sharing knowledge can reduce dependencies and can contribute to sustainable digital development enabled through ICT, which further delivers insights into the role of ICT in digital ecosystems (Sect. 2.5.3). Shifting from the technical perspective on knowledge (ICT) to the development perspective (D), knowledge is likewise crucial in development cooperation (Sect. 2.1), where knowledge for instance includes scientific and technical expertise and implicit knowledge gained from implementing development interventions and working in local contexts (e.g., Keijzer et al. 2018; Janus et al. 2014). More importantly for this dissertation's focus on sustainable digital development, Stuermer et al. (2017) highlight the sustainability potential of knowledge and the impact of technology on the creation and use of knowledge since "[...] knowledge has to be seen as a resource that itself should be sustainable, to preserve its value for society and ensure that it can permanently contribute to the goals of sustainable development" (Stuermer et al. 2017:247).

Against this background, Adler (2001) for instance further distinguishes increasing knowledge-intensity between "the rising education level of the workforce (living or subjective knowledge) and the growing scientific and technical knowledge materialized in new equipment and new products (embodied or objectified knowledge)" (Adler 2001:216). In the ICT4D context of this dissertation, digital literacy and skills here are essential forms of knowledge. Their role for sustainable digital development is further discussed with regard to multi-stakeholders in Sect. 2.5.2. Arguably, technical knowledge is especially required in dealing with the digital transformation. With regard to state capacity, new questions arise: Is there a correlation between digital development and state capacity and/or state fragility? Who possesses knowledge, how is knowledge shared, how is knowledge used? These are questions that will be addressed with regard to this dissertation's case study of Kenya (Chap. 4). In the context of ICT4D it is especially important to distinguish here between endogenous and exogenous knowledge and innovation similarly to Mansell (2011) and their impact on sustainable digital development in the context of the discussed dependencies and imbalances. This leads over to the next challenge to sustainable digital development.

Digital Colonialism

These power and knowledge asymmetries are further discussed in the reviewed literature in terms of concepts such as 'digital colonialism' and 'digital imperialism' (e.g., Kwet 2019; Young 2019; Hill 2014) or 'technocolonialism' (e.g., Madianou 2019) have emerged to highlight digital development gone 'wrong'. Similarly, the term 'data colonialism' has received attention (e.g., Thatcher, O'Sullivan and Mahmoudi 2016; Couldry and Mejias 2019b,a), emphasizing data as a form of capitalist expropriation or also in terms of surveillance capitalism (e.g., Zuboff 2019). Hence, 'digital colonialism' combines colonialism with digital development. According to Kwet (2019), instead of the colonial conquest of land however, digital technologies are colonised. Thereby digital colonialism is seen as a "twenty-first century form of colonisation" expressed in an "[a]ssimilation into the tech products, models, and ideologies of foreign powers" (Kwet 2019:4). This reflects elaborated critique in the reviewed literature of ICT4D (Sect. 2.3).

Digital colonialism includes both foreign donors as well as foreign private sector companies as the 'new colonializers', especially with regard to infrastructure control, dispossession of valuable resources and ownership by colonial powers. The importance of (digital) infrastructure for digital development as been highlighted in this dissertation with regard to sustainable digital development (Sect. 2.1 and Sect. 2.3.1). On the other hand however, authors supporting the concept of digital colonialism see self-beneficiary/self-serving motives behind infrastructure development by foreign powers that is also described 'privatised forms of governance' on infrastructure development in the Global South (Kwet 2019:7). Besides infrastructure, as the basis of access to digital technologies, software and hardware are likewise discussed with regard to digital colonialism (Kwet 2019) (as also addressed in Sect. 2.3.6 and Sect. 2.2.4). As discussed in Sect. 2.1.4, domination as a form of power asymmetries (see also Sect. 2.3.6) are asserted through centralized ownership of software, hardware and network connectivity, mostly with regard to the US (Kwet 2019). According to Kwet (2019), firms in the Global South cannot compete with established tech giants, for instance due to network effects creating barriers to entry for competition and economies of scale posing barriers to entry so that digital colonialism can refer to both the firm but also government/country level.

Especially in the context of (digital) infrastructure, other stakeholders than from 'the West' are involved such as SSC actors. As elaborated in Sect. 2.1.4, SSC actors are important players in development cooperation, but also in digital development. China is arguably one of the most influential actors in digital development worldwide, hence also in Sub-Saharan Africa, with the Chinese 'Belt and Road Initiative' encompassing a specific 'Belt and Road Digital Economy International Cooperation Initiative' as one example. China not only promotes infrastructure (Sect. 2.1.4), but due to engagement of Chinese digital tech companies, China is also one important actor in digital ecosystems in Sub-Sahara Africa and therefore is a member of the ecosystem. Hence, digital colonialism is not only attributed to 'the West' but elements may also be found in SSC, which requires further research.

In conclusion, if tech hegemony prevails in terms of technology dominance but also in terms of dominance of actors creating a non-inclusive digital ecosystem in developing countries, there is a danger of dependency (Sect. 2.3.6), which rather increases inequalities. Hence, these circumstances are factors contributing to possible digital colonialism (e.g., Wakunuma 2019; Kwet 2019).

Lock-in Effects

Synthesizing challenges emerging from the digital transformation as well as from development theory, this dissertation argues that dependencies in form of lock-ins and as a result of asymmetries (Sect. 2.3.6), are a crucial challenge for sustainable digital development. While lock-ins are cross-referenced among the various disciplines, here lock-ins in technological and political sense are of interest. On the one hand, lock-ins can occur of whole technological system due to economies of scale and influence over the political system (Ayres 1991:255). Expanding on the technological angle of lock-ins, a certain technology can have an initial advantage over competitors, which creates a path dependence (Sect. 2.5.2) due to switching costs that hence lock out alternative solutions (Cecere, Corrocher, Gossart and Ozman 2014). This impacts a chosen path in that with increasing returns to adoption, one path becomes dominant due to self-reinforcing processes and absorbing states, and this causes the economy to lock itself in to an outcome (Cecere et al. 2014). This is also known as 'network externalities' (e.g., Katz and Shapiro 1986; Shapiro and Varian 1999) describing that the value of adopting a technology increases with the number of other users that join the network. This is specifically for this dissertation's analysis of implications of ICT as the presence of such network externalities often leads to the emergence of lock-ins, "[...] because of technical interrelatedness (complementarity/modularity) among hardware and software components, system scale economies and quasi irreversibility of the investments because of high switching costs" (Cecere et al. 2014:1041). Such interdependencies and dependencies of actors here specifically refer to hardware and software providers (especially multinational tech companies as for instance illustrated in The World Bank 2016a:44).

On the other hand, the political component of lock-ins is further essential for this dissertation's focus on sustainable digital development, since "[a]s much as organizations can be locked-in and experience inertia, so can institutions" (Cecere et al. 2014:1043). More importantly for this dissertation, lock-ins relate to coordination effects at the level of organizations, policy makers and society—particularly to the role of institutions and policies in fostering or hindering the emergence and diffusion of innovations (e.g., Rogers 2003). Network externalities not only lead to lock-ins but also to the creation of standards, which in ICT4D is closely linked to the formation and transformation of institutions specifically with regard to codification and *standardization* processes (e.g., the responsible organization ITU). The role of standards for coordination in digital ecosystems for sustainable development will be discussed in more detail in Sect. 4.4.6. More importantly for this dissertation's (network) governance approach, networks and standards represent a new hybrid legal-institutional form of governance, which is developing across a wide range of geographies and public policy arenas (Cohen 2020). This dissertation will therefore analyze lock-ins on both the technological level and the political level, especially in terms of development cooperation, as it is a factor explaining interdependencies, power asymmetries and dependencies on protocols and actor requirements (Sect. 2.1.5, Sect. 2.3.6, Sect. 2.5.1).

After discussing possible benefits and challenges of digital technologies for development, the next section will introduce and analyze modes for governing the digital transformation in the context of ICT4D and digital development.

2.4 Solution (Micro Level): Coordination of Digital Ecosystem Activities through Network Governance for Sustainable Digital Development

Synthesizing the previously discussed 'D' in ICT4D, 'ICT' in ICT4D and barriers to ICT4D, this dissertation emphasizes the role of governance for achieving sustainable digital development in digital networks (ecosystems). It argues that network governance delivers prospective solutions for sustainable digital development of digital ecosystems in developing countries, which can be achieved through network governance. The interlinkage between sustainable digital development and governance is for instance highlighted by Meuleman (2019) assessing sustainable development as both "a laboratory for governance innovation" as well as "for policy innovation" (ibid.:7). This perception highlights governance as not an end, but a means to an end, which likewise has been elaborated in Sect. 2.2 with digital transformation as both a goal and a means itself. More specifically, this dissertation asks whether and how digital ecosystems can be governed for sustainable digital development. For this purpose it is important to highlight that network governance here is chosen as a coordination mechanism since the network is understood as a governance approach treating networks as the unit of analysis. Similarly, networks are also a mechanism of coordination (Provan and Kenis 2008:232).

This section will briefly elaborate on governance research with special regard to relevant concepts of governance in the digital transformation in addition to ICT4D literature (Sect. 2.1.2). These governance concepts offer insights into the analysis of political implications of the digital transformation. However, these digital governance concepts will only be discussed briefly, since the focus of this dissertation is more on the relation and interaction between multi-stakeholders in digital ecosystems in developing countries rather than on the inner workings of individual organizations. This likewise applies to the related concept of sustainability governance, which will be discussed briefly to approach sustainable (digital) development. The interplay of these governance theory approaches will be discussed and synthesized for the sustainable digital development perspective in the following.

2.4.1 Forms of Governance related to the Digital Transformation

Theoretical approaches to an analysis of the digital transformation and political implications are manifold. Given their scope and the research design of this dissertation, these will only be addressed briefly. To begin with, governance as *New Public Management* focuses on how to improve the performance of government by shifting the role of implementation to non-state actors (e.g., Dunleavy, Margetts, Bastow and Tinkler 2006; Dunleavy and Carrera 2013). This dissertation however does not aim at an analysis of internal public administration processes influenced by the digital transformation, but on network governance that focuses on the complex processes taking place in networks of public and non-public actors (Provan and Kenis 2008).

Secondly, the concept of *Digital-Era Governance* has emerged from the public administration perspective. Adding to the logic of New Public Management (NPM), authors such as Margetts and Dunleavy (2013); Clarke (2018); Dunleavy et al. (2006) have established understandings of digital-era governance. Thirdly, e-government is conceptually related with e-governance, e-participation and e-democracy, focussing further on citizen participation in public governance for instance via ICTs (Choi and Chandler 2020). E-government addresses the implications of (digital) innovations on the efficiency of public sector organizations and quality of their services (see e.g., Choi and Chandler (2020)).

Internet Governance (e.g., DeNardis 2014; Carr 2015) is possibly the most prominent of the digital governance concepts, especially with regard to multi-stakeholder internet governance (e.g., Hofmann 2016). Not least since the establishment of the IGF has Internet Governance been known to a wider audience.

Here, the debate for instance revolves around the internet as a public good and responsibilities for providing it (The World Bank 2016a).

2.4.2 Sustainability Governance of Digital Transformation

Shifting the governance focus from digital governance concepts to this dissertation's focus on sustainable digital development, *sustainability governance* is proposed in the literature for addressing sustainability challenges as conceptualized in the SDGs (e.g., Meuleman 2019). By transferring sustainability governance perspectives to the digital transformation (in the Global South), this dissertation aims at contributing to the current research gap as "[c]urrent models fall short of capturing such disruptive technological advances" (TWI2050 2018:76). As the TWI2050 (2018) highlights, reforms are needed for governing the transformations towards sustainability.

Moreover, *digital sustainability* is of importance in ICT4D literature, more specifically with regard to the sustainability of digital systems and the impact of digital systems on the sustainability of other systems (Heeks 2016:48–49). Hence, sustainability governance offers a further theoretical perspective of governing the digital transformation for a sustainable digital development in developing countries. This dissertation provides insights into the role of governing the digital transformation as a goal as well as an enabler of sustainable development. Fig. 2.5 demonstrates this interlinkages of the SDGs and governance.

Fig. 2.5 Interlinkage of SDGs and governance (own illustration based on Monkelbaan 2019:14)

More specifically, sustainability governance addresses global fragmentation especially from a stakeholder lens on the increasing number of public and private actors involved which "may be evaluated as problematic" (Glasbergen 2011:191). This is highlighted in this dissertation as one major barrier to sustainable digital development (Sect. 2.1.5) and the number of actors hence derived as one precondition for coordination through network governance Sect. 2.5.1.

Achieving the SDGs hence calls for "pluralistic approach, partnerships and 'bottom-up' action" (Monkelbaan 2019:28) instead of only a hierarchical approach (specifically of public sector actors). Here, governance play a crucial role: "Governance for the SDGs requires a variety and dynamic mixture of approaches, ranging from centralized to decentralized; from legally binding to voluntary; and from hierarchical to network and market styles of governance." (Monkelbaan 2019:28–29) As a consequence, this dissertation will analyze this proposed partnership formats in form of instruments PPPs and MSPs in the context of the digital transformation and sustainability from a governance perspective (see Sect. 2.5.4).

2.4.3 Network Governance

As illustrated in Sect. 2.1.2 and Sect. 2.2, networks play a crucial role in digital transformation, development cooperation but also in governance research. Firstly, the study of networks is not new in academic research and has increasingly attracted attention (e.g., Börzel 1998; Klijn, Koppenjan and Termeer 1995; Klijn 2008; Klijn and Koppenjan 2016; Isett, Mergek, LeRoux, Mischen and Rethemeyer 2011; Molin and Masella 2016b; Provan and Milward 2001; Provan and Kenis 2008; Sørensen and Torfing 2005; Sørensen and Torfing 2017). Generally, networks can be defined as a definable set of elements or nodes with one or more (social) relations (Jansen 2017). Depending on the theoretical lens, networks can be further distinguished for instance between formal and informal networks (e.g., Isett et al. 2011; Sørensen and Torfing 2009) or in terms of their density (strong or weak ties). This level of network analysis is for instance subject of ANT or network research, where methods include approaches to describe networks in terms of their density, distance or centrality (see e.g., Stegbauer and Häußling 2010). In development research, ANT has so far been hardly used (Heeks and Stanforth 2015), but more so in ICT4D and IS theory as ANT increasingly gains attention and influence (e.g., Avgerou 2017; Bon and Akkermans 2019; Walsham 2017). ANT for instance focuses on the way in which actors (both human and nonhuman such as technology) seek to build and maintain networks. While ANT similarly offers approaches of describing networks rather than offering insights into coordination or governance of these interactions, the only argument used from ANT is the argument of recognizing technology itself as 'actor' (e.g., Avgerou 2017) as it feeds into possible governance roles (Sect. 2.5.3) and formats (Sect. 2.5.4) for ecosystem coordination. Against this background, this dissertation analyses networks (here ecosystems) on a macro analysis level where it argues that the form of ecosystems has an influence on coordination for sustainable digital development. Rather than analyzing the network nodes themselves, this dissertation derives determinants for governance in digital ecosystems from network governance (e.g., Sect. 2.5.1).

Secondly, network governance offers the possibility to synthesize sustainability research, ICT4D and governance for answering this dissertation's research question. Networks form an important structure in the relation between digital and development as well as its governance. More specifically, the internet and digital infrastructure itself follow a network structure: "Because the digital infrastructure is networked, there is a natural affinity between ICTs and real-world network structures" (Heeks 2020a:7). This is further reflected for instance in 'knowledge networks' (e.g., Clegg et al. 2016) or 'knowledge society' and 'innovation networks' (e.g., Clegg et al. 2016; Desmarchelier et al. 2019).[8] Thereby this dissertation follows the current discourse in ICT4D and governance research concerning networks and offers (social science) insights into the conceptual role of networks in ICT4D: "For ICT this is even more obvious as it is a network technology itself, but evidently it is not limited to such cases." (Bon and Akkermans 2019) In specific reference to networks and network governance in the present *digital development context*, the emergence and availability of new (digital) technologies is ascribed the potential of changing perceptions, power relations, or institutional structures in governance networks and the interaction processes within these networks (Klijn and Koppenjan 2016, as further discussed in Sect. 2.3.6).

Another crucial factor for governing networks especially in the digital transformation are network effects describing that the value of networks increases with the number of users. In ICT4D this for instance applies to digital platforms (see e.g., Mansell 2015b; Allen 2017; Bass et al. 2013; Foster and Heeks 2013; Koskinen et al. 2019; Heeks 2020a) and more generally to ICT as "ICT's role in development includes its ability to create network effects, generating a network of knowledge that can be shared and is open to a larger stakeholder group" (Bass et al. 2013:22). On the contrary however, network effects can likewise prevent actors from entry (here into digital ecosystems), create user lock-in and dis-/re-intermediation (Heeks 2020a).

---

[8] For further insights into technological research on networks, Camarinha-Matos, Afsarmanesh and Antonelli (2019) for instance demonstrate network collaboration from technical IT perspective in fields of industry 4.0, big data analytics, open innovation, services, food and agribusiness and value creation networks.

Those are further examples of potentially inhibiting sustainable digital development. More specifically with regard to this dissertation's focus on coordination in digital ecosystems, ICTs are discussed as driving or enabling every type of network and changing the role of hierarchies in coordination: "ICTs have levelled a part of the playing field: hierarchies were traditionally better than self-organized networks at tasks such as coordination or focusing resources, and, with ICTs in place, this is no longer the case" (Heeks 2020a:8 based on Castells 2000). This will be further discussed with regard to governance formats in the context of digital ecosystems in Sect. 2.5.4. Moreover, what is hence crucial for the effectiveness and the success of networks, is *collaboration* between network participants i.e., multi-stakeholders. Medaglia et al. (2018) highlights the importance of collaboration by arguing that "[...] in these collaborations, it seems that the governance structure is one of the key issues to solve, particularly understanding how governance works in a particular context." (ibid.:9) As indicated in Sect. 2.4, in order to use the driving potential of digital technologies and (digital) innovation for development, collaboration in the sense of shared knowledge has been derived from the literature as crucial for governing the digital transformation sustainably. The value of collaboration in the digital transformation is further highlighted in Sect. 2.2 for instance with regard to networks and collaborative innovation (e.g., Desmarchelier et al. 2019) or collaborative platforms (Ansell and Gash 2018), which this dissertation identifies as one potential governance format for sustainable digital development (Sect. 2.5.4).

Thirdly, besides the network character of ICTs, this dissertation argues that in addition to the natural affinity between networks and ICT, networks likewise occur through collaborating sets of stakeholders. As multiple stakeholders form a network (e.g., Bevir 2011, as discussed in Sect. 2.1.4), effective stakeholder management is likewise important for ensuring sustainable (digital) development. This is for instance highlighted as "[e]ffective identification, understanding and management of stakeholders, their triggers and their expectations will improve the ability to reduce risk, tailor mitigation measures and achieve the SDGs" (Monkelbaan 2019:122). Hence network governance further serves as a mechanism for achieving coordination in the ecosystem from an actor perceptive on networks as digital ecosystems likewise form networks as elaborated in Sect. 2.2.5. Conclusively, this dissertation includes a focus on networks being formed by stakeholders (Sect. 2.1.4).

Fourthly, there are various theoretical approaches to networks in the literature, specifically in order to "conceptualise the trend towards governing through and in networks" (Klijn and Edelenbos 2007:199). Generally, networks can be defined as a "stable articulation of mutually dependent, but operationally autonomous actors [...], who interact through conflict-ridden negotiations that take place within an institutionalized framework of rules, norms, shared knowledge and social imaginaries [...] and contribute to the production of public values" (Sørensen and Torfing 2009:236). From a network governance perspective and most importantly for this dissertation's research question, the emergence of networks can be further explained by enhancing integration and coordination between services as summarized in Molin and Masella (2016a). The literature mostly distinguishes between policy networks, network governance and network management (e.g., Atkinson and Coleman 1992; Klijn et al. 1995; Klijn and Koppenjan 2016; Molin and Masella 2016b; Provan and Kenis 2008; Rhodes 1997; Molin and Masella 2016b), which often are used synonymously. What these three have in common is that they all analyze reasons for the development of networks (Molin and Masella 2016b). Governance research further highlights that these network-related explanatory and research strategies can be combined with other theoretical approaches (Jansen and Wald 2007). More specifically with regard to this dissertation's focus on networks in the digital development context, innovation transfer and underlying information and knowledge sharing take place through networks (e.g., Ciesielska and Janowski 2019, as further highlighted in Sect. 2.3.6). Hence, networks in which interdependent actors collaborate, such as digital ecosystems, are seen as a "valuable framework for enhancing collaborative innovation" (Sørensen 2014:1).

On the one hand, the network literature in public administration has examined the roles of interaction, collaboration, trust (e.g., Agranoff 2007; Provan and Milward 2001; Provan and Kenis 2008; Klijn and Koppenjan 2012), resource dependencies (Sect. 2.1.1) and power dynamics in the public and nonprofit sectors as similarly argued by Whetsell, Leiblein and Wagner (2020). More importantly for this dissertation, power constellations are dissected as crucial factors from the literature review of both development cooperation and the digital transformation (Sect. 2.3.6). This is another factor explaining this research project's choice of network governance since in network governance "the relevant value is the allocation of power between the governing actors, allowing them to act as a legitimated controllers over the others" (Molin and Masella 2016b:831). This further includes traditional authoritative power such as resources allocation, laws definition, but also skills and capacities (e.g., Sørensen and Torfing 2005). Hence, this perspective provides theoretical contributions towards this dissertation research question of how digital networks can be coordinated for sustainable digital development. Given their relevance in network governance and in ICT4D, the factors trust, resources and power relations among others have been derived as important determinants for this dissertation's framework of governing digital ecosystems (Sect. 2.5).

On the other hand, a specific network governance approach to digital ecosystems is lacking, especially in the synthesis with sustainability and development research. Hence this dissertation has developed a framework for governing digital ecosystems and its multi-stakeholder activities in the context of developing countries for sustainable digital development, which will be discussed in the next section while synthesizing the previous findings of development context (Sect. 2.1), digital technologies as an influencing factor (Sect. 2.2) and challenges to ICT4D (Sect. 2.3). Additionally, network governance will be further expanded with regard to specific governance roles of actors in the ecosystem (Sect. 2.5.3) as well as further governance formats for coordination (Sect. 2.5.4).

2.5 Framework for Coordination of/in Digital Ecosystems for Sustainable Digital Development

This dissertation not only uniquely synthesizes development, sustainability and governance research, but also develops a framework for coordination of and in digital ecosystems in developing countries for achieving sustainable digital development. From the network governance perspective (Sect. 2.4.3), the emergence and functionality of networks as a form of coordination is dependent on both the characteristics of networks but also on the ecosystem's participants with related embeddings such as trust or resources (see e.g., Jansen and Wald 2007). The framework incorporates four main determinants debated within network governance literature identified by Molin and Masella (2016a): 1) starting conditions, 2) form of network governance, 3) meta-governance and 4) the outcome of network governance. These four have been expanded and adapted by this dissertation to the digital development context. More specifically, preconditions for network governance (determinant 1, Sect. 2.5.1) will be elaborated in the following with regard to digital development. These are not only derived from network governance (Sect. 2.4.3), but also from the previous chapters on the context (Sect. 2.1) and digital transformation potentials (Sect. 2.2). As a second part of this framework, challenges for coordination of digital ecosystems (Sect. 2.5.2) are introduced incorporating both network governance challenges but more specifically ICT4D challenges (Sect. 2.3). Thirdly, governance roles (Sect. 2.5.3) are elaborated that mainly originate from network governance literature but are supplemented by ICT4D perspective on actors' roles in digital ecosystem and possibly in governing digital ecosystems. Fourthly, governance formats for coordination of digital ecosystems are introduced both based on governance literature and with regard to development research. These ultimately influence sustainable digital development (determinant 5/outcome, Sect. 2.5.5). The framework is further illustrated in Fig. 2.6. Ultimately, this framework serves as analytical basis for analyzing the digital ecosystem of Kenya (Chap. 4, Sect. 4.4).

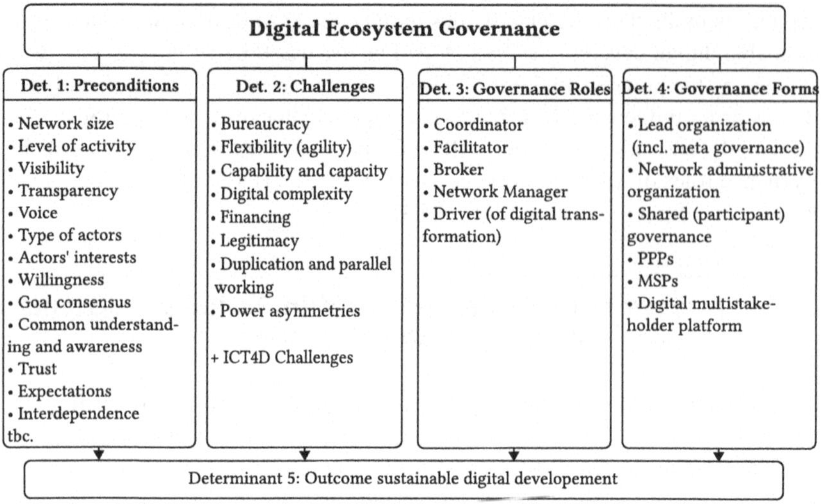

Fig. 2.6 Theoretical framework: Digital ecosystem governance for sustainable development (own illustration)

2.5.1 Determinant 1: Preconditions for Network Governance

As demonstrated, networks are an important structure for governing digital development. The here derived preconditions can be drivers or enablers i.e., their absence can cause failure and their presence can cause success. The success of network governance is recognized to be largely dependent upon the form and functioning of the relations between different organizations in the network. Hence, interaction of multi-stakeholders in the network (here ecosystem) leads to 'success' if interactions are "[...] free from conflict and power struggles, and based on continuous co-ordination among all relevant and affected actors" (Wilson, Morrison and Everingham 2017:502–503 based on Sørensen and Torfing 2005). However, as shown in Sect. 2.3.6, settings with power and knowledge asymmetries prevail and hinder the success of networks, which is why they build a factor of determinant 2 (challenges).

Against this background, the following preconditions or dimensions are derived from the literature and are adapted to digital ecosystems. These preconditions are similar to the literature review of (Molin and Masella 2016b,a) that includes trust, size of the network, goal consensus, joint action and motivation, leadership and commitment as well as embeddedness and diversity. While determinants such as size of network usually are usually analyzed quantitatively (as discussed in the previous section), this dissertation however follows the argument of Klijn and Koppenjan (2016) that these are equally necessary for qualitative understanding actor relations. Based on the previous sections, especially on network governance and network analysis literature (Sect. 2.4 and ICT4D), this dissertation derives the following preconditions for governance of and in digital ecosystem for sustainable digital development:

- Network size (number of actors)
- Level of activity
- Visibility
- Transparency
- Voice
- Type of actors
- Actor's interest
- Willingness (also motivation)
- Goal consensus
- Perceptions: Common understanding and awareness
- Trust
- Expectations
- Mutual dependency (here interdependence)
- Frequency of interactions
- Duration of actors' relationships
- Ease of entry (inclusive vs. exclusive ecosystem)
- Intensity (closeness and distance)
- Degree of centralization (also embeddedness)
- Network's rationale
- Network effectiveness and efficiency

This list likewise determines the structure of this subsection, where each of these preconditions will be analyzed individually in the following. From the ICT4D, differing interests of actors (Sect. 2.3.3), common understanding (Sect. 2.2.1, Sect. 2.1.1) and (inter-)dependencies (Sect. 2.1.5, Sect. 2.3.6) have been derived as ICT4D challenges and hence function as both preconditions (from network governance perspective) and as challenges (from ICT4D perspective).

Size (Number of Actors)

As a first precondition, the network size determined by the number of actors in the network (here digital ecosystem) is crucial for coordination through network governance. More specifically, the size of the ecosystem not only refers to the amount of multi-stakeholders in the ecosystem, but also to their possible activities and connections (Provan and Kenis 2008). Against this background, when the number of actors increases, their activities and interactions increases exponentially, which affects coordination of activities in adding complexity to governance of activities within the network (e.g., Provan and Kenis 2008:238, Molin and Masella 2016a:497). Thereby also complexity increases since "[...] the needs and activities of multiple organizations must be accommodated and coordinated." (Provan and Kenis 2008:238) Additionally, this leads to a "trade-off between the size of the group of actors and the added value that can be achieved" (Klijn and Koppenjan 2016:162) because one the one hand, with the increasing number of actors also increase costs and risks of the interactions, but on the other hand, a small number of actors leads to the risk of exclusion for instance of actors or resources. Hence, the size of the network has implications on the performance of network governance depending on which governance form is chosen. Academic research here revolves around the vault lines of formalization degrees of governance (high to low degree of formalization or top-down to bottom up), as in networks with a high number of actors the need for formalization or top-down governance roles increases (Span, Luijkx, Schols and Schalk 2012:193). The correlation between network size and network governance form will be discussed in more detail in Sect. 2.5.4. Most importantly for this dissertation's analysis of the (Kenyan) digital ecosystem in the context of sustainable digital development, the increasing number of actors in the digital ecosystem poses a specific importance. Here the network governance perspective on the network size is supplemented by findings of Sect. 2.1.5 and Sect. 2.1.4 with an increasing number and types of actors in digital development. Moreover, the number of multi-stakeholders involved is important for sustainable digital development in terms of inclusion and even a possible enlargement of multi-stakeholders that influences coordination (e.g., Keijzer et al. 2018).

Level of Activity

Closely linked to the number of actors in the ecosystem, is their level of activity which also is an important factor that needs to be assessed for network analysis (Klijn and Koppenjan 2016:263). Moreover, the activity level of actors is likewise part of the definition of actors, and here (multi-)stakeholders, in that actors are characterized by actively influencing for instance content, course and outcomes of social interaction processes (Scharpf 1997; Klijn and Koppenjan 2016). Nonetheless actors "[...] who are currently not actively acting and who are not present in arenas of decision making, may decide to do so in the future" (Klijn and Koppenjan 2016:73) and are still affected by processes or polices even without participating. This classifies them as passive (rather than active) actors. Especially in ICT4D there can be various reasons why actors choose or are not able to participate, for instance due to lack of access and ease of entry (Sect. 2.5.1), interests (Sect. 2.3.3), asymmetries in (digital) knowledge and power (Sect. 2.5.2). The activity level of multi-stakeholders in this dissertation likewise follows a methodological goal as it is used as one dimension of a proposed matrix for analyzing the correlation between activity and influence in digital ecosystems (Table 3.3), which will be further completed with empirical data from the Kenyan case (Table 4.1).

Visibility of Actors

This dissertation argues that the level of activity in the digital ecosystem can likewise be determined by the visibility of actors, especially against the background of development cooperation and also digital transformation (Sect. 2.2.3). Visibility here refers to the visibility of actors, not necessarily the visibility of innovations or ICT per se. At a first glance, it appears that only the actors who are visible in the ecosystem are part of the ecosystem and possibly are the (more) active ones. More specifically, in the network governance literature, actors from the periphery of the network or ecosystem for instance are assessed as less visible and under less pressure to conform to established norms and routines (Jansen 2017:35). This emphasizes the interlinkage between centralization (Sect. 2.5.1) and visibility of actors.

From a network governance perspective, visibility is further crucial for identifying actors in the ecosystem. Moreover, this includes perspectives of development in so far as visibility (and transparency of action) is a crucial aspect of both aid effectiveness (Sect. 2.1.5) and sustainability and inclusivity. The angle on visibility from a transparency perspective on interactions will be discussed in the following subsection. More importantly, the development landscape is characterized by "visibility pressure" (Lundsgaarde and Keijzer 2019:230), especially of donor organizations. Visibility can hence be understood in terms of accountability as a motivational factor (Sect. 2.5.1), actor proliferation and participation in 'hype' sector (Sect. 2.2.3, Sect. 2.1.6) or interest of actors (Sect. 2.3.3). With regard to public sector actors, this for instance includes visibility of donor programs (Klingebiel, Mahn and Negre 2016). For private sectors visible activities include CSR activities among others (Sect. 2.3.3).

Transparency

Closely linked to visibility of actors is visibility of their actions in digital development—or else: transparency. With regard to the D in ICT4D, transparency is one essential theme of development cooperation, especially in the context of sustainable development, as discussed in Sect. 2.1.6 and Sect. 2.1.7. Transparency of actions, especially of public sector actors, is a further determinant for political stability as discussed in Sect. 2.1. Additionally, ICT are ascribed the potential for increased transparency for instance through open data or open source (Sect. 2.2.4).

Synthesizing these perspectives with transparency in network governance, transparency is argued to be "[...] key to building trust and mitigating power- asymmetries and should thus be promoted by engaging the members in open and transparent communication, decision-making, and evaluation" (Pattberg and Widerberg 2016:47). Against the background of the diverse set of multi-stakeholders in the digital ecosystem as well as for analyzing their activities in digital transformation and digital development, this dissertation analyzes transparency differences also with regard to the role of transparency for the involved actors. From a network perspective and referring to Sect. 2.1.4, differences in terms of transparency are observed, especially between public and private actors: "Regarding information, public parties will emphasize the need for transparency and openness. This is not self-evident to private actors. They may fear that company-sensitive information or innovative ideas will be leaked to competitors, or that information provided to public parties will be used by them to propose unfavourable regulations or contract conditions" (Klijn and Koppenjan 2016:167). Thus, transparency has an impact on the interactions and relations between multi-stakeholders and hence for coordination of their actions so that this is identified as an essential precondition for governing digital ecosystems.

Voice

Levels of activity and visibility can be further determined by the factor of voice. From a network governance perspective, "[v]oice is complex because many actors are involved and clear rules are often lacking" (Klijn and Koppenjan 2016:220). Generally, voice (especially of citizens) is an essential part of democracy. More importantly for this dissertation, voice is also part of governance indicators such as accountability and political stability (Sect. 2.1.2). The digital transformation likewise influences 'voice', most prominently with regard to digitally-enabled social movements that also give voice to alternative politics. In this dissertation's framework, voice however is neither understood technically (e.g., digitally enabled communication via voice message or voice control) nor understood from a democratic perspective based on the voices of citizens. Instead, voice here is operationalized on the organizational level of analyzing multi-stakeholder organization's activities in the digital ecosystem in terms of expressing activity and exerting power, for instance determined by participating in the ecosystem (and e.g., in activities such as policy processes). Additionally, it is essential not only to determine who exerts their voice in digital development processes in the ecosystem, but also whose voices is listened to (determinant of inclusion as part of sustainable digital development).

In particular, the actor group of NGO are attributed as a counterbalancing power (Sect. 2.1.4). Similarly to the activity level, voice further follows a methodological goal as it serves the second indicator for assessing influence of multi-stakeholders in digital ecosystems (see further Table 3.3), which will be further completed based on empirical data from the analysis of Kenya.

(Diverse) Type of Actors

Following up on this interconnectedness of the number and also the types of actors in the analysis of digital ecosystems, the type of actors in the network /ecosystem and their diversity are a further important precondition for network governance (e.g., Span et al. 2012; Molin and Masella 2016b,a). Differences between the actors are for instance manifested in the nature of the organization (governmental or non-governmental), operational values, skills, and knowledge, which are relevant to how the network can best be governed (Span et al. 2012:193). The diversity of actors is especially crucial for this dissertation's analysis of digital development or ICT4D context which involves a range of multi-stakeholders, and also especially actor groups of donors and private tech actors (Sect. 2.1.4, Sect. 2.1.5). Hence, the type of actors range from governmental to non-governmental actors—both being essential for addressing complex and fragmented 'wicked' problems or social needs (here: needs associated with digital transformation and development).

Coordination in form of network governance here is especially crucial since a), the different involved actors in the ecosystem can have conflicting interests that need to be balanced for the network's efficiency and effectiveness (Molin and Masella 2016b:833) and b), the different actors come with different expertise, skills, and other diversity-increasing attributes, especially in the context of the digital transformation, development cooperation and ICT4D. These factors around the type of actors likewise have an impact on the governance form (top-down or bottom-up, as further discussed Sect. 2.5.4) (Span et al. 2012): The governance of a heterogeneous network is discussed on the network literature, with Span et al. (2012) for instance highlighting that in these heterogeneous networks it is more suitable to achieve network goals if it follows a bottom-up governance approach. Here, not individual actors are addressed in network governance, but organizations that represent single interests or points of view (Molin and Masella 2016b:833). A diverse set of stakeholders in the ecosystem is especially needed in settings of very complex and wicked issues (Raeymaeckers, Vermeiren, Noël, Van Puyvelde and Willems 2020), in this case issues such as aid fragmentation and imbalances of power and knowledge in ICT4D.

As an antidote to diversity serves the emphasis on similarity or homophily in the network literature, with the purpose for instance of explaining why certain actors are attracted to others and hence possibly interact more (Provan and Kenis 2008:239). This will be further discussed in more detail under 'degree of closeness and distance' (Sect. 2.5.1).

Interests of Actors

As alluded in the previous subsection on the type of actors, the various multi-stakeholder groups likewise follow differing, often conflicting interests (e.g., McGuire and Agranoff 2011; Klijn and Koppenjan 2016; Molin and Masella 2016b; Sørensen and Torfing 2017). Actors' interests are highlighted in network governance literature as one barrier to achieving coordination of actors' activities or processes (Klijn and Koppenjan 2016:89). This impacts the performance of the network and requires governance for instance through identifying compromises (Molin and Masella 2016b:833). The implications of differing interests between the multi-stakeholder in digital development or ICT4D have been discussed in more detail in Sect. 2.3.3, but due to their importance are here argued to be a further essential determinant for governing digital ecosystems. From the ICT4D literature analysis, interests of actors have likewise been identified as ICT4D barriers to sustainable digital development (Sect. 2.3.3).

Willingness

Closely linked to the network actors' interests is the precondition of willingness. Willingness here is understood with regard to motivation, commitment, and also interest of actors as mentioned beforehand. Molin and Masella (2016b) for instance highlights commitment or motivation as a precondition which is enhanced by goal consensus (goal consensus will be discussed in more detail in the next subsection). Motivation (or willingness) has an impact on network governance as it enhances the quality and strength of relationships. Willingness is further preconditional for network governance since the actors' activities are both shaped by "[...] the extent to which they are passionate about the problem or the solution, and their willingness to use their resources" (Klijn and Koppenjan 2016:270). Hence, if actors are unwilling to participate or to invest in the process or if there is a conflict, this leads to impasses or blockages (Klijn and Koppenjan 2016:84). In sum, combined willingness, capability, and resources of partners are stated as factors for successful networked governance (Pattberg and Widerberg 2016:46). With regard to governance forms, there is debate in the literature concerning top-down or bottom-up governance forms. As a brief insight, Molin and Masella (2016b) argues (in reference to Provan and Kenis (2008)) that in networks with low commitment "[...] shared form of network governance could perform better since decision-making power is diffused among participants" (Molin and Masella 2016b:502), while Klijn and Koppenjan (2016) provides further insights that actors are only willing to interact and collaborate under the threat of hierarchical interventions. More specifically to this dissertation's analysis of digital development, the willingness of actors to participate in the digital ecosystem and essentially to participate in the digital transformation is a further issue for this dissertation with regard to public sector actors and political contexts (Sect. 2.1), which will be discussed in more detail in Sect. 2.5.2.

Goal Consensus

Closely linked to actors interests (Sect. 2.3.3) is the next precondition for coordinating networks in terms of goal consensus, as identified for instance by Molin and Masella (2016a). Goals here include both organizational goals (although in this dissertation these are more subsumed under actors interests) and network-level goals such as attracting network-wide funding, but also processes-orientated goals such as reducing competition or conflict among participants (Provan and Kenis 2008:239). According to Provan and Kenis (2008), for reaching goals the goals of network members do not necessarily have to be similar, since "[...] similarity of purpose can result in difficulties in working together, especially when competitive pressures make network organizations reluctant to cooperate and share information [...]" (Provan and Kenis 2008:239). A lack of alignment of goals (here of organizational goals with network goals) can lead to difficulties in establishing goal consensus and to conflicts (Raeymaeckers et al. 2020:1039). This constitutes alignment of goals (and to the political context) as one indicator for assessing network governance effectivity (Sørensen and Torfing 2009). Hence, goal consensus impacts the network performance as the actors are more likely to collaborate and to commit to the network (Raeymaeckers et al. 2020:1039). Raeymaeckers et al. (2020) further argue that an absence of goal consensus and difficulties in collaboration of actors can lead to the failure of the network while Provan and Kenis (2008) argues that networks can still be effective with only moderate levels of goal consensus. Goal consensus among actors in the network is at least argued to be easier the more similar actors' perceptions are (Klijn and Koppenjan 2016:103). The correlation between goal consensus and perceptions (here especially common language and understanding) will be discussed further in Sect. 2.5.1.

With regard to this dissertation's focus on coordination through network governance, in networks where actors can agree on network goal, self-governed governance forms are most likely to be effective since they can collaborate while still attaining their own goals (Provan and Kenis 2008:240). This will be discussed in more detail in Sect. 2.5.4. The authors further highlight the correlation between goal consensus and the before mentioned trust, which are not necessarily related to each other since "[t]rust is based on reputation and past interaction experience, whereas consensus is based on goal similarity" (Provan and Kenis 2008:240). On the contrary, in networks where goal consensus is lacking, governance by lead organizations or Network Administrative Organization (NAO) are more effective, however more on a short-term basis (Provan and Kenis 2008:240). For reaching goal consensus, Raeymaeckers et al. (2020) emphasize the role of the network governor or coordinating actor while simultaneously considering the diversity among the network actors. The governance roles will be discussed in more detail in Sect. 2.5.3.

Perceptions: Common Understanding and Awareness

Further linked to interests of actors in the digital ecosystem are their perceptions. In line with Klijn and Koppenjan (2016), perceptions in this theoretical framework are defined as "[...] different understandings of the content (nature of the problem, desired solutions, and so forth) and of the other actors (for instance, actors' power and competence)" (ibid.:104), which are created through their experiences and interactions. For this dissertation both perspectives are crucial and will hence be distinguished between *common understanding* of multi-stakeholders in the digital transformation and their *awareness* of each other as well as of each others activities. Thus, networks can be further characterized by the degree of shared perceptions between actors (Klijn and Koppenjan 2016:104) since outcomes of network governance for instance imply a certain degree of joint perceptions and a shared language, among others (Klijn and Koppenjan 2016:95).

Moreover, common language and understanding has been further identified by this dissertation as a important determinant for achieving coordination for sustainable digital development. This is further highlighted for instance in the "need to develop a common language for understanding and describing the outcomes and impact of activities related to private sector engagement" (OECD 2018b:10), and in "[...] at least at a certain level a joint understanding of the situation is needed" (Klijn and Koppenjan 2016:104). These are required for solving problems and finding solutions, especially since awareness of a problem can be perceived in various ways (Klijn and Koppenjan 2016:64–65). This likewise can lead to wicked governance challenges due to a fragmented knowledge base and a lack of consensus among stakeholders as to how the underlying problems should be defined and solved. This dissertation argues that this is especially relevant for ICT4D considering the wide range of actors (donors, private tech companies, NGOs etc.) that all have a different background and hence draw on different bases of knowledge (Mulder 2020). Given each actor's limited views and limited capacities to influence outcomes, "[e]ffective network governance requires interactive learning by and between all of these stakeholders" (Monkelbaan 2019:34). However, against this background, actors "will not always decide to collaborate and coordinate their strategies in order to arrive at joint policies for co-produced services" (Klijn and Koppenjan 2016:152). Hence, the downside of joint actions in ecosystems relates to interaction costs (investment of time and money) or transaction costs (Klijn and Koppenjan 2016:152). This dissertation will deliver empirical insights into this trade-off of the role of common perceptions and joint actions with regard to the Kenyan digital ecosystem (Sect. 4.4.4).

In sum, this dissertation argues that an awareness as well as common understandings of the implications of the digital transformation on the development of developing countries are essential for sustainable digital development and for coordination. This is especially relevant considering the complexity of the digital transformation (Sect. 2.5.2, Sect. 2.2), knowledge and power asymmetries that foster asymmetries in understanding and awareness (Sect. 2.3.6) and the discourse around hype (Sect. 2.2.3). Common language as a precondition for network governance is likewise derived from ICT4D discourse, but is identified there as a challenge (Sect. 2.2.1, Sect. 2.3).

Trust

Another important precondition for network governance and understanding network-level interactions of digital ecosystems is trust among network members (Provan and Kenis 2008; Molin and Masella 2016a). Hence networks can be characterized by the degree of trust between actors in the ecosystem (Klijn and Koppenjan 2016:116). Trust can for instance be defined as the perception of the good intentions of other actors (Klijn and Koppenjan 2016:115). Trust in this dissertation is especially important as it not only has a (positive) effect on cooperation in facilitating interaction, but also is an important condition for realizing innovation (e.g., Klijn and Koppenjan 2016). More importantly with regard to this research question of coordination for sustainable digital development, trust can be "[...] a possible coordination mechanism for complex decision making in networks and supports the need for innovative solutions in decision making" (Klijn and Koppenjan 2016:119). Therefore trust is not only essential for coordination, but likewise enables (digital) innovations as it reduces uncertainty about opportunistic behavior (ibid.). Similarly, uncertainty in the digital innovation setting can arise due to knowledge and power asymmetries between multi-stakeholders in the digital ecosystem as discussed in Sect. 2.3.6.

On the one hand, trust has implications for the *performance* of the ecosystem for instance in terms of reducing transaction costs (Klijn and Koppenjan 2016:116), enhancing network performance (Klijn and Koppenjan 2016:117) and network effectiveness (Molin and Masella 2016b), or improving knowledge exchange and learning (Klijn and Koppenjan 2016:118)—all of which are highlighted as essential for sustainable digital development (Sect. 2.3.6). Nonetheless, the absence of trust likewise affects the network performance in the case of complex environments or processes as well as hostility between actors that hinder coordination of their actions (Klijn and Koppenjan 2016:97).

On the other hand, trust is essential for network performance and *sustainability* (Provan and Kenis 2008:237). Sustainability in this dissertation revolves around sustainable digital development, which is one area of implementation pf the SDGs as discussed in Sect. 2.1.3. Here, "[t]rust is thought to be especially important when dealing with the complexity involved in implementing the SDGs because vertical means of control are difficult given the interdependencies between actors" (Monkelbaan 2019:136). Moreover, trust is the basis for institutionalized governance formats such as MSP or PPPs, especially in digital development, which will be discussed in more detail in Sect. 2.5.4 and Sect. 2.5.4. In networks where trust is absent, "[...] shared governance will not be effective since there will be little basis for collaboration among network members" so that "[...] network governance is likely to be brokered, either through a lead organization or through an NAO" (Provan and Kenis 2008:238). This will be discussed in more detail in Sect. 2.5.4.

Expectations

Closely linked to the precondition of trust between network actors are their expectations, since trust implies "[...] an expectation about the intention of another actor and entails the expectation that the other actor will respect the interests of the 'trusting' actor" (Klijn and Koppenjan 2016:115). Molin and Masella (2016b) (based on Vangen and Huxham 2003) similarly highlights the interrelation between trust and expectations arguing that a cycle of trust, expectation and risks strengthens collaboration between network actors. Moreover, the needs and expectations of networks actors need to be satisfied by the network (Provan and Milward 2001:417). Therefore expectations are another important factor for coordination through network governance which involves expectation management of multi-stakeholders in the network. This dissertation argues that expectations of ecosystem members is especially crucial for sustainable digital development regarding the discussed expectations aligned to the digital transformation (Sect. 2.2), and specifically expectations around what ICTs can do and what not (Sect. 2.2.3). Expectations thereby are further linked to (digital) knowledge and interests.

Interdependence (Mutual Dependency)

Furthermore, mutual dependency or interdependence between the actors in the network (ecosystem) are essential for coordination through network governance. Mutual dependency (Molin and Masella 2016b) and interdependence are both used interchangeably, while this dissertation will follow the terminology of interdependence (e.g., Klijn and Koppenjan 2016; Monkelbaan 2019; Provan and Kenis 2008; Sørensen and Torfing 2009). In the reviewed literature, interdependence is not only seen as the starting point for any networked governance process (Monkelbaan 2019:137), a measure to assess network governance effectivity (Sørensen and Torfing 2009), but also an incentive for collaboration (Monkelbaan 2019:137). Hence the network further comprises of a web of relationships between autonomous yet interdependent government, business, and civil society actors (Klijn and Koppenjan 2016:11). Interdependencies are especially manifested with regard to resources, since resources necessary for solving problems are owned by different actors (Klijn and Koppenjan 2016). With regard to this dissertation's focus on ICT4D, ICTs are assessed to even strengthen interdependencies (e.g., Jansen 2017, Sect. 2.3). This dissertation argues that these interdependencies between actors in the digital ecosystem are especially high due to the resources required for participating in the digital transformation (Sect. 2.1.1), particularly financial resources and (digital) know-how (Sect. 2.3.6), which likewise contribute to power constellations (Sect. 2.3.6). Demonstrating the interdependencies between actors, Klijn and Koppenjan (2016) have developed a typology of dependency relations as illustrated in Table 2.1.

Table 2.1 Typology of dependency relations between actors and resources (own illustration based on Klijn and Koppenjan 2016:75)

	Substitutability of the resource	
Importance of the resource	High	Low
Great	Moderate dependency	High dependency
Snall	Low dependency	Moderate dependency

Table 2.1 shows that the greater the importance of the resource and the lower the substitutability of the resource, the higher is the mutual dependence between actors. This can be transferred to this dissertation in that digital technologies and services (with respective financial and human capital resources as discussed e.g., in Sect. 2.5.2) are crucial for digital development and are not easy to substitute, for instance given user-lock ins (Sect. 2.3.6) and unclear suitability of open source as an alternative (Sect. 2.2.4). According to Klijn and Koppenjan (2016), based on Alrich (1979), "[o]rganizations can try to avoid or influence interdependency by 1) acquiring crucial resources, 2) finding alternative resources, and 3) acquiring authoritative powers to coerce other actors or changing ambitions and goals" (Klijn and Koppenjan 2016:30). Interdependence of multi-stakeholders not only poses a chal-

lenge to sustainable digital development (Sect. 2.3.6), but likewise to coordination. Here, Klijn and Koppenjan (2016) distinguish between coordination (governance) formats depending on pooled interdependence (governance form: standardization), sequential interdependence (dependence input on output form others, governance form: coordination 'by plan') and a mutual dependency relationship (outputs constitute inputs for others; governance form: coordination by mutual adjustment) (Klijn and Koppenjan 2016:31). Governing formats for interdependencies in digital ecosystems are discussed in this dissertation's Sect. 2.5.4.

This dissertation further follows Klijn and Koppenjan (2016) assessment of interdependence as closely linked to frequency of interactions through which stable interaction patterns are created: "These patterns are often quite logical in the sense that actors who interact frequently with one another are also dependent on one another's resources to achieve their goals." (Klijn and Koppenjan 2016:103). Synthesizing the previous determinants, the authors argue that "dependencies between actors can lead to more or less durable interaction patterns, shared perceptions (more specifically trust) and the emergence of institutional rules that guide behaviour" (Klijn and Koppenjan 2016:292). Duration and frequency of interactions will hence be further discussed in the following.

Frequency of Interactions

Expanding on the correlation between interdependence and frequency of interactions, frequency as a further precondition for network governance is mostly used to describe the structure of the network or interaction patterns between organizations (among other relational factors such as intensity, centrality, which will be discussed in the following) (Klijn and Koppenjan 2016:24). More specifically, frequency demonstrates how often organizations have contact with one another ranging from very frequent to sporadic (Klijn and Koppenjan 2016:281). Another dimension of frequency refers to frequency of interaction with regard to other actors (with whom): "For instance, actors can interact frequently with a very limited set of actors in the network, but they can also interact infrequently with a large array of other actors (which can be organizations or individuals for instance)" (Klijn and Koppenjan 2016:281). Hence frequency is not directly a precondition for inclusiveness (Sect. 2.1.3), but can be an indicator of exclusiveness or silos (Sect. 2.5.1) in cases where actors only interact with a certain actor group.

Nonetheless, frequency in the network governance literature is an indicator for stability in contacts, that is for instance influenced by interdependence and repetitiveness of interaction in networks: "[...] actors who interact frequently with one another are also dependent on one another's resources to achieve their goals." (Klijn and Koppenjan 2016:103) Thereby decreasing or even non-existent interactions are a challenge for coordination since they can potentially fuel social impasses, especially when factoring in the diversity of actors (Sect. 2.5.1) and their conflicting interests (Klijn and Koppenjan 2016:276, see further Sect. 2.3.3). Another factor that influences the frequency of interactions in a network is the size of the network as previously discussed in Sect. 2.5.1 since "[i]n a small network it is also easier to maintain contact with organizations on a frequent basis as well as to evaluate the contributions of all parties involved" (Span et al. 2012:192), which potentially decreased transaction costs.

Duration of Multi-stakeholders' Interactions

Closely linked to frequency of actors' interactions in the ecosystem, is the duration of their interactions and relations. This is especially relevant for this dissertation's focus on sustainable digital development, where long-lasting impact is one dimension of sustainability (Sect. 2.1.3). More specifically, networks/ecosystems can be characterized for instance as short- or long-lived (Sørensen and Torfing 2009). Moreover, the relationships between multi-stakeholders in the ecosystem are directed at service delivery in the digital transformation and hence are thought to be permanent (e.g., Sørensen and Torfing 2009; Isett et al. 2011; Molin and Masella 2016b.

The impact of the duration on network governance performance and as a precondition for network governance is discussed critically since

> "[t]he duration of a round or a process does not tell us a lot about the success or the quality of a process by the way. A fast process may result in a substantive solution, policy, or service that does not meet standards. A prolonged process may produce products of high quality." (Klijn and Koppenjan 2016:95)

This is for instance criticized with regard to the long duration of decision-making processes (Klijn and Koppenjan 2016:131) and hence serves as one indicator for process evaluation. Duration of multi-stakeholder activities is likewise addressed with regard to bureaucracies and their implications for sustainable digital development in Sect. 2.5.2.

Ease of Entering (Exclusivity vs. Inclusivity)

Another relational precondition identified in the literature review of network governance is the degree of inclusiveness or exclusiveness of the ecosystem (e.g., Provan and Milward 2001), or also known as open vs. closed networks (Sørensen and Torfing 2009). In this dissertation, inclusion of actors in the ecosystem is analyzed for instance with regard to their inclusion and consideration of their interests, but also as an angle for explaining to which degree entrance in the ecosystem is possible or not. More specifically, the network (ecosystem) can be characterized as inclusive in terms of including a variety of actor (groups), which however can be challenging (Klijn and Koppenjan 2016:226) as also discussed in Sect. 2.5.1. Moreover, networks as vehicles for solving policy problems or arranging service delivery are discussed as being relatively closed since only actors whose resources are considered critical are included (Rhodes 1997). Nonetheless, networks are also perceived as becoming more accessible to new actors (Klijn and Koppenjan 2016:227). In this dissertation, these 'problems' revolve around the implication of the digital transformation on governance structures and government action in developing countries. Inclusiveness and exclusiveness in this research project hence are further indicators for (un-)sustainable digital development asserting inclusion as part of sustainability (Sect. 2.1.3). In addition, inclusiveness and exclusiveness are further determinants for coordination such as horizontal or vertical coordination with regard to legitimacy (Fawcett and Daugbjerg 2012). Legitimacy will be discussed in more detail in Sect. 2.5.2.

This dissertation argues that the ease of entering determined by levels of inclusivity and exclusivity in digital ecosystem can likewise be assessed through dimensions of access and openness. Access in this dissertation has been emphasized as essential for sustainable digital development (Sect. 2.3.4, Sect. 2.3.1, Sect. 2.1.3). Against this background, access not only entails access to ICTs as prerequisite for participation in digital transformation, but also for entering the digital ecosystem. Thereby it is closely linked to resources for entering (access to resources highlighted by Raeymaeckers et al. 2020). Limitations to access further relate to the fact that on the one hand, certain issues are not allowed onto the agenda in decision making processes, which would contribute to the network as closed especially when costs of participating are too high or because existing actors more or less deliberately limit new actors' access (Klijn and Koppenjan 2016:29). On the other hand, actors make decisions in various arenas, often inaccessible to others likewise challenging coordination of activities (Klijn and Koppenjan 2016:66). This dissertation argues that these two factors are especially relevant to sustainable digital development in digital ecosystems, especially against the background of diverse set of actors with power and knowledge asymmetries in the digital transformation (see e.g., Sect. 2.3.6).

Degree of Closeness and Distance

Expanding on the discussion around exclusive or inclusive networks, the degree of closeness (and distance) between actors in the digital ecosystem are further determinants identified in the context of coordination for sustainable digital development. The *closeness* as a relational indicator between actors are also referred to as close ties (Raeymaeckers and Kenis 2016) or intensity of interactions as an interaction pattern (Klijn and Koppenjan 2016:103). Closeness as a precondition for network governance here is understood in line with Klijn and Koppenjan (2016) for determining which actors have the most contact with one another, which actors are central and which are rather peripheral to the network (ibid.:103). Moreover, networks demonstrate different characteristics that are both relevant for the entry of the network as well as for the actual network interaction. More specifically, networks can be distinguished according to strong or weak relations or ties (e.g., Jansen and Wald 2007). Networks with strong ties imply a high level of effort in maintaining relations or contacts while networks with weak ties often follow hierarchical patterns or patterns of center-periphery (Jansen and Wald 2007, similarly to development theory as discussed in Sect. 2.1.1). The potential level of coordination here lies in the differentiation between actors in different network positions. This is of special relevance for this dissertation as it investigates if and how digital networks in developing countries can be governed for sustainable development.

Opposed to closeness between actors in the ecosystem is the degree of *distance*. The complexity of governance network processes is likewise highlighted since interactions between multi-stakeholders in the network can occur in more than one area, i.e., some actors participate in some areas while other arenas are difficult to enter for them as illustrated in Fig. 2.7. The ease of entry has been further discussed in

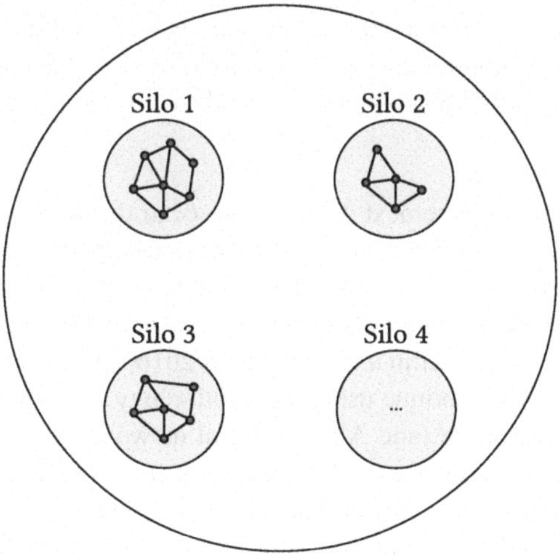

Fig. 2.7 Siloed governance network processes (own illustration based on arenas in Koppenjan and Klijn 2004:57)

the subsection above. Possible interactions between multi-stakeholders in different arenas leads to a fragmentation of interaction that makes influencing or governing these interactions difficult (Klijn and Koppenjan 2016:90), as similarly discussed with regard to aid fragmentation (Sect. 2.1.5). These areas can be further characterized as subsystems, closed sector policy networks (Klijn and Koppenjan 2016:26) or *silos* (e.g., Hertting and Vedung 2012). Silos are one form of institutional complexity where certain actor groups are unable or unwilling to "[...] cope with some broader and intractable problems" (Hertting and Vedung 2012:30). Silos here are understood as closed arenas in which actors collaborate in a closed group and only between themselves instead of interactions with others. This is for instance observed in digital development activities following a siloed approach. The World Bank (2016a) for instance emphasizes silo mentalities between ministries and departments, for instance as they compete against one another for scarce budgetary resources, further demonstrating a resistance to horizontal collaboration (The World Bank 2016a:179). This is similarly discussed by Sørensen and Torfing (2011) arguing that "[...] public sector is commonly associated with rule-bound, bureaucratic silos characterized by red tape, inertia, and stalemate" (ibid.:846). Siloed approaches are further discussed in the literature with regard to donor organizations (e.g., Klingebiel, Mahn and Negre 2016) or academia (e.g., Walsham 2017), among others. In order to address silos, networking and cooperative structures in the form of networks are called for, thus underlining this dissertation's research question of coordination through networks for sustainable and inclusive interactions in digital development efforts. Effects of such siloed approaches that are here identified as challenges to ICT4D and to coordination in the digital ecosystem are analyzed in more detail in Sect. 2.5.2.

Degree of Centralization

This directly leads over to the next determinant: centralization. The degree of centralization in the network is again as part of network design process, similarly to the discussed frequency and intensity (e.g., Klijn and Koppenjan 2016 and Provan and Kenis 2008). Moreover, the centrality of actors in the network is also defined as 'network embeddedness' (e.g., Klijn and Koppenjan 2016; Molin and Masella 2016a), but this dissertation will continue using the terminology of centrality or centralization. In network governance (and ANT or social network analysis), centralization describes the position of the actors in the network (in the center or periphery, see e.g., Jansen 2017). However, centrality here is understood in a physical sense, meaning locally centralized or decentralized. Since this dissertation does not use social network analysis but governance analysis, the centrality of actors in the network in the sense of "prominence, popularity, or power" (Whetsell et al. 2020:82) or connections to other network actors is not the focus here. Opposed to the perspective of centrality of actors in the network, centrality in this research project is understood with regard to the context (macro) factors of the network, such as whether the digital ecosystem is centralized on specific areas such as urban areas instead or rural areas or different factors that explain a (digital) divide. This allows insights into the impact on (un)-sustainable digital development. Moreover it is thereby closely linked to the precondition of ease of entry into the ecosystem due to physical barriers (Sect. 2.5.1, Sect. 2.5.1). This degree or dualism of centralization vs. decentralization is likewise essential to governance formats for coordination, which will be discussed in more detail in Sect. 2.5.4.

Network Rationale

A further precondition identified in the literature review of network governance and ICT4D is the rationale of the network or ecosystem. According to Molin and Masella (2016b), the "[n]etwork rationale refers to the reasons behind the implementation of the network—i.e. to define public policies or deliver public services [...]" (ibid.:831). This definition emphasizes that the network rationale is closely linked to actors interests, willingness and motivation as discussed at the beginning of this framework's first determinant. Network rationale in network governance is likewise discussed for instance in McGuire and Agranoff (2011); Sørensen and Torfing (2009); Torfing (2005). This dissertation understands such network rationales with regard to sustainable digital development as the ultimate goal (as universally highlighted by the SDG). Hence this precondition is referred to identically with Sect. 2.1.3 and is summarized under determinant 5 (outcome: sustainable digital development, Sect. 2.5.5).

Network Efficiency and Effectiveness

The network rationale as the goal of achieving or fostering sustainable digital development is further closely associated with efficiency and effectiveness of the network. Network effectiveness can for instance be defined as the achievement of collective outcomes otherwise not possible by individual organizations alone (Provan and Kenis 2008:230). For this dissertation's focus on ICT4D, this can be further synthesized with regard to aid effectiveness and aid efficiency as discussed in Sect. 2.1.7. Adding the development perspective on efficiency in the digital transformation context, ICTs are for instance ascribed the potential for increasing efficiency of processes, including public sector services for instance in the context of E-government effectiveness and efficiency (see e.g., Choi and Chandler 2020). Efficiency in the ICT4D context has likewise been discussed in Sect. 2.2.1 and Sect. 2.2.2.

Similarly to efficiency of aid as the desire outcome in the (digital) development context, efficiency is a further measure for network governance in terms of outputs over inputs (Provan and Kenis 2008:242). However, as indicated in the previous preconditions, efficiency is likewise affected by the other preconditions. With regard to actors interests and inclusion of (all) diverse multi-stakeholders for instance, "[i]n networks, the primary tension regarding efficiency is between the need for administrative efficiency in network governance and the need for member involvement, through inclusive decision making" (Provan and Kenis 2008:242). More specifically with regard to this dissertation's analysis of the multi-stakeholder interactions in the digital ecosystem, collaboration is "seldom an efficient endeavor, however" (Provan and Kenis 2008:242), for instance due to intensity of time and resources. As potential governance formats for achieving network efficiency (in digital ecosystems) especially in terms of inclusiveness, serves for instance the NAO mode of governance which "[...] is likely to provide a greater balance than either of the other two forms regarding the tension between the need for efficient operation and inclusive decision making" (Provan and Kenis 2008:242). This will be discussed in more detail in Sect. 2.5.4. Hence, network efficiency and effectiveness in this framework are understood as a precondition for network governance (performance), and simultaneously as a goal to be achieved (closely related to the goal of sustainable digital development).

2.5.2 Determinant 2: Specific Sustainable Digital Development Challenges for Ecosystem Governance

The second determinant proposed in this framework for governance of digital ecosystems arises from governance challenges in ICT4D context of digital ecosystems. The focus here lies on governacne challenges, although it has to be noted that these are to be complemented by ICT4D challenges (Sect. 2.3). Determinant 2 synthesizes these challenges and argues that these are not only specific challenges that influence coordination of digital development activities in ecosystems, but likewise are preconditions for a sustainable digital development if preconditions (Sect. 2.5.1) are fulfilled. Using Provan and Kenis (2008)'s term of network tensions, this dissertation complements this theoretical network governance perspective through a digital development angle. This subsection differs from the previous one (Sect. 2.5.1) in that these challenges are derived from the previous sections in specifically applying to ICT4D (Sect. 2.3) and thereby pose a challenge to governance or coordination and a (digital) development challenge. These challenges are further illustrated in

Fig. 2.6. More specifically, the following digital ecosystem governance challenges will be explored in more detail:

- Bureaucracy
- Flexibility (agility)
- Capability and capacity
- Digital complexity
- Financial resources
- Legitimacy
- Duplication and parallel working
- Power asymmetries

Bureaucracy and 'Traditional' Governance Structures

Beginning with the first factor, this framework argues that bureaucracy in the sense of 'traditional' governance structures are a challenge for coordination in digital ecosystems and to sustainable digital development. Firstly, sustainable digital development can likewise be interpreted with regard to stability (Sect. 2.1, Sect. 2.1.3) and durability (Sect. 2.5.1). Network governance as this dissertation's theoretical angle suggests that "[t]he most obvious mechanism for maintaining stability is the formation of a formal hierarchy" (Provan and Kenis 2008:245) exerted by governments. More importantly regarding the digital transformation, governance through hierarchy or so-called 'bureaucracies' face a trade-off between high flexibility or adaptability, legitimacy and efficiency (Provan and Kenis 2008:245). Here the interests of each actor group need to be highlighted (Sect. 2.5.1) and their legitimacy, which will be discussed in the next subsection. Moreover, this alludes to the discussion around the role of the state in (digital) development (Sect. 2.1), for instance in executing tasks such as tax collection, social security, law and criminal justice, and other public services that "[...] are all often organized in classical bureaucracies with all the typical characteristics of those bureaucracies" (Klijn and Koppenjan 2016:7).

Secondly, bureaucracies here a understood as a challenge to coordination for sustainable digital development due to the nature of political-administrative organizations in the context of high transaction costs, slow processes, inefficiencies etc., as discussed in Sect. 2.1.4 and Sect. 2.1.5. Moreover, bureaucratic barriers between institutions are likewise mentioned as a challenge for sustainable (digital) development (e.g., Lang, Wiek, Bergmann, Stauffacher, Martens, Moll, Swilling and Thomas 2012; Cochrane and Cundill 2018) and also with regard to *intergovernmental coordination* especially in the cross-cutting context of digital transformation (Sect. 2.1.2, Sect. 2.2.1, Sect. 2.3.5).

Additionally, coordination, complementarity, learning and expectation effects of actors, especially with regard to inter-governmental interaction in digital development (Sect. 2.5.1), are likewise discussed in the context of path-dependent development (Sydow 2017:47). In this dissertation's framework, path dependencies are further analyzed in the context of lock-ins in the digital transformation (Sect. 2.3.6) and in terms of interdependence (Sect. 2.5.1). Path dependency however has a further development theory angle since dependency on (internal) protocols and donor requirement as part of aid fragmentation Sect. 2.1.5 are further argued as challenges for sustainable digital developmental and (donor) coordination (Sect. 2.1.6).

Flexibility

Thirdly, (digital) innovations are characterized by a high degree of flexibility (as discussed in Sect. 2.2). With the innovative power in digital transformation being mostly associated with the private sector, flexibility is especially ascribed to this actor group (see e.g., Sect. 2.3.1). Moreover, the ability to adapt (e.g., technologies) is also where a potential for leapfrogging lies (Sect. 2.2.2). Regarding governance approaches in the digital transformation, "[...] each government is responsible for the identification of the best digitalization governance strategy based on the country culture, customs, and economic needs" (Linkov et al. 2018:5). However, the digital transformation requires governments to employ flexible and adaptive governance approaches allowing involved stakeholders (Sect. 2.1.4) to equally adjust their practices for deriving an inclusive and sustainable digital development. Hence, multi-stakeholders, especially public sector actors such as partner governments and bilateral donors, are required "to iteratively improve and adapt their policies that govern the digital economy to balance technological gains against the potential for social and environmental disruption and externalities" (Linkov et al. 2018:5).

More specifically, flexibility in this dissertation is further understood specifically in the digital context with rapidly changing digital innovations and agile approaches (e.g., Lindgren and Van Veenstra 2018). This however stands in contrast to the public sector nature, which can potentially be "cumbersome and bureaucratic" (Provan and Kenis 2008:244), as discussed in the paragraph above. Hence, this underlines the value of coordination in the digital transformation through networks, where organizations can collaborate for achieving specific goals that require combined resources and expertise that hierarchies alone could not readily accomplish (Provan and Kenis 2008:244): "This flexibility allows networked organizations to respond quickly to competition and other environmental threats, as well as to opportunities." (ibid.). Moreover it is important to highlight that flexibility here is not understood as antagonal to stability, which is subsumed under durability for sustainable digital development (Sect. 2.5.1).

Duplications and Parallel Working

From the variety and amount of other multi-stakeholders as well as the discussed challenge of Sect. 2.2.3, duplications of activities and parallel (isolated) working are here argued as important challenges to coordination for sustainable digital development. Especially with regard to the development cooperation context, this dissertation has highlighted the importance of aid fragmentation (Sect. 2.1.5) that poses the danger of duplication or parallel (siloed) activities. It has been argued that aid fragmentation is even fostered in the digital transformation sector due to setting of hypes (Sect. 2.2.3). Similarly, the competitive nature of private sector actors likewise can have duplication as a consequence.

These determinants of duplications and parallel working in this framework are hence understood as one indicator of distance in relations (Sect. 2.5.1) and unsustainable digital development highlighting the importance of coordination in the ecosystem. Hence, *alignment* of activities (and strategies) can be a crucial factors for achieving sustainable digital development through coordination. From a network governance perspective, awareness or perceptions (Sect. 2.5.1) in the ecosystem as well as goal consensus are important (Klijn and Koppenjan 2016:270). Alignment here refers to the actors' awareness enabling them to co-exist in the network without necessarily sharing the same perceptions (ibid.:128).

Digital Complexity: Capabilities and Capacities

This is closely linked to a further challenge applying to the capability and capacity of multi-stakeholders in the digital transformation, especially of public sector actors and the mentioned 'bureaucracies'. This determinant is derived from Sect. 2.3.5 and is understood as not only a challenge, but also a precondition that is necessary for achieving coordination for sustainable digital development. Moreover capabilities and capacities of actors in the ecosystem for governing the digital transformation are used interchangeably with competencies. Competencies here "concern the formal authority to make certain decisions and to take responsibility for the realization of specific solutions" (Klijn and Koppenjan 2016:267–268) that are assigned to specific actors, which likewise constitutes their role in the network. While capabilities and competencies are preconditions for governance, especially in form of MSPs (Pattberg and Widerberg 2016), this dissertation argues that digital capabilities of multi-stakeholders are necessary for coordination and influence the quality and performance of governance roles (Sect. 2.5.3, Sect. 2.5.4). Digital capabilities and capacities likewise refer to knowledge (Sect. 2.3.6), especially digital literacy and digital skills. This is one area that is emphasized in ICT4D literature as crucial for sustainable digital development (Sect. 2.2.1, Sect. 2.3).

More importantly, in this framework digital capability of actors are understood as challenge for actors conducting sustainable digital development activities that is aligned with the challenge of *digital complexity*. Complexity here not only refers to the technological aspect of digital complexity with regard to the functioning of digital technologies (Sect. 2.2), but also regarding the especially complex actor setting in digital ecosystems of developing countries (as discussed in Sect. 2.1.4, Sect. 2.1.5 and Sect. 2.2.1). Here this dissertation argues that governance in the setting of (digital) complexity is crucial for achieving sustainable digital development in terms of governance networks as the answer to complexity (Klijn and Koppenjan 2016:1). Moreover, this stresses the applied ecosystem logic since complexity requires interaction and network relations, for instance due to interdependencies (Sect. 2.5.1) and power asymmetries (Sect. 2.3.6), but also because interaction is detrimental for achieving effective outcomes (Klijn and Koppenjan 2016:10).

Financial Resources

Closely linked to ICT4D challenges of capacities and capabilities, especially in developing countries, are financial resources (in addition to the before mentioned resource of knowledge or know-how). Financial resources are understood as available financial assets and budgets, which enable to realize (digital) solutions and to cover (transaction) costs attached to participation in complex interaction processes (Klijn and Koppenjan 2016:267–268), here the digital transformation. Thereby financial resources are not only an essential part of exchanges within a network (Sect. 2.3.5, Sect. 2.3.6), but also of interdependencies and, most importantly for digital development, relate to affordability of digital infrastructure or devices (Sect. 2.5.1). Hence, these resources are crucial for this framework in governing the digital transformation for sustainable digital development on both the interaction level (micro), but also on the national (macro) level considering the countries' resources. This is similar to the logic of resource exchanges within the network/ecosystem and between the ecosystem and its environment (Castells 2004:36).

Firstly, affordability of ICT services, infrastructure or hardware are essential for achieving sustainable digital development (Sect. 2.1.3), especially in the context of the digital divide (Sect. 2.3.4). Moreover, financial resources are a determinant for participation in digital transformation and in the ecosystem (Sect. 2.5.1). Secondly, (financial) resources and (inter-)dependencies are closely linked. The more important a resource is for the realization of network objectives and the more difficult it is to acquire the resource elsewhere, the more dependent one organization is upon another (Klijn and Koppenjan 2016:268, as further discussed in Sect. 2.5.1).

With regard to the role of financial resources ins digital ecosystems, an imbalance between financial resources of actors are hindering sustainable digital development as they manifest in one form of power asymmetries, especially with regard to the private sector vs. the public sector (similar to other resources such as know-how, as further discussed in Sect. 2.1.4 and Sect. 2.3.6). Similarly, financial resources, among others, are an important factor for the performance of ICT4D projects. This is one example where many ICT4D projects are dependent on donors, multilateral institutions and private sector organizations with regard to financial resources and technology expertise (Delponte et al. 2015, Sect. 2.3.1).

Legitimacy

In a nation state system, legitimacy is an essential part of government action. Elected political bodies are actor groups that "have the ability to give legitimacy to, or withhold legitimacy from, decisions made to solve difficult problems" (Klijn and Koppenjan 2016:74). This includes legitimacy for certain solutions, "[...] but also for the process followed and the work methods used to accomplish that, and for the actors involved in that process" (Klijn and Koppenjan 2016:267–268). As argued in Sect. 2.1.4, in the digital transformation the private (tech) sector is a further essential actor group that actively influences the digital development of a country (and its ecosystem). Here, similarly to the discussion around the involvement of private sector in development cooperation (Sect. 2.1.4), the question of legitimacy of action arises. This dissertation takes an actor perspective on legitimacy with regard to the internal or external legitimacy of networks as proposed by Provan and Kenis (2008). Therefore, this dissertation follows legitimacy in terms of legitimacy of multi-stakeholders' activities, with a special focus on the government since "[t]he governmental apparatus is professionalized and aimed at the production of effective policies and services doing justice to the principles of equality, legitimacy, and legality" (Klijn and Koppenjan 2016:5). Moreover, this dissertation highlights that legitimacy of action, also in terms of accountability, is an essential part of sustainability and sustainable digital development (see Sect. 2.1.2). The question is what the implications of a (lack of) legitimacy of actions are in the digital ecosystem on sustainable digital development, to which the empirical analysis will deliver further insights (Chap. 4).

Power Asymmetries

Synthesizing the findings of the previous sections (Sect. 2.3, Sect. 2.3.6, power asymmetries in this dissertation are argued to be a major challenge for both coordination through network governance in digital ecosystems and for sustainable digital development. Power asymmetries are derived as existing on the actor level (who are powerful and less powerful actors) and how power asymmetry is exerted. As for the first perspective on actors, power asymmetries especially between private tech and public sector actors have been discussed in Sect. 2.3.6. This secondly refers to the perspective of 'how': Here dependencies are a crucial factor in explaining the existence and implications of power asymmetries. Dependence in ICT4D and digital development literature is often linked to and compared with aid dependence (Sect. 2.1), which is likewise relevant for network governance since power is a central concept in the resource dependency model (Klijn and Koppenjan 2016): "Networks emerge because of resource dependency between actors [...]" (Klijn and Koppenjan 2016:33). Especially asymmetrical interdependencies that create power differences between actors (ibid.) are a challenge for coordination, which is further discussed in Sect. 2.5.1. Hence power both refers to traditional authoritative power (e.g., resources allocation, laws), but also to resources, skills and capacities that can give authority to specific actors (e.g., Torfing 2005). More specifically for this dissertation, power here both includes financial power but also know-how or knowledge and similar knowledge asymmetries (Sect. 2.3.6).

This dissertation's framework further proposes 'solutions' as to how these power asymmetries in the digital ecosystem can be governed for sustainable digital development in Sect. 2.5.3 and Sect. 2.5.4. More importantly with regard to its research question, sustainability here is further challenged by the degree of exclusion (Sect. 2.5.1) arising from power asymmetries where "[...] certain actors may be denied access, and points out that problems and agendas tend to include certain definitions or subjects and exclude other" (Klijn and Koppenjan 2016:37). This is further linked to the veto power of powerful actors that can hinder (joint) decision-making or goal consensus (e.g. see Tsebelis 1990, Sect. 2.5.1).

2.5.3 Determinant 3: Governance Roles in the Network

As a third determinant, governance roles are proposed in this framework as to how digital networks can be governed. The following roles have been identified in the reviewed literature mostly based on network governance but supplemented by ICT4D perspectives. These network governance roles are specifically relevant to this dissertation as they provide further insights into how networks can be coordinated, because all of these roles are coordinating activities in the network. This is in line with the network (governance) literature, where "governance refers to actors' conscious attempts to influence processes in networks" (Klijn and Koppenjan 2016:36). More importantly, governance roles are understood as collections of activities and rules for achieving the goals of a network (Brass, Galaskiewicz, Greve and Tsai 2004; Span et al. 2012), in this case sustainable digital development. From the literature review, the following governance roles have been derived in the context of coordination for sustainable digital development while noting no claim for completeness and further highlighting that these roles are often used interchangeably or overlapping. A brief overview is provided in Fig. 2.6. Each governance role, as further listed below, will be elaborated in the following with references to the relevant multi-stakeholder groups in digital ecosystems.

- Coordinator
- Facilitator
- Broker
- Network manager
- Driver (of digital transformation)

From the D in ICT4D perspective, the role of the government is crucial for the development of the state, hence this framework pays special attention to the governance roles of the government for sustainable digital development in digital ecosystems. While noting that these governance roles in digital ecosystems may apply to all governments regardless of development background, this dissertation shines special light on governments in developing countries and their specific roles in the digital ecosystem since they face further challenges as discussed in Sect. 2.1. More specifically, since these governance roles are derived from the synthesis of network governance and ICT4D, the role of governmental organizations in developing countries and the role of ICT in the digital ecosystem are highlighted in line with discussed research calls for an holistic (analysis) approach (Sect. 2.2.1). Hence, on the one hand, the government is especially essential for the network governance analysis as it fulfills two roles, on a macro level and micro level (which simultaneously mirrors the logic of this dissertation's design and framework, Fig. 1.1). More specifically, at a macro level the government takes a role as a facilitator (Sect. 2.5.3) in defining shared visions and goals for multi-stakeholders as well as through establishing communication and accountability mechanisms (Molin and Masella 2016b). On a micro level it interacts with the ecosystem through its public agencies and hence is a network member itself (ibid.). The exact role of the government here however is contested in the literature, which will be discussed in more detail in the following.

Besides the essential role of multi-stakeholders for coordination, for this dissertation's research purpose of sustainable digital development another force is essential: ICTs. As discussed in Sect. 2.2.1, the ICT4D literature debates whether ICT are a means to an end, or an actor in digital ecosystem themselves. Hence, ICTs further impact governance since they can either be a new form of network governance, for instance with regard to platforms (Sect. 2.5.4, an entity that influences stakeholders' interactions or an additional actor whose actions require coordination.

Coordinator

As a first governance role, and most importantly for this dissertation's research question, this framework analyses the governance role of a coordinator that coordinates multi-stakeholder digital development activities in the digital ecosystem. A coordinator here is understood as coordinating multi-stakeholders' activities, not in the sense of technical coordinators (as for instance discussed in computer science research with regard to software ecosystem, see e.g., Jansen, Cusumano and Brinkkemper 2013). The role as a coordinator can be exerted by multi-stakeholder organizations or by different governance formats, which will be discussed in Sect. 2.5.4. Regardless of the specific form of the coordinator (as a separate organization or an existing actor), the coordinator takes an important role in governing networks (Provan and Kenis 2008; Edelenbos, Van Buuren and Klijn 2013; Raeymaeckers et al. 2020).

This dissertation understands the coordinator role as actively coordinating, even to the extent proposed by Raeymaeckers et al. (2020) that "[...] the coordinator can make all decisions without consulting the network partners" (Raeymaeckers et al. 2020:1040). Although the coordinator role can take different features that here qualify as individual governance roles such as a facilitator, co-producer and commissioner (as proposed by Raeymaeckers et al. 2020:1038), this dissertation analyzes the roles of a coordinator and a facilitator separately. This allows an in-depth analysis of both governance roles with regard to multi-stakeholders in the digital ecosystem, but also the role of ICT as influential factors.

Network Manager

Closely linked to the role of a coordinator is the role of a network manager (e.g., Klijn and Koppenjan 2016:163). Responsibilities of a network manager include for instance the identification of key resources (here specifically financial resources and knowledge, see Sect. 2.5.2, Sect. 2.5.2), the management of the involved interdependent actors and responding to tensions (e.g., arising from conflicting interests) for achieving common goals (e.g., Agranoff 2007; Provan and Kenis 2008; Molin and Masella 2016b; Keast, Mandell, Brown and Woolcock 2004; Keast and Brown 2010, see further Sect. 2.5.1). In doing so, and in line with this dissertation's focus on the implications of the digital transformation on governance structures and government action, Keast and Brown (2010) argue that managing the network is specifically challenging due to balancing autonomy and fostering innovation. According to Goldsmith and Eggers (2004), further responsibilities of a network manager include identifying possible patterns, bringing stakeholders to the table, analyzing current operations, determining and communicating expectations to members, devising strategies to maintain the network, activating the network and coordinating activities between network participants for knowledge sharing, overcoming cultural differences and building trusting relations. If the governance role of a network manger is lacking or not fulfilled entirely, the literature suggests different roles of governance such as in the form of a facilitator or broker (e.g., Molin and Masella 2016b; Span et al. 2012; Raeymaeckers et al. 2020). These roles will be further discussed in the following.

Facilitator

After discussing these two governance roles, the roles of a facilitator and broker are likewise highlighted in the reviewed literature—often used interchangeably (e.g., Provan and Kenis 2008; Raeymaeckers et al. 2020; Span et al. 2012; Klijn and Koppenjan 2016; Molin and Masella 2016b). Moreover, opposed to the rather active role of a coordinator, the following roles of a facilitator and broker are further understood as more passive than active due to indirectly engaging in the network. While Klijn and Koppenjan (2016) associates facilitators with process mangers (ibid.:174), Raeymaeckers et al. (2020) and Span et al. (2012) further distinguish between commissioner, co-producer and facilitator. For this research project, only the role of facilitator is pursued for research pragmatic reasons. Nonetheless, this dissertation follows Raeymaeckers et al. (2020)'s argument of the facilitator being a bottom-up format where the decision making power lies in the hands of the participating actors (ibid.:1040).

The governance role of a facilitator is especially relevant to this dissertation which argues that the digital transformation and emergence of digital networks is highly complex (Sect. 2.3.6, Sect. 2.5.2), emphasizing the need of a facilitating role. Moreover, the role of a network facilitator aims at maintaining and building trust among network members (Provan and Kenis 2008:243). This is closely linked to the previously discussed role of trust in (governance of) digital ecosystems (Sect. 2.5.1). With regard to the correlation between the governance role of facilitator and governance formats (Sect. 2.5.4), a network facilitator or broker can take the form of a NAO or a formal organization (Provan and Kenis 2008:236). This will be discussed in more detail under determinant 4 (Sect. 2.5.4).

Broker

As discussed above, the governance roles of facilitator and broker are often used interchangeably. The role of a broker here is understood similarly to the responsibilities of the facilitator. The role of a broker is further discussed by Börzel and Risse (2010); Klijn and Koppenjan (2016); Kenis and Provan (2009). With special regard to this dissertation's focus on implications of the digital transformation on governance structures and government action, when one or more actors overtake the role of a broker, deadlocks or stagnation can occur, for instance in cases of innovative public polices and services realized (Klijn and Koppenjan 2016:306). More specifically, in the ICT4D literature the role of 'innovation intermediaries', for instance in diffusing and scaling innovations, are discussed (e.g., Heeks et al. 2013, Sect. 2.2.1). Such intermediaries enable innovation through knowledge brokering, especially between the knowledgeable ones (who can redesign technology and those who understand how the technology needs to be used) and through inclusion in technology or program development (Sanner and Nielsen 2019; Foster and Heeks 2013).

Similarly to the previously discussed roles, the question around the role design, whether it is top-down, bottom-up or by single organization, likewise applies to the role of a broker. According to Provan and Kenis (2008) for instance, if the network broker is a NAO, this organization plays a key role in coordinating and sustaining the network (ibid.:236). This dissertation here further includes the development cooperation perspective with respective actors, which is especially important against the background of actors potentially seeing themselves as brokers, especially donors. Here, Otoo, Agapitova and Behrens (2009) for instance refer to donors and their experts acting as catalysts, facilitators, and brokers of knowledge and technique. Expanding the notion of a catalyst, Whetsell et al. (2020) sees the government in the role as a catalyst, for instance with regard to "cooperation between competitors on scientific and technological landscapes" (ibid.:81). Hence, the role of a network broker is likewise associated with the role of a catalyst. Catalyst in this present framework likewise accommodates the function of catalyzing the digital development in the digital ecosystem, which hence is closely linked to the next governance role of a driver of digitalization.

Driver

A specific governance role in digital ecosystems (networks) are drivers of the digital transformation as argued by this dissertation. More specifically, two perspectives on drivers for coordinating in the digital ecosystem are emphasized here: One, the actor perspective (who is driving) and second, what or where digital technologies are driving development. With regard to the latter perspective, Sect. 2.2.1 demonstrates that ICTs are discussed as drivers (for development) among the ICT4D literature. They are seen as drivers for development (economic, social and political) also in terms of leapfrogging (Sect. 2.2.2). Driver is hence not understood in terms of a role leading the network, which is instead expressed in the governance role of the network manager (Sect. 2.5.3).

Secondly, the governance role of driver of digital transformation aims at analyzing the interactions of multi-stakeholders and are essential to this dissertation's research question. This is closely linked to ICTs since they are discussed to be actors themselves (Sect. 2.2.1). Emphasizing the actor perspective of driving the digital transformation is here similarly understood with regard to levels of activity (Sect. 2.5.1) that indicate who are the active and less active actors in the digital ecosystem. Hence the question here revolves around which actor in the ecosystem overtakes this role of actively driving digital development. The identified actor groups are discussed individually with regard to implications if they are overtaking the driving role (Sect. 2.1.4). This dissertation argues that the governance role as a driver of digital transformation is closely linked to not only activity and motivation (Sect. 2.5.1, Sect. 2.5.1), but also actors' interests (Sect. 2.3.3), capacities (Sect. 2.5.2) and respective power (Sect. 2.5.2, Sect. 2.3.6). Of special importance for this dissertation's research focus on government action is the question whether or how governments here are in the 'driver's seat' (Keast, Mandell and Brown 2006). In the network governance literature, Molin and Masella (2016b) provide an insight into this debate arguing that the state is no longer seen in this position "[...] since it needs the technical expertise and skills of non-governmental actors, giving life to a complex multilateral dependency between all the actors involved and their resources" (Molin and Masella 2016b:843). The empirical analysis of this dissertation will deliver further insights into these governance roles and the role of the multi-stakeholders in shaping the digital ecosystem for sustainable digital development.

2.5.4 Determinant 4: Governance Formats for Coordination

A further determinant of this dissertation's framework are governance formats for coordination in digital ecosystems (see Fig. 2.6). Besides the before mentioned governance roles, governance formats are a further approach to achieve coordination for sustainable digital development through network governance incorporating all of the previous three determinants (Sect. 2.5.1, Sect. 2.5.2, Sect. 2.5.3). Especially with regard to Sect. 2.5.1, depending on which of these preconditions are fulfilled in the digital ecosystem, a different governance format is assessed as suitable for the desired outcome in the reviewed network governance literature. In the literature, the question of how networks, and hence their internal activities, can be governed is highly debated (Raeymaeckers et al. 2020:1040). Here, the approach of Provan and Kenis (2008), and similarly of Kenis and Provan 2009, is most frequently cited and hence likewise serves as the basis of this research project's framework. Kenis and Provan (2009) propose three different forms of network governance: lead organization governance, NAO and a shared participant governance form as illustrated in Fig. 2.8.

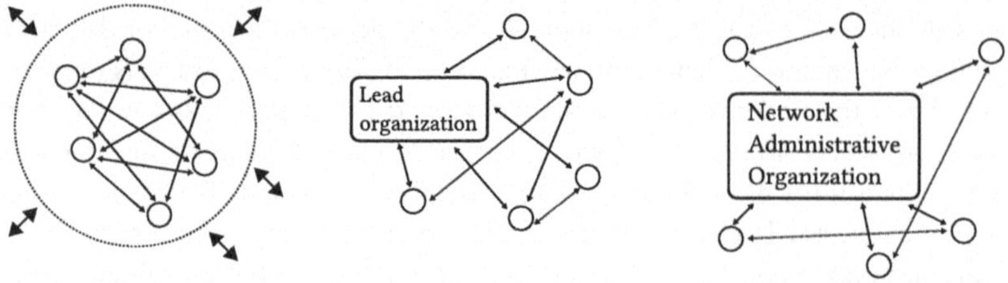

(a) Shared governance network **(b) Lead organization network** **(c) Network administrative organization**

Fig. 2.8 Forms of network governance, (a) illustrates shared governance form, (b) illustrates lead organization network and (c) illustrates network administrative organization (own illustration based on Kenis and Provan 2009:447)

Moreover, network governance forms can be further distinguished in so far as they "may or may not be brokered" (Provan and Kenis 2008:233). While this reminds of the previously discussed governance role of a broker (Sect. 2.5.3), here not the dynamic of interactions and actor roles are of interest, but the role of institutions for coordinating or governing the ecosystem are at the center of attention. Hence, brokered or non-brokered forms as proposed by Provan and Kenis (2008) are used for explaining degrees of formal intervention that are reflected in either a (new) organization (lead organization or NAO, see Fig. 2.8b and 2.8c) or (already existing) network members (shared participant or also self-governance, see Fig. 2.8a). This approach is reflected in this dissertation's approach since "[...] governance networks are not only dynamic decision and interaction systems, but also have an institutional dimension" (Klijn and Koppenjan 2016:102). With regard to the network literature, Raeymaeckers et al. (2020) notes that most research has focused on lead organization types of governance with only some work being conducted on shared participant types of governance where the network is regulated by all network actors (ibid.:1040).

Fig. 2.9 illustrates the network governance forms in coordinating digital ecosystems derived by this dissertation. The forms' positions on the continuum of formalization will be explained in the following. The three network governance forms of lead organization, NAO and shared forms will be discussed in more detail in the following as individual sections with regard to coordinating digital ecosystems, while incorporating the closely related meta governance in order to deliver insights into implications of a meta organization governing the digital ecosystem (Sect. 2.5.4). In order to also incorporate the digital research perspective, digital (collaboration) platforms will be discussed as one organizational governance form similar to the NAO (Sect. 2.5.4). Further incorporating the development cooperation perspective, MSPs and PPPs are derived from the development focus in ICT4D as possible governance formats. These governance formats can be further supplemented by network governance forms focusing on collaboration, such as alliances (Sect. 2.5.4). All of these follow a multi-stakeholder approach that is argued to be essential for sustainable and inclusive digital development, although to different extents. The empirical analysis will deliver further insights into the probability of these governance formats for sustainable digital development from the Kenyan case.

Fig. 2.9 Network governance forms according to degrees of formalization (own illustration)

Shared Form or Self-governance

As a first governance form, the network governance literature suggests the shared form or self-governance as demonstrated for instance by Raeymaeckers and Kenis (2016) and the most cited Provan and Kenis (2008). In the shared governance form, the network (here ecosystem) is governed by network members with various interactions (Provan and Kenis 2008:233–234). Adding to the before mentioned preconditions (e.g., size and type of actors) in participant driven governance, "[...] governance becomes extremely complex" (Provan and Kenis 2008:238), so that this governance form is "[...] best suited to small networks of organizations" (ibid.). With regard to this dissertation's focus on interactions, also from a dependency perspective in ICT4D (Sect. 2.1.1), in this governance format power lies with the network participants. Hence, shared self-governance is often seen as highly desirable by network participants, since they can retain full control over the direction of the network. However, the downside of self-governing forms is that "[...] processes get blocked or stagnate, because of conflicting interests, perceptions, strategies, or institutional barriers" (Klijn and Koppenjan 2016:8) highlighting the importance of cooperation.

In this dissertation's framework, self-governance and self-regulation of activities in the network are understood as interchangeably where "[g]overnance refers to self-regulation of actors within networks; the 'networking' of these actors" (Klijn and Koppenjan 2012:194). For this purpose the concept of meta-governance has been introduced in the governance literature to address this ambiguity (ibid.), which will be further discussed in Sect. 2.5.4. Consequently, in the governance format of shared or self-governance there is decentralized instead of centralized coordination of action and is one option how digital ecosystem can be governed. This dissertation further emphasizes that this governance approach is similarly mirrored in ICT4D literature. Stuermer et al. (2017) for instance argues that: "Good governance means the digital artifact and its ecosystem is not controlled by a single individual or organization, but governed decentralized by its contributors and other stakeholders." (ibid.:253), as further discussed in Sect. 2.1.2.

As a summary, the following list demonstrates for which networks shared governance is efficient or to be chosen:

- *Trust*: As identified by Provan and Kenis (2008) shared governance is most likely to be an effective form when trust is pervasive throughout the network providing a dense web of trust-based ties between network participants (ibid.:238; high density of trust).
- *Size*: Shared self-governance is best suited to *small* networks of organizations, e.g., with fewer than six to eight organizations (Provan and Kenis 2008:238–239)
- *Type*: When diversity of actors (different actor types) is limited, a shared governance model can be implemented with actors coordinating their activities themselves, enhancing trust and mutual monitoring (Molin and Masella 2016a)
- *Goal Consensus*: Self-governed forms are most likely to be effective when participants can generally agree on network-level goals (High level of goal consensus) (Provan and Kenis 2008:239)
- *Willingness*: When commitment is relatively low, shared form of network governance could perform better since decision-making power is diffused among participants (Molin and Masella 2016a:502)

This dissertation argues that due to the fact that in this governance format of self-governance "[...] networks depend exclusively on the involvement and commitment of all, or a significant subset of, the organizations that comprise the network" (Raeymaeckers and Kenis 2016:405) and "are themselves responsible for managing internal network relationships and operations as well as external relations" (ibid.), this may be difficult with regard to a sustainable and inclusive digital development as emphasized in determinants 1 and 3 (Sect. 2.5.1, Sect. 2.5.3). Consequently in this dissertation's framework, determinants 1 and 3 are understood as explanatory factors that influence (and might hinder) governance forms such as (shared) self-governance.

Alliances and Working Groups

Similar to the self-governance governance formats are alliances and working groups as forms of cross-institutional and interorganizational interactions to coordinate organizational fragmentation and environmental complexity (Keast and Brown 2010:8). Related types include task forces, coalitions, or ad hoc committees (Raeymaeckers and Kenis 2016). These governance formats, especially alliances, are less formalized than the other discussed governance formats as "[...] coordination is achieved by negotiated rules" (Klijn and Koppenjan 2016:31). Governance here is exerted through regular meetings between involved multi-stakeholders (respective representatives of the organizations) (Raeymaeckers and Kenis 2016) and hence is limited in time mostly to the occurrence of multi-stakeholder meetings. This is specifically relevant with regard to *working groups* for top-level consultations involving multi-stakeholders, which are mostly restricted to information exchange due to limited time capacities of actors (Klijn and Koppenjan 2016:165). Moreover, working groups are mostly coordinated by a steering group in the role of a process manager (ibid.). Additionally, they are characterized by "[...] variety, a mix of collaboration and competition, loose coupling, and differentiation and co-specialization within a structure" (Jansen 2017:22), which is seen as one reason for their increasing emergence, particularly in R&D and joint ventures in new technologies. Alliances in network governance are formed for the purpose of information sharing and other resource exchange, capacity building, problem solving and service delivery (Agranoff 2007; Raeymaeckers and Kenis 2016) as well as for strategic purposes, especially from a business view (Whetsell et al. 2020). Due to the fact that these ties between multi-stakeholders in these formats are voluntary and do not rely on formalized agreements (such as contracts in the case of PPP), they can be seen as further forms of shared participant types of governance or self-governance (e.g., Raeymaeckers and Kenis 2016; Foster-Fishman, Salem, Allen and Fahrbach 2001; Isett et al. 2011).

Multi-stakeholder Partnerships for Sustainable Digital Development

A further governance form are multi-stakeholder approaches for instance in the form of MSPs, as most common for instance in development cooperation and internet governance (Sect. 2.4.1). With regard to this dissertation's network governance approach, MSPs can be defined as "manifestations of networked governance" (Pattberg and Widerberg 2016:46). This dissertation emphasizes MSPs as a crucial governance format for coordinating digital development activities in digital ecosystems since they are tools in both development cooperation and in the digital sphere. Thereby they offer the potential of synthesis for analyzing digital (development) ecosystems. MSPs are formed on both the international level but also the national level and are one prominent tool of *Multi-stakeholder Governance*.

Emphasizing the relation of networked governance and MSPs, success factors for MSPs include combined willingness, capability, and resources of partners, among others (Pattberg and Widerberg 2016:46)—similarly to the here identified preconditions for network governance of digital ecosystems (Sect. 2.5.1). Pattberg and Widerberg (2016) have established nine conditions[9] as part of a literature review on both the levels of actors (here: micro level), process (here: meso level) and context (here: macro level). The role of MSPs in ICT4D and in sustainable development will be discussed in more detail in the following while highlighting its coordination features.

MSPs in International Development Cooperation and Sustainable (Digital) Development MSPs are not only part of network governance, but more importantly for this dissertation, instruments in international development cooperation and increasingly in digital development. MSPs are also called 'multi-actor partnerships' (see e.g., Wehrmann 2018; TWI2050 2018) or 'multi-partner' partnerships (e.g., Cochrane and Cundill 2018). Generally, MSPs seek to combine assets of public and private actors with initiatives varying according to stakeholder composition, scope, function, and thematic focus (Beisheim and Simon 2016) and are based on voluntary commitment (Lundsgaarde and Keijzer 2019). This is especially relevant to the (digital) development context, since there is a "[...] heightened emphasis on multistakeholder development approaches including partnerships between public and private actors reflects both an acceptance of the importance of actors [...]." (Lundsgaarde and Keijzer 2019:218). As elaborated in Sect. 2.1, the international development landscape becomes increasingly complex due to the increase of (diverse) actors. With the number of actors increasing, MSPs likewise increase as they are "[...] supposed to promote public policy objectives in a network rather than hierarchy setup" (Keijzer et al. 2018:50). Hence, MSPs are seen as "potential arenas for coordination and the various actors involved" (Lundsgaarde and Keij-

[9] Optimal partner mix; effective leadership; stringent goal-setting; sustained funding; professional process management; regular monitoring, reporting, and evaluation to support organizational learning; active meta-governance; favorable political and social context; and fit to problem-structure (Pattberg and Widerberg 2016:46).

zer 2019:221). Following this argument, this dissertation elaborates on MSPs as a coordination format in digital ecosystems.

More importantly for this dissertation's research question, MSPs are a governance format for achieving sustainable (digital) development in line with the SDGs, especially goal 17 ('Global Partnerships'). The emergence of 'partnership' discourse, "reflects a realization that global challenges are too complex to rely only on national governments to achieve collective goals" (Lundsgaarde and Keijzer 2019:218). As highlighted in the literature, "[s]ustainable development requires dedicated efforts to work collaboratively between diverse stakeholders", so that the rising attention to MSPs is seen as "an expression of a changing approach to addressing development goals" (Lundsgaarde and Keijzer 2019:218). More specifically, MSPs are likewise part of sustainability research, also with regard to MSPs for sustainable development (e.g., Bäckstrand 2006.) Under the principles set out in the 2030 Agenda, MSPs should also be designed to be inclusive and to leave no-one behind (Ellersiek and Beisheim 2017:7–8). As mentioned, inclusiveness and transparency are success factors for network governance, including the governance form of MSPs. Inclusiveness in the (sustainable) development context is especially important, which is highlighted for instance in articles 84 and 89 of the 2030 Agenda that call on partnerships and other non-state actors to report their activities (Ellersiek and Beisheim 2017:7–8). Moreover, partnership formats such as MSPs highlight the relation between public and private actors, especially in knowledge and resource exchange: "Partnerships involving private-sector actors have thus attracted interest as a vehicle for mobilizing resources by providing incentives such as financial guarantees or accompanying technical assistance, while promising to transfer knowledge from the business world" (Lundsgaarde and Keijzer 2019:218). Hence, MSP are a potential vehicle for overcoming ICT power, knowledge and resource imbalances (Sect. 2.3.6) by collaborating and pooling resources.

Hence this dissertation analyzes MSPs as a governance format in the sense of a 'new' coordination agenda, shifting from top-down state-centric approaches to multi-stakeholder formats. However, "[a]longside other examples of diversifying cooperation models, global multi-stakeholder initiatives create new coordination challenges at multiple governance levels" (Lundsgaarde and Keijzer 2019:224). Table 2.2 summarizes the main shifts underlying a reorientation towards MSP in the debate around coordination in development cooperation:

Table 2.2 Comparing main characteristics of past and present efforts to promoting coordination in development cooperation (own illustration based on Lundsgaarde and Keijzer 2019:225)

State-centric coordination	Multi-stakeholder coordination
Negative framing, emphasizing the costs of non-coordination	Positive framing, emphasizing the gains of joint action
Limited to official actors: Governments and their bi- and multilateral development partners	Multi-stakeholder orientation, and explicit ambition to got beyond official actors
Emphasis on top-down planning and centralized decision-making	Focus on enabling as opposed to directing coordination
Country-specific "across the board" effects	Cross-country and sector-centric efforts

MSPs in ICT4D /Digital Development Moreover, MSPs are likewise discussed as a governance format in ICT4D. Here, MSPs have become a framework reflecting the before mentioned transformations in development cooperation (Uimonen and Hellström 2015; Lundsgaarde and Keijzer 2019). Due to the blurring of the private and public sphere in ICT4D, it is also referred to as a "hybrid domain" (Aoyama and Parthasarathy 2017:258). Calls for multi-stakeholder approaches in ICT4D are recently growing, especially in terms of AI and gender equality as highlighted in the IGF (2019) report. The most common examples of MSPs (and PPPs) in the ICT context are infrastructure (Nederhand and Klijn 2019), e-government and national e-IDs (Medaglia et al. 2018). Following the public-private relation of governments and the private sector in ICT4D, partnerships and cooperation are addressed on various levels. With special regard to the donor level, "[m]any donors contribute to and participant in D4D-relevant fora such as the eCommerce Week, the WSIS Forum and the Internet Governance Forum" (UNCTAD 2019:38). The latter Internet Governance Forum has been previously discussed in Sect. 2.4.1. Despite theses advances, there is still no specific forum for facilitating exchange, for instance between donors "in matters of support to the SDGs through digital economies in developing countries" (UNCTAD 2019:38). This dissertation aims at contributing to research around governance formats for facilitating exchange but also coordination.

As illustrated, national governments often consult the private sector and civil society organizations. As private tech companies constitute the majority of the internet's backbone, they also "enact policy not only in carrying out their core functions but also as actors responding to events on a larger political stage" (DeNardis 2014:12). Hence, for-profit actors are often regarded as important actors who are able to provide much-needed knowledge, innovations and investments (Wehrmann 2018). More specifically with regard to developing countries, governments consult a broad range of national stakeholders in designing national development strategies: "Almost all partner country governments (93 percent) report they consulted three or more stakeholders when designing their national development strategy." (OECD/UNDP 2019:59). Moreover, "[p]artner country governments and the private sector demonstrate promising mutual trust and a willingness to engage with each other. Governments express sincere interest in engaging the private sector. Private sector stakeholders demonstrate shared optimism in this area" (OECD/UNDP 2019:72). This again highlights the importance of identified preconditions for coordination such as trust and willingness of multi-stakeholders (Sect. 2.5.1).

Public-Private Partnerships for Sustainable Digital Development

Similar to MSPs, PPPs aim at cooperation between public and private actors, here with a focus on the context of digital development activities. Klijn and Teisman (2003), for instance define PPP as a "[...] cooperation between public and private actors with a durable character in which actors develop mutual products and/ or services and in which risks, costs and benefits are shared" (Klijn and Teisman 2003:137). Similar to the international attention to and promotion of partnerships such as MSPs and PPPs (e.g., SDGs)), these formats are strongly supported by donor organizations, national governments and other organizations. This is emphasized for instance in the PPP Knowledge Lab initiative[10] or the develoPPP initiative on a multilateral level (Bayliss and Van Waeyenberge 2018). Similarly, Nederhand and Klijn (2019) argue that PPP are seen as 'the' governance strategy among governments, especially with regard to improving service delivery and infrastructure projects, including the ones in developing countries (see also Bayliss and Van Waeyenberge 2018).

While MSPs mostly address cooperation and partnership on an international level, PPPs are mostly implemented on a national level (e.g., Börzel and Risse 2010). More importantly for this dissertation's analysis of the interaction between multi-stakeholders in digital ecosystems and their impact on sustainable digital development, PPPs can be distinguished from traditional public procurement through contracts between public and private sector actors (Bayliss and Van Waeyenberge 2018): They "usually take the form of a concession-type contract with variations in the requirements from the private partner in the configurations of Design, Finance, Build, Operate, Maintain and Transfer of the infrastructure asset" (Bayliss and Van Waeyenberge 2018:577). Consequently, governments are similarly accounta-

[10] see https://pppknowledgelab.org

ble for services provided and funding derived from either public or private sector or both (Bayliss and Van Waeyenberge 2018). However, the literature states that in most PPPs, public actors take the role of a buyer while private sector agencies are suppliers (e.g., Medaglia et al. 2018). This dissertation understands this as one example for (institutionalized) power asymmetries and inherent dependencies, while likewise emphasizing the degree of closeness between the public and private actors as rather distant (Sect. 2.5.1) due to its contractual relation (even though it is labeled as a 'partnership'). More importantly for this dissertation, there is debate as to whether the power asymmetry between actors might be more balanced or more equal in ICT projects such as e-ID projects because of the interdependence of the different actors' resources (Medaglia et al. 2018:2). Nonetheless, stakeholder involvement in PPPs is under-researched (Nederhand and Klijn 2019). More specifically, the authors argue that PPPs as "a part of a larger network where various stakeholders are involved is often neglected in PPP literature" (Nederhand and Klijn 2019:1002). This dissertation will deliver further insights into the role of partnership formats such as PPPs as a governance format for sustainable digital development in the empirical analysis of the Kenyan digital ecosystem (Sect. 4.4.6). As this dissertation highlights, governance of relations between the digitally involves stakeholders is key, especially with regard to governance of PPPs (Klievink and Janssen 2014; Klievink, Bharosa and Tan 2016; Medaglia et al. 2018).

Advantageous implications of PPPs for digital development are for instance seen with regard to the innovativeness of PPP projects (Nederhand and Klijn 2019), for instance due to (multi-)stakeholder involvement in improving content and innovativeness or enhancing democracy (see also e.g., Klijn and Koppenjan 2016; Sørensen and Torfing 2007; Nederhand and Klijn 2019). Besides possible fostering of innovations or innovativeness through PPPs, another possible advantage is seen with regard to efficiency and finance in terms of returns to investors on the one hand and service or infrastructure delivery on the other (Bayliss and Van Waeyenberge 2018). For public sector or governmental actors in the partnership, (digital) innovation can be encouraged through "striking the right balance between maintaining the control needed to secure public value and enabling autonomy that can encourage innovation" (Medaglia et al. 2018:3). This however poses a challenge, especially with regard to PPPs in ICT4D, regarding PPP's design and human capital (Medaglia et al. 2018; Ojo and Mellouli 2016). Consequently, this dissertation aims at contributing to the current research since there is a need for guidance and learning regarding the design and management of multi-partner models (Cochrane and Cundill 2018; Macharia 2016).

Challenges of MSPs and PPPs for Sustainable Digital Development
Besides these potential benefits of partnerships for digital development, the relation between public and private actors and their cooperation, faces challenges. Firstly, financial and technical resources as well as capacities, which are needed to address capacity constraints that limit participation in partnership formats as highlighted in Sect. 2.5.2 and further emphasized in OECD/UNDP (2019). As one example, the OECD/UNDP (2019) report states that "[p]ublic and private actors are willing to participate in dialogue, but have limited capacity to engage effectively" (OECD/UNDP 2019:72).

Secondly, effectiveness of partnership formats is a further challenge for sustainable (digital) development. The OECD study for instance emphasizes that consultations in the PPD "could be more effective, inclusive and transparent" (OECD/UNDP 2019:65), which this dissertation highlights as one factor hindering sustainable digital development. In this context, also common language and understanding (Sect. 2.5.1) come into play as "the views of governments and private sector stakeholders differ on the quality of their PPD" (OECD/UNDP 2019:71). This is closely linked to challenges arising from often poor designs and weak monitoring systems of partnership formats such as PPP and MSP (OECD 2018b). Hence Keijzer et al. (2018) state that "[e]fforts to monitoring the engagement of private sector actors in multi-stakeholder platforms need to strike a balance between joint learning and doing justice to the different forms and contexts of various initiatives" (Keijzer et al. 2018:51), which currently remains a challenge. Moreover, PPPs have been characterized by mixed success (Aldasoro et al. 2010; Beisheim et al. 2017) and even very high failure rates of PPPs in ICT4D are attested (e.g., Mirza 2013, similarly to Sect. 2.3.1).

Besides these exemplified advantages of PPPs, disadvantages of this instrument are likewise debated in the literature. For instance, "PPPs remain a highly contested vehicle for infrastructure financing and delivery" (Bayliss and Van Waeyenberge 2018:577). Similar to any project, including ICT4D projects (Sect. 2.3), MSPs likewise are not immune to failure in delivering sustainable digital development. As stated, private sector actors are the main actors in the development and deployment of ICT, especially infrastructure (Uimonen and Hellström 2015). This puts them in a position of power, with power asymmetries arising (Sect. 2.3.6). As the private sector is bound by a market logic, inequalities can be commercially valuable: "The deployment of ICT generally follows patterns of economic inequality, the most lucrative markets being the best served." (Uimonen and Hellström 2015:8) This has a negative impact on PPPs as companies aim at selling their products and services, so that "their approach to public-private partnerships often translates into businesses selling goods and services to governments" (Uimonen and Hellström

2015:8). Here the divergence of interests between multi-stakeholders as a challenge for both coordination and sustainable digital development becomes obvious (Sect. 2.3.3, Sect. 2.1.4). Ismail et al. (2018) further conclude concerns about private sector domination in PPPs and MSPs as well as about private sector domination in ICT4D partnerships in many developing countries.

Nonetheless, the literature has elaborated on the importance of overcoming theses challenges of partnerships such as MSPs and PPPs. Pattberg and Widerberg (2016) for instance identify conditions for improving collaborative settings such as partnerships in terms of "a) the processes (goal setting, funding, management and monitoring), b) contexts (meta-governance, problem structure, and socio-political contexts) and c) the actors (leadership and partners)" (Cochrane and Cundill 2018:923–924). This is similarly reflected in this dissertation's design (Fig. 1.1).

Lead Form of Network (Ecosystem) Governance

Opposed to the logic of self- or shared-governance, an additional governance format for coordinating digital ecosystems are lead forms of networks as proposed by Provan and Kenis (2008). Here, network governance is conducted by and through a single organization (lead organization), which acts as a centralized network broker that coordinates all network-level activities (Provan and Kenis 2008:234–235). While the authors understand the lead form of governance in the role of being a network broker, this dissertation analyzes the role of a broker separately from any governance format (Sect. 2.5.3). The lead organization format for governing as a centralized format is further characterized by asymmetrical distribution of power and by being governed internally by a single participating network member (Provan and Kenis 2008:235–236). In this research project, the lead organization is understood as a member of the digital ecosystem on a national level (within a country) or one outside it (e.g., international level). The implications of which organization participating in the ecosystem takes a governance role is discussed in more detail in Sect. 2.5.3.

As a summary, the following list demonstrates for which networks shared governance is efficient or chosen:

- *Trust*: In networks with low-density trust levels, network governance is likely to be brokered, either through a lead organization or through an NAO, for instance due to threats of opportunistic behavior of network members (*Low density*, highly centralized) (Molin and Masella 2016a; Provan and Kenis 2008)
- *Size*: With increasing size and increasing complexity to govern increased number of (diverse) actors, a lead organization or NAO governance form could better manage the number of actors (than self-governance) (Provan and Kenis

2008:239) where a "separate administrative entity could ensure the goals' achievement and activities coordination" (Molin and Masella 2016a:502) (Moderate number of actors, Provan and Kenis 2008:237)
- *Type of actors / diversity*: High degree of diversity is better governed through lead organization or NAO model, "in which a single organization has the power of coordinating the others' activities, behaviors and goals, avoiding opportunistic conducts" (Molin and Masella 2016a:502)
- *Goal Consensus*: In networks with *moderately low* goal consensus, governance through lead organizations is preferred (Provan and Kenis 2008:240). In low levels of goal consensus, the presence of external factors can avoid the risk that participants pursue their own interests (Molin and Masella 2016a:502) (relevant for both lead organization and NAO form)
- *Willingness*: Where commitment (here willingness) of network members is strong, lead organization or NAO are claimed as suitable for governing the network as the role of leader is clearly defined (Molin and Masella 2016a:502)

Network Administrative Organization

As a third form of network governance proposed by Provan and Kenis (2008) functions the NAO model. This governance form is similar to the before mentioned lead organization form as the network is likewise governed by a separate organization for coordinating the network activities. In contrast to the lead organization, in the NAO model the network is externally governed and established outside the network's boundaries for the purpose of network governance through mandate or members themselves, so that it may stem from a variety of actors groups (public or private) (Provan and Kenis 2008:236). In this, it fulfills similar roles as discussed in Sect. 2.5.3, since the NAO is responsible for facilitating or brokering the network. The discussion likewise reminds of the discussed role of external actors in digital ecosystems such as multinational tech companies or donor organizations (Sect. 2.1.4) similarly to the development theory perceptions of development through endogenous or exogenous factors as discussed in Sect. 2.1.1.

As a summary, the following list demonstrates for which networks shared governance is efficient or chosen:

- *Trust*: Moderate density (Provan and Kenis 2008:237)
- *Size*: Moderate number to many actors (Provan and Kenis 2008:237))
- *Type of actors / diversity*: High degree of diversity is better governed through Lead Organization or NAO model, "in which a single organization has the power of coordinating the others' activities, behaviors and goals, avoiding opportunistic conducts" (Molin and Masella 2016a:502)

- *Goal Consensus*: Moderately high (Provan and Kenis 2008:237)
- *Willingness*: see section lead organization

In this dissertation the NAO governance form is combined with meta governance with regard to the governance format of a 'meta-governor' or 'meta-coordinator' of digital development actions in the digital ecosystem. Metagovernance here is understood as NAO governance format as an organization that 'stands above' the ecosystem and it will deliver insights into the role of the establishment of a new external organization.

Meta Governance as Network Governance Form

As a further governance form, this research project uses meta governance research as closely linked to the previously mentioned network governance formats of self-governance, lead organization and NAO. Meta governance is a concept describing "governance of governance" (Meuleman 2019:15). With regard to this dissertation's focus on governing digital ecosystems and especially the role of multi-stakeholders, meta governance is chosen for this framework as it focuses on the efforts of public authorities in steering networks through different strategies and the way in which networks are empirically governed (Molin and Masella 2016a; Baker, Edwards and Doidge 2012). More importantly for this dissertation, parallels can be seen between sustainable development aiming at creating balances between economic, political, environmental and social dimensions as a 'meta-policy' (Meadowcroft 2011) and meta-governance that is "[...] about creating balanced combinations of tools from three different governance styles, notably hierarchical, network and market governance" (Meuleman 2019:5). Hence, meta-governance is seen as suitable for addressing the numerous 'wicked' problems that the SDGs aim to address, for instance by Monkelbaan (2019).

In the literature, meta governance is further seen as an approach, foremost in governance research and sustainability research, for *network management* (Klijn and Edelenbos 2007; Sørensen and Torfing 2007; Jessop 2002). Hence it is chosen by this dissertation for analyzing digital ecosystems in developing countries as it incorporates both sustainability with governance perspectives. However, in the reviewed literature, only limited references to ICTs or ICT4D have been found. In this dissertation metagovernance will be understood as "a means by which to produce some degree of coordinated governance, by designing and managing sound combinations of hierarchical, market and network governance, to achieve the best possible outcomes [...]" (Meuleman 2008:68). Meta governance from the network governance angle is especially relevant for this research project since it provides not only insights into governance structures but also into government action with regard to the role of the government in network governance.

Firstly, while governments are seen as responsible for steering different sets of organizations, it is argued in (meta) governance literature that such traditional (hierarchical) governance mechanisms are to be adapted to networks where governments act as meta-governors of networks exerting network design, framing, management and participation (e.g., Keast et al. 2004; Sørensen and Torfing 2009; Molin and Masella 2016b). The degree of intervention on behalf of the government in the role of a meta-governor spans from limited to strong intervention (Sørensen 2014). Daugbjerg and Fawcett (2017) for instance established a heuristic continuum ranging from high to low for assessing the state's capacity to metagovern. Similarly, Meuleman (2008) argues that regarding the performance of public sector organizations those responsible see public managers as 'metagovernors' (ibid.:75). This dissertation will elaborate on whether this is the case in digital development networks in the Global South or if it rather applies to actors from the private sector. Especially since it is important for the actor governing a network to be inclusive and pool assets, to foster collaboration in the network, to stimulate innovation and to diffuse knowledge and innovation within but also beyond the network (Sørensen 2014).

Secondly, focusing on the state's capacity to metagovern, the state's role is necessary for input and output legitimacy (see also Sect. 2.5.2). In terms of input legitimacy "it has the broader legitimacy necessary to make decisions about who to include and exclude from the network" (Daugbjerg and Fawcett 2017:1232). Critical here is its flexibility, which is "constrained by the broader structural and ideational context of the policy area in question" (Daugbjerg and Fawcett 2017:1232) as similarly discussed in Sect. 2.5.2. Regarding output legitimacy, the state's role is crucial "because non-state actors are unlikely to command all of the necessary resources and authority to deliver high levels of output legitimacy. They are also unlikely to be under the same obligation as the state to deliver collective welfare functions and may adopt rent-seeking behaviors." (Daugbjerg and Fawcett 2017:1232–1233) This is especially relevant in governing the digital transformation where non-state actors, especially the private (IT) sector, undertake functions previously performed by the state. This applies also to situations of limited statehood, where non-state actors are increasingly present (see e.g., Daugbjerg and Fawcett 2017 as discussed also in Sect. 2.1. In these cases, if the state only has weak capacity to metagovern, the responsibility falls to non-state actors with "necessary interest, resources, authority, and legitimacy to perform this role" (Daugbjerg and Fawcett 2017:1232). This will be discussed in more detail with regard to Kenya in Sect. 4.3.4 and Sect. 4.4.5.

Thirdly, one proposed solution addressing this trade-off between public and private governance, lies in strengthening government responsibilities and capabilities

(Glasbergen 2011). Similarly to the logic of (good) governance, "[...] states are the ultimate and exclusive authoritative agents in public affairs" (Glasbergen 2011:191). In contrast to governance and coordination via the market, which "only delivers sustainable progress if the market is embedded in politically designed frameworks", the state needs to set the 'rules of the game' within which boundaries private forms of governance need to act (Glasbergen 2011). Hence, one argument in favor of a strong position of the state in meta-governing networks comes from its democratic credentials given private regulation's "[...] negative influences on state power, and their inability to induce the necessary sustainable improvements" (Glasbergen 2011:191). More importantly with regard to (sustainable) digital development, it is to be highlighted that "[b]usinesses are not run on democratic lines" (Meadowcroft 2007:19). Hence, Glasbergen (2011) questions the implications of relying on the role of states in meta-governance of networks. The author raises concern about the role of the state in sustainability issues in global networks. Due to the complexity of these problems, "[..] a central steering role for states very unlikely" (Glasbergen 2011:194). Moreover, the question of who governs these networks by referring to the fact that there is no supranational authority setting parameters is highlighted. Whoever it may be, Provan and Kenis (2008) claim that it is important for coordinating a network of service providers or stakeholders to have a central or lead organization. Additionally, the question whether any institutions can restrict or prevent private regulations is raised considering "the validity of the assumption of self-serving activities of private governance arrangements" (Glasbergen 2011:194). With regard to sustainable development, private regulations therefore are seen as unable "to induce the necessary sustainable improvement" (Glasbergen 2011:191). These arguments are counterarguments towards the state and hierarchy as the 'metagovenor' in policy networks.

Fourthly, meta governance is widely referred to as a means to overcome fragmentation. The overarching challenges of aid fragmentation, aid effectiveness and donor coordination have been established in Sect. 2.1. As further established in Sect. 2.5.4, alternative governance arrangements such as MSPs are an indicator for the emerging need for addressing fragmentation that often leads to functional overlaps and competition (see e.g., Pattberg and Widerberg 2016). This is further in line with this dissertation's findings in Sect. 2.1.5 and Sect. 2.5.2. Meta governance offers "provisions for *increasing aid effectiveness* and to *facilitate PPPs*" (Beisheim et al. 2017:107, emphasis added in the original). However, little is known in research "about the effect of meta-governance frameworks on the performance of partnerships" (Beisheim et al. 2017:107). As the authors assert, there is little research on national-level meta-governance for MSPs. While Beisheim et al. (2017) have analyzed meta governance of two water partnerships in Kenya for sustainable development, this dissertation aims at contributing to filling this research gap by analyzing the role of partnerships in governing the digital transformation (also using Kenya). As a consequence of only little research, the authors suggest meta-governance for fulfilling functions of enabling and ensuring partnership's effectiveness (Beisheim et al. 2017; Beisheim and Simon 2016). Pattberg and Widerberg (2016) further argue that for mitigating the risk of fragmentation, MSPs should consider meta-governance.

Fifthly, and most importantly for this dissertation's focus on digital ecosystems, the literature around meta-governance in the digital context is scarce. Most recently as a rare example, meta-governance in the context of IoT is discussed by (Sutherland 2020). Similarly, platforms are argued to play the possible role of 'meta-governance' or 'orchestration' (Ansell and Gash 2018), as further discussed in Sect. 2.2.5. Platforms are here seen as 'meta-organizations' coordinating actors (Ansell and Miura 2020) and providing resources (such as information) for its users. Against this background, this dissertation's framework proposes a digital platform acting as a (meta-)governor for sustainable digital development of digital ecosystems with a focus on developing countries in the next section.

Digital Multi-stakeholder Platform for Ecosystem Governance

As elaborated in the previous section, platforms can serve the purpose of meta-governors. Moreover, as further discussed in Sect. 2.2.5, platforms are one of the current 'hype' technologies and with regard to the current trend of platformisation in the digital transformation context. Moreover, platform ecosystems are used interchangeably with (other) forms of digital ecosystems (Sect. 2.2.5). With regard to this dissertation's focus on ICT4D, it needs to be noted that an understanding of what digital platforms are, what their main features are and how they generate value are lacking (de Reuver, Sørensen and Basole 2018).[11] This is also due to the fact that digital platforms have only been addressed in ICT4D research in the last years (Bonina et al. 2021:881). Following Constantinides, Henfridsson and Parker (2018) for instance, digital platforms can be defined as sets of digital resources (e.g., digital services or content) that facilitate interactions between its participants. In this dissertation, participants or users of a digital development platform for coordination or governance would originate from the multi-stakeholder actor groups (donor organizations, governmental organizations, private sector, NGOs and academia). Platforms can serve "as the organizing nexus for multiple networks rather than for a single network" (Ansell and Miura 2020:262) and thereby fit into the here underlying goal of sustainability and inclusivity.

[11] Bonina et al. (2021) provide an overview and literature review on digital platforms for development.

This dissertation proposes the use of digital platforms as a governance format for coordinating digital development activities. Platforms here are not analyzed with an economic focus as it aims at analyzing potentials of platforms for governance and thereby follows Ansell and Miura (2020)'s argument that governance platforms do not build on market transactions. According to Ansell and Miura (2020), *governance platforms* can be defined as "an institution that strategically deploys its architecture to leverage, catalyse, and harness distributed social action for the purpose of achieving certain governance objectives" (Ansell and Miura 2020:264). This dissertation follows Ansell and Miura (2020)'s approach in that digital platforms are analyzed with regard to whether platforms and their organizing logic of connecting, intermediating and mobilizing can be harnessed for public administration and governance (ibid.:262). So far, platforms in and for governance still remain "largely experimental" (Ansell and Miura 2020:262), which highlights the timely relevance of this dissertation. This is especially underlined with regard to the only recent focus of ICT4D research on platforms and more importantly with regard to the context of digital ecosystems. Bonina et al. (2021) for instance argue that "there is little attempt to examine the influence of the wider ecosystem members and stakeholders such as regulators, governments and alike on governance regimes" (ibid.:880). Synthesizing governance research with digital platforms for governance demonstrates potentials for achieving important governance roles (Ansell and Miura 2020). This is similarly confirmed by Bonina et al. (2021)'s claim that categorizations of platforms such as in form of transaction platforms are "[...] equally suited to grasp the broader context of achieving developmental goals" (ibid.:874). This is especially relevant with regard to the aim of platforms for information sharing and emphasizing the role of platforms as a facilitator: "[...] transaction platforms are largely intermediaries or online marketplaces that enable seemingly distant people and organizations to share information, to sell, buy or access a variety of services" (Cusumano et al. 2019:19).

This leads over the question how digital platforms themselves can be governed as similarly reflected in the previous discussion around brokered or non-brokered governance formats in the beginning of this section. While the digital platform literature reviewed in Bonina et al. (2021) tends towards a platform owner coordinating and governing the wider platform ecosystem and its members (Iansiti and Levien 2004; Ghazawneh and Henfridsson 2013; Wareham, Fox and Cano Giner 2014), the question of governance here is also a technical one as addressed for instance in Sect. 2.2.5. However, this plays only a subordinate role for this research project as the focus here is on what digital platforms can offer for coordination of multi-stakeholders activities in the digital ecosystem. The identified preconditions in Sect. 2.5.1 here likewise apply and need to be included for platforms' functioning as a governance form. Moreover, the empirical analysis will deliver further insights into digital platforms for sustainable digital development (Sect. 4.5.3).

2.5.5 Determinant 5: Outcome of Governance for Coordination

As a last section of the proposed framework serves the outcome of network governance of digital ecosystems in developing countries for sustainable digital development. Sustainable digital development in this dissertation is not only the goal to be achieved (Sect. 2.1.3), but also the network rationale and therefore the pursued governance outcome. Hence determinant 5 likewise functions as an interim conclusion of the theoretical background as it incorporates all of the previous findings, which is further illustrated in Fig. 2.6.

In the previous sections, it has been elaborated that effective digital development promotion needs to be inclusive for a sustainable development. More specifically, ICT4D and digital development has been studied "in terms of its intended purpose, process and impact under labels such as inclusive, frugal, grassroots, pro-poor and bottom-of- the-pyramid innovation" (Sanner and Nielsen 2019:219). Sustainable development and inclusivity is a crucial theme in ICT4D literature that gains importance (e.g., Sanner and Nielsen 2019; Nan and Markus 2019; Heeks et al. 2014; 2013). On an international level, the digital transformation is discussed as potentially facilitating international economic inclusion through sectors like the gig economy (Sect. 2.2.5). Digital inclusion further entails both a socio-economic, international as well as a political dimension and spans from the digital innovations itself to processes around digital inclusion (e.g., Sanner and Nielsen 2019). For instance, digital technologies contribute to intra-national inclusion of previously marginalized parts of the population, especially through mobile banking (Sect. 2.2). With regard to political inclusion, digital technologies potentially contribute to increased transparency of action and participation (Sect. 2.4.1, Sect. 2.5.1). What becomes obvious is that inclusivity is crucial for sustainable digital development on various levels. Dimensions of inclusion for sustainable digital development are for instance summarized by Sanner and Nielsen 2019:219:

- Participatory dimension: bridge traditional dichotomies between users and producers (based on Foster and Heeks 2014; Chataway, Hanlin and Kaplinsky 2014)
- Supply side and the demand side (based on Foster and Heeks 2013)
- Align top-down standardization with local innovation (based on Foster and Heeks 2014; Heeks et al. 2014)

- Degree of inclusion: inclusion in consumption and use of innovations; inclusion in innovation development processes; inclusion in defining and shaping the very structures through which innovations are produced (based on Heeks et al. 2013)

Besides these potential contributions of digital technologies to development, as highlighted in the TWI2050 (2018) report, if the digital transformation is "not shaped appropriately and geared to the SDGs of the 2030 Agenda, then digitalization could multiply already existing problems in many societies" (ibid.:125). These include both economic as well as political inequalities for instance in the labor market, education, data sovereignty and civic rights with further eroding governance capacities of public organizations (ibid.). Hence, while the advantages and possible positive driving features of digital technologies have likewise been analyzed (Sect. 2.2), there are still very few instances of successful processes for shaping and governing digitalization around the world (TWI2050 2018; WBGU 2018). This dissertation provides insights into how digital development can be shaped in order to be sustainable and inclusive as opposed to what implications it has for sustainable digital development if it is shaped 'wrong' i.e., unsustainably.

Against this background, this dissertation argues that one essential challenge for sustainable digital development is the inclusion of multi-stakeholders in the digital ecosystem. Sect. 2.1.4 for instance addressed the crucial role of non-state actors such as the private sector as intermediaries in digital development promotion, especially in ICT4D projects. Sect. 2.3.1 emphasized that ICT4D programs and projects of bilateral and multilateral donors among others often aim at solving a particular problem based on programmatic solutions. Such processes, however, are often inclusive in outcome, but not in process (Sanner and Nielsen 2019). Hence, inclusiveness of digital development is crucial on various levels of action, not only in terms of outcome but from the process beginning on. Additionally, it has been elaborated that in order to 'do digital development right', sharing and scaling-up of knowledge and innovations among others in the network of stakeholders and across settings with variable resources and capacities needs to be coordinated. Only if these relations are governed inclusively and sustainably (Sect. 2.4), can they contribute to sustainable digital development. Otherwise, dependencies and asymmetries between the multi-stakeholders will even increase (Sect. 2.3.6) and dangers of not only increasing the digital divide (Sect. 2.3.4), but also notions of digital colonialism (Sect. 23) might be (unsustainable) consequences. For this purpose, various governance roles in digital ecosystems have been derived from the synthesis of network governance and ICT4D literature and discussed with regard to the various multi-stakeholders (Sect. 2.5.3) including coordinators, drivers of digitalization, facilitators, brokers and network managers.

This dissertation proposes a framework for sustainable digital development in digital ecosystems in developing countries using network governance in combination with ICT4D literature for analyzing coordination. From a development research angle, coordination has been highlighted for addressing aid fragmentation challenges for instance in terms of reduced administrative capacity of partner governments, reduced transaction costs and transparency through information and knowledge sharing (Sect. 2.1.5, Sect. 2.1.6, Sect. 2.1.7, Sect. 2.3.6). More specifically, aid and donor fragmentation has been identified with regard to *administrative capacity burdens*, for both donors and recipients, which "could be reduced if the aid coming from numerous donors was well coordinated, predictable, and used joint approaches to reduce the management burden" (Davies and Klasen 2019:245). Secondly, coordination can reduce *transaction costs* associated with aid delivery by limiting the administrative burden on recipient governments (Lundsgaarde and Keijzer 2019) through better donor coordination (Bigsten 2006; Nunnenkamp, Sotirova and Thiele 2016). *Information and knowledge sharing* is a third advantage which can possibly be achieved through coordination beyond it merely being based on information sharing among donors (Lundsgaarde and Keijzer 2019) and other multi-stakeholders. In terms of aid transparency and coordination:

> "Available research on initiatives to improve aid transparency illustrates how information availability and information sharing may enable coordinated action. Efforts to increase the scope and quality of aid data such as the International Aid Transparency Initiative (IATI) have expanded donor reporting, however researchers have questioned the impact of these global information sharing platforms [...]" (Lundsgaarde and Keijzer 2019:22)

This is further discussed in Sect. 2.2.5 and Sect. 2.5.4. Most importantly for this dissertation, inclusiveness, fairness, accountability, and transparency are not only identified in the context of aid effectiveness (Sect. 2.1.7), but also for sustainable digital development. Sustainable and inclusive digital development is hence not only the goal to be achieved (Sect. 2.1.3) and the network rationale of the digital ecosystem (Sect. 2.5.1), but also the intended outcome of coordination through network governance.

Ultimately, the theory of network governance is chosen for analyzing how digital ecosystems can be coordinated for sustainable digital development as it offers 'solutions' (Sect. 2.5.4) for overcoming ICT4D challenges (Sect. 2.5.2). In line with Molin and Masella (2016b)'s argument that "[...] we do not actually know how to define and conceptualize the outcome of network governance and, therefore, how to evaluate and measure it" (ibid.:836), this dissertation understands the outcome of network governance of digital ecosystems as sustainable digital development.

Consistent with the multi-stakeholder approach of this dissertation, network outcomes can be evaluated on three different levels (Provan and Milward 2001): the individual organization (i.e., the single organization), the network itself and the community of services users (i.e., local area served by the network). In this dissertation the analysis level of network outcome refers to the community or rather multi-stakeholder interaction level. Further evaluation outcomes in network governance usually include effectivity and efficiency (Sect. 2.5.1) or democracy (Sørensen and Torfing 2005), which here are analyzed from the aid effectiveness and legitimacy of action angle (Sect. 2.1.7, Sect. 2.5.2). These are especially important for sustainable digital development since input and output legitimacy in network governance both include inclusiveness, fairness, accountability, and transparency (Klijn and Edelenbos 2007) (input legitimacy) as well as the effectiveness of networks, including their capacity to solve problems and deliver better policy outcomes (Daugbjerg and Fawcett 2017) (output legitimacy). This underlines the relation between aid effectiveness, sustainable digital development and network legitimacy. Thus, the assumption can be made that if a network/ecosystem is effective, aid effectiveness of involved stakeholders in the network would be improved as well. Against this background, not only governance roles of multi-stakeholders in the digital ecosystem are discussed (Sect. 2.5.3), but also possible governance formats, including digital platforms as governance formats (Sect. 2.5.4).

Conclusively, the proposed framework (Sect. 2.5) serves as a theoretical insight into governing digital ecosystems for sustainable digital development by providing preconditions for governance (Sect. 2.5.1), specific challenges to sustainable digital development (Sect. 2.5.2) before introducing possible governance approaches to coordination in terms of governance roles (Sect. 2.5.3) and governance formats (Sect. 2.5.4). Before this will be transferred to the empirical case of the Kenyan digital ecosystem (Chap. 4), methods as to how the research goal is pursued will be discussed in the following.

Research Design and Methodology

3

This chapter elaborates on this dissertation's research design and methodology for gathering empirical insights into how digital ecosystems can be governed for sustainable digital development using the case study of Kenya. As introduced in Chap. 1, a qualitative and inductive research design is applied for this research purpose using expert interviews for the data gathering process and qualitative content analysis for the data analysis. More specifically, this chapter will firstly discuss the methodological positioning of this research project before analyzing the fieldwork preparation process including sampling, preparation of the interview guideline, interview pretest and field access. The data collection through guideline-based expert interviews will be further discussed in addition to the data analytic process of transcription and interpretation of verbal data using qualitative content analysis and qualitative network analysis.

3.1 Methodological Positioning

Firstly, the methodological positioning of this dissertations follows a qualitative research design. The individual work steps of the research project will be explained in the following sections with regard to the consistency of the methodological assumptions. A qualitative approach to the implications of the digital transformation for governance of digital ecosystems in developing countries is chosen after assessing the fit with the exploratory nature of the research goal and the target group of experts in the field (Przyborski and Wohlrab-Sahr 2014a). Hereby, following the pragmatic position on the design of Creswell and Creswell (2018), one important factor is that the phenomenon of governing the digital transformation in developing countries is assessed as a phenomenon that applies to a circumstance that "[...]

C. Stötzel, *The Digital Ecosystem of Kenya*,
https://doi.org/10.1007/978-3-658-45553-8_3

needs to be explored and understood because little research has been done on it [and] because it involves an understudied sample" (ibid.:19). These factors merit the qualitative approach. In addition to the need for exploring a previously under-researched problem, which is also expressed by the lack of appropriate theories that are able to adequately grasp the empirical reality, qualitative access to the dissertation is appropriate if the research project's objective is a detailed understanding of the complexity of the investigation problem. In contrast to a quantitative study, a qualitative design also allows to explore the context conditions and should be used especially if—as in the mentioned dissertation project—interactions between (network) actors should be mapped:

> "We use qualitative research to develop theories when partial or inadequate theories exist for certain populations and samples or existing theories do not adequately capture the complexity of the problem we are examining. We also use qualitative research because quantitative measures and the statistical analyses simply do not fit the problem. Interactions among people, for example, are difficult to capture with existing measures, and these measures may not be sensitive to issues such as gender differences, race, economic status, and individual differences." (Creswell 2013:39–40).

Here, the context of developing countries and especially of Kenya is crucial for this research project. The theoretical (and empirical) evidences have highlighted the importance of contextual factors for the implications of the digital transformation (Sect. 2.1). Moreover, interactions among multi-stakeholders in a digital ecosystem are the unit of analysis in this dissertation. Against this background, the theoretical background (Chap. 2) simultaneously serves the purpose of a literature review.

Moreover, the methodological positioning also includes a reflected approach to the significance of theoretical concepts and hypotheses in one's own qualitative research project. While theoretical concepts may serve as a guide for researchers to explore which aspects of empirical reality are to be investigated, in qualitative studies, however, theories are usually not tested. Instead, they serve the purpose of object-related theory extraction, whereby the analysis of the data is done inductively (Creswell 2013). In this dissertation a framework for governing digital ecosystems in developing countries is developed through a literature review (no theory test). This is further underlined with regard to the classification of the relationship between qualitative research and theory formation into an explorative, explanatory or constructive function (Lamnek 2010). According to Lamnek (2010), in cases where qualitative research is given an explorative function in the development of theory, then it is reduced to a purely preliminary exploration of the field of research. Thereby qualitative research can also have the function of formulating, modifying and checking theories. This is the case in this present research project where theories stemming from development theory / ICT4D and network governance are modified and transferred to the digital transformation context of digital ecosystems.

This is further relevant for this dissertation since qualitative research is field research and the investigation process cannot be defined in advance, in contrast to controlled laboratory experiments, since the research process for qualitative researchers is emergent and hence may change or shift after entering the field. This enables the researcher to learn about problems or issues from participants and adapt the research accordingly (Creswell 2013:39). This openness and flexibility is especially crucial for this dissertation's analysis of the current and highly dynamic digital transformation. Despite this openness, reflection in the planning of the research project and understanding the research as communication or interaction between the researcher and the object under investigation are required, as they will have an influence on the results of the data collection (Lamnek 2010:20). Thereby, the possibility of objectivity is denied, since the researcher is not understood as an uninvolved third party, but as a research tool, so that qualitative research is interpretative research: "Qualitative research is a form of inquiry in which researchers make an interpretation of what they see, hear, and understand. The researchers' interpretations cannot be separated from their own background, history, context, and prior understandings" (Creswell 2013:39). Although this research project aims at recognizing objectivity, the subjective background of the researcher and hence possible non-intended or unconscious biases cannot be fully denied. Since the methodological positioning has a significant influence on the criteria of the scientific quality of the investigation, the research project must fulfill different quality criteria. These are discussed in more detail in Sect. 3.1.3 as to how these are addressed in this dissertation.

3.1.1 Ethics

The scientific quality of a research project is further related to the quality criteria, which refers to working in a methodically correct way. Basically, every social research is accompanied by ethical implications, because "research involves collecting data from people, about people" (Creswell and Creswell 2018:88). This gives rise to an ethical responsibility to avoid harm to the respondents or those involved, which occurs when members of the field are directly physically, psychologically or emotionally harmed by research, the necessary consent is not obtained or confidentiality is violated (ibid.). This dissertation has formulated a detailed information document for participants, which was sent as an attachment of each interview request, also informing about possible disadvantages of research participation. Since this dissertation conducted research with experts in the field representing international organizations, national organizations, private sector and NGOs, they do not account for vulnerable groups. Moreover, each interview participant has been given a declaration of consent for using their verbal data, with different options to choose from. Furthermore, this document highlighted that participation in the research is voluntarily. As a requirement for conducting research in Kenya, and further emphasizing this dissertation's ethical research approach, this research project was also assessed by the *Research Ethics Committee of the University of Passau* with the declaration "justifiable".

3.1.2 Cross-cultural Data Collection in Field Research

As this dissertation analyses governance of the digital transformation, more specifically of digital ecosystems, with a focus on Kenya, this research project operates in a cross-cultural context. Although the intended field work in Kenya could unfortunately not be pursued locally due to the COVID19 pandemic, participants from Kenya and other countries served as interview participants. Hence, the implications of cross-cultural data collection is of special relevance for this research project. The position of the research here is manifold: On the one hand, the position of the researcher in the field is essentially that of an outsider, for instance in terms of nationality, local and expert knowledge (Smith 2018). Hence, research in Kenya poses challenges for the researcher regarding linguistic barriers, socio-economic and cultural difference. Nonetheless, it is worth mentioning here that in this dissertation not all interviews were of a cross-cultural nature, since some interview partners were of the same nationality as the researcher. The different positionality of the researcher however leads to an emergence of (perceived) power relations in the interview process as "[a]n interviewer's education, university affiliation, race or ethnicity may be intimidating to the interviewee" (Sands, Bourjolly and Roer-Strier 2007:355–356). Since the research participants in this context were experts in their field, barriers relating to education however can in this instance be assessed as limited.

Moreover, as a minimum requirement to conduct culturally sensitive research, serves the acquisition of knowledge from the cultural context of field research (Liamputtong 2010). Since the actual field work trip to Kenya could not be pursued, this only applies to the actual communication process in the interviews, which due to the digital format, only has limited implications for this research project. However, linguistic barriers were anticipated for this research project, since English is not the first language of most interviewees: "Different languages construct different ways of seeing social life, which poses methodological and epistemological challenges for the researcher" (Larkin, Dierckx de Casterlé and Schotmans 2007:468). It shall be noted at this point that the collected data in this research project is "regarded as preproduced, culturally stored and independent of the interviewer-interviewee relationship" (Ryen 2002:336). Further observed challenges due to the cross-cultural data collection for this research project will be further discussed in Chap. 6.

3.1.3 Quality Criteria

As previously indicated, the quality criteria of any research project are essential. Regarding the question which quality criteria should be applied to qualitative research, three basic positions can be distinguished depending on the scientific-theoretic localization: 1. the transfer of quantitative quality criteria to qualitative research, 2. the formulation of superior criteria for both research paradigms and 3. the preparation of a catalog of genuinely qualitative quality criteria (Steinke 2004). This dissertation follows the quality criteria logic number 2, especially the 'Four Criteria of Trustworthiness' by Lincoln and Guba (1985). The Lincoln criteria catalog by Lincoln and Guba (1985) enjoys universal recognition and most citations in the literature (Döring and Bortz 2016:108–110). Due to the fact that by parallelizing their criteria to the respective quantitative counterparts, a higher connection to the previous discourse on scientific goodness could be established. Hence the authors formulated independent criteria for qualitative research, which are based on the methodological rigor of the quantitative criteria and are intended to have an equivalent effect in terms of their function. More specifically, according to Lincoln and Guba (1985), "good" qualitative research must meet the upper criterion of credibility (trustworthiness), i.e., convincing the audience that the results of the respective qualitative study are meaningful. The four criteria for guaranteeing trustworthiness are *credibility*, *transferability*, *dependability* and *confirmability* (Lincoln and Guba 1985).

Beginning with the first criterion *credibility*, this dissertation has conducted a comprehensive data collection through an extensive time period in the field (from October 2020 to December 2021). Moreover, the data was debriefed by peers in an expert round table and through a pretest (Sect. 3.2.3) and additionally credited through the method of expert interviews. Secondly, *transferability* is assured in this research project due to its case study design, which allows for transfers to other case studies (e.g., other countries). The respective contexts of the countries (see e.g., Sect. 2.1) here is important for transferability. Moreover, transferability is ensured through the expert interviewees who function as representatives of their organizations and hence conclusions for the respective actor group (such as government, tech company or donors) can be drawn. Thirdly, an inquiry audit was conducted in this research project in form of the participant information, which was sent to the expert interviewees prior to the interview, where the researcher explained the study's approach. In the expert round table, this approach was also discussed. Hence, *dependability* of the research is secured. Fourthly and lastly, *confirmability* is achieved by this research project since conducted measures to ensure ethical correctness of the research was confirmed by the Ethics Committee of the University of Passau. Moreover, the participants of the interview have been fully informed in the participant information document prior to the research and they confirmed the use of their data in a chosen way. Additionally, the questionnaire was derived from the literature review in the theoretical background thereby referencing peer-reviewed sources (Döring and Bortz 2016; Lincoln and Guba 1985; Schou, Høstrup, Lyngsø, Larsen and Poulsen 2012).

In addition to these quality criteria, the conducted pretests (Sect. 3.2.3) were further used as assurance procedures for the survey instrument and the entire survey design, hereby offering an additional quality criterion (Weichbold 2014:299). *Triangulation* could not be conducted in this dissertation through the usual research observation due to the digital setting of data gathering (for instance research was not in the field in the physical sense, most participants switched their camera off, etc.).

3.1.4 Case Study

Case studies are characterized by their flexibility in terms of their design as they can follow a qualitative or quantitative paradigm or correspond to a mixed-method design. In addition, different methods of data collection and analysis can be combined (Döring and Bortz 2016:215). According to Simons (2009), "[t]he primary purpose is to generate in-depth understanding of a specific topic [...], programme, policy, institution or system to generate knowledge, and/or inform policy development, professional practice and civil or community action" (ibid.:21). In this dissertation, it is aimed to gather more insights into the digital ecosystem of Kenya while indirectly generating knowledge of and for involved multi-stakeholders, especially from the public sector.

Firstly, it is debated in the literature whether a case study qualifies as an independent research method or even a research methodology. Simons (2014) and Stakes (2005) for instance argue that the case study is not characterized by a research process that is characteristic of it, but by its research object, namely the case. For the most part, however, the case study is classified as a separate research methodology (e.g., Creswell 2013). This dissertation follows the definition of Yin (2014) who defines a case study as an "[...] empirical inquiry that investigates a contemporary phenomenon (the 'case') in depth and within its real-world context, especially when the boundaries between phenomenon and context may not be clearly evident." (Yin 2014:16) A case study as a research design is particularly useful if it is assumed that the context conditions are relevant for understanding the case, because the case study also collects data on the social environment of the case. In addition, the case itself often cannot be investigated separately from its environment (Yin 2014:16–17). Hence, this is suitable for this dissertation, which analyses the rapidly changing context of the digital transformation and its implications for development (Sect. 2.2), where boundaries are not entirely clear and also for using the real-world context of Kenya (Sect. 2.1).

Secondly, and nonetheless, the case needs to be defined and restricted. Gerring (2017) for instance defines a case as follows:

> "[…] [A] case connotes a spatially and temporally delimited phenomenon of theoretical significance. Cases may be compromised of states or state-like entities (e.g. empires, nation-states, regions, municipalities), organizations (e.g. firms, non-governmental organizations, political parties, schools), social groups (defined e.g. by ethnicity, race, class, gender or sexuality), events (e.g. foreign policy crisis, revolution, democratic transition, decision-point), or individuals (e.g. a biography, case history). However a case is defined, it must compromise the phenomena that an argument attempts to describe or explain." (Gerring 2017:27)

In this dissertation the case is comprised of the state of Kenya, more specifically the stakeholders (organizations) present in the Kenyan digital ecosystem. This therefore addresses the argument that the case definition also depends on the sample of (Yin 2014:31–32). Sample in this case are representatives of the multi-stakeholders in the digital ecosystem. Narrowing the case down can be challenging, especially when it comes to determining the time span, which is often arbitrarily designed, especially in terms of events such as revolutions or legislative acts (Gerring 2017:27). This similarly applies to this dissertation, since the digital transformation is a gradual process with no exact start date (other than possibly the emergence of the internet). With regard to the Kenyan digital ecosystem, the start date of the case study could be referred to in terms of the era of Bitange Ndemo as then Permanent Secretary in the Ministry of Information and Communication (2005 to 2013) at which point digitalization found its way on the agenda of the Kenyan government (see e.g., Sect. 4.1).

Thirdly, case studies can also be differentiated according to the type of case in terms of its relationship to the research object: a typical, an atypical, a critical, or a new yet undescribed case (Yin 2014:52). While the study of Kenya in ICT4D literature is not new and on the contrary even often quoted as a 'best-practice' example, the study of the Kenyan digital ecosystem from a governance, sustainability and development theory perspective is new. Hence this case could classify as both classical and new case. Kenya has been chosen against this background and more specifically since it is assessed as one of the fastest-moving digital economies with a narrative driven around "Africa Tech Rising" (Osiakwan 2017:60).

Fourthly, (qualitative) data collection in case studies is of further interest in this research project. The most important principle of data collection in a case study is data triangulation, because it allows statements to be made that exceed the scope of the specific case:

> "A major rationale for using multiple sources of evidence in case study research relates to the basic motive of doing a case study in the first place: to do an in-depth study of a phenomenon in its real-world context. Being both in-depth and contextual—a context that potentially includes events over a period of time—means collecting a variety of relevant data and hence relying on multiple sources." (Yin 2018:127)

According to the author, these multiple sources can stem from up to six different sources (Yin 2018:113–135):

1. Any kind of written documents (emails, media coverage, official publications, scientific studies etc.)
2. Archived documents (public such as statistics, etc.)
3. Interviews
4. Direct observation (both as a stand-alone data source and as a by-product of interview situations)
5. Participative observation (researcher as an active part in the surroundings)
6. Physical objects (art, but also technical equipment)

In this dissertation, all of these data sources are employed expect for the participative observation (5). The reasons for this have already been established due to the digital field setting. However, this data source is likewise assessed as sensitive since it can pose the danger of data distortion (Yin 2018:113–135). The first data source (written documents) is used throughout this dissertation. Scientific studies are referred to in the theoretical background (Chap. 2) as well as in the empirical part (Chap. 4) and discussion chapter (Chap. 5). Secondly, archived documents are used in the empirical part, for instance with regard to public statistics. Interviews (3) are the major data course of this research project. Direct observation in this dissertation, albeit in digital setting, is a data source in terms of a by-product of interview situations, which simultaneously had an impact on the interview data gathering (such as lack of internet coverage; see Chap. 6). This is closely linked to data source 6, where technical infrastructure is of interest for this dissertation regarding the influence of the digital transformation, and hence, digital (information and communication) technologies. How the field for gathering interview data is designed and furthermore prepared in this dissertation will be explained in the next subsection.

3.2 Fieldwork Preparation

The question of how to gain access to the field of research plays an essential role in qualitative research. Wolff (2004) highlights the importance of field access as follows:

> "A preoccupation with the way into the field serves not only methodological or research-pragmatic purposes, it also yields insights into structures and sequences in the research as a social event, and into the field of action that is under investigation. The trial paths, detours and false trails that researchers often complain about and feel to be burdensome, and even the failed attempts at gaining access—which are normally carefully suppressed—all then become 'critical events', the analysis of which opens up chances of making discoveries." (Wolff 2004:196).

The field in this dissertation is determined as actors (individuals of selected institutions) present in the digital ecosystem in Kenya (especially Nairobi). The exact sample and sampling strategies will be discussed in the next subsection (Sect. 3.2.1).

Prior to entering the field via conducting interview, various preparatory steps needed to be undertaken. Field work preparations of this dissertation started by the end of 2019 until the actual entry of the field in October 2020. Initial fieldwork preparations were influenced by the global COVID19 pandemic. In this phase, field work was planned for the period of August to October 2020 with data collection during a research stay in Nairobi. Conducting research in Kenya requires a research permit provided by the National Commission For Science, Technology and Innovation (NACOSTI). This permit requires an affiliation with an approved institution. This dissertation has chosen an affiliation with the British Institute in Eastern Africa (BIEA), which was provided after successful application. For this purpose, a 'Letter of Ethical Clearance' was further required. This was provided by the *Research Ethics Committee of the University of Passau*, as discussed above. Subsequently, requirements for the research permit application with NACOSTI were hereby fulfilled and the research permit was granted. Simultaneously, the researcher has applied for a Deutscher Akademischer Austauschdienst / German Academic Exchange Service

(DAAD) scholarship for funding field work associated costs, which was granted in September 2020. Due to the COVID19 pandemic still prevailing in 2021, the research trip to Kenya unfortunately had to be canceled. In agreement with the research supervisor (and DAAD), the research was hence conducted digitally. As another crucial element of fieldwork preparation, the researcher has conducted extensive research into the relevant stakeholders which were then classified according to the stakeholder groups of donors, government, NGO and tech companies (Sect. 2.1.4). As a next step, relevant employees of the institutions were researched (or at least attempted), which will be discussed in more detail in the following. Implications of this fieldwork approach on the actual research outcome will be further addressed in Chap. 6.

3.2.1 Sample and Sampling Strategy

Additionally, the validity of the empirical data depends on which case selection or sample it was generated. The selection of cases and the process of sampling will hence provide a preliminary decision on how and if the results of the study can be generalized (Kruse 2014). Although there are no requirements in qualitative research in terms of sample size, certain criteria regarding the composition of the sample should be met. For this research project, sampling methods aiming at gaining as heterogeneous a sample as possible in order to maximize the variability have been chosen (Schreier 2013). More specifically, sampling here followed the top-down approach (instead of bottom-up) (Przyborski and Wohlrab-Sahr 2014b), since criteria for the composition of the sample were already established, in this case by the stakeholders' presence in the digital ecosystem of Kenya.

Firstly, the three most relevant non-probabilistic top-down sampling techniques are qualitative sampling, deliberate sampling of specific case types, and snowballing (Schreier 2013:196–197). This research project follows the sampling technique of snowballing. Firstly, a intentional sampling was established by defining ex ante criteria determining the inclusion or exclusion in the sample, including the choice of respondents based on their expertise (Daniel 2012:88, 91). Since the method for data collection of this dissertation are guideline-based expert interviews, this approach was followed. The defined criteria referred to experts present in the digital ecosystem derived from both empirical and theoretical research. Secondly, the snowball process recruits potential interviewees by establishing contact with respondents. This method has been chosen since in the research preparation phase, there was insufficient knowledge of the sample with access being obtained in a previously unknown field, and the respondents being experts in foreign countries posed high access barriers (Schreier 2013:198). Hence, this method is useful for studying (social) networks in better determining the relevance of actors in the field (Daniel 2012:112–113). However, this snowball sampling does not cover an entirely appropriate detection of the field, which is neither the claim nor goal of this dissertation and is also exceeding the scope of this dissertation. This potential limitation is compensated through the selection of a case study.

Against this background of sampling strategies and with regard to the actual sample of this research, important stakeholder groups were identified in the research process and fieldwork preparation. As a result, the stakeholder groups of bilateral and multilateral donor organizations, NGOs (both nationally in Kenya as well as internationally), international and national tech companies, national hubs and national governmental institutions have been identified due to being active in either/or digital transformation and development cooperation. For each group, organizations were identified and the attempt was undertaken to identify actual experts in form of representatives of the institutions. In the first sample research process, there was no constraint as to how many organizations per group. In order to achieve possible comparisons between the stakeholder groups, for the analysis process an equal amount of experts was intended. In the following, the exact sample composition shall be explained. This composition is closely linked to the stakeholder description in Sect. 2.1.4.

Secondly, after elaborating the sample strategies and composition, the sample size is to be further elaborated. As stated in the beginning of this section, there is no fix rule concerning sample size in qualitative research. According to Flick (2009) for instance, the sample size depends on the decision for a particular research strategy, which implies criteria for the generalizability of the results (similarly in Baker et al. 2012). Since this dissertation conducts an exploratory design with expert interviews, the sample size is thereby already restricted (elite of experts, more information in Sect. 3.3.1). Moreover, it is aimed at analyzing the various multi-stakeholders present in the digital ecosystem of Kenya. Since it would be impossible to interview every participant in the digital ecosystem of a whole country (in addition to its vastly changing composition and development) within the scope of a single dissertation, this research project chose to further divide the sample into stakeholder groups as representatives of their organization in the specific sector (public, private, national, international). For the purpose of comparability, a sample of at least one expert was chosen for each group. Due to the here applied snowballing strategy, the number of representatives per group varies. For exploratory studies, 20 to 150 participants are listed for instance by Daniel (2012). With 23 interviews, this dissertation thereby fulfills this requirement. Moreover, it shall be emphasized that this dissertation not only gathers data through the method of interviews, but in addition uses other data sources due to the case study design (Sect. 3.1.4). The sample of this dissertation is comprised of the following organizations. Table 3.1 provides an overview about the numeric distributions of the sample which will be discussed in more detail in Sect. 3.3.2.

Table 3.1 Sample group and sample size (own illustration)

Sample group	Sample size
Bilateral Donor Organizations	6
Multilateral Donor Organizations	5
NGOs	5
Private Tech Sector	4
Startups	2
Academia	1
Kenyan Government	0

Due to privacy choices and data protection reasons, the specific organizations' names are not listed, but the actor fields and classifications of Kenyan or international organizations will be elaborated. *Bilateral donor organizations* have been chosen and contacted according to their activities in digital development and their ODA contributions to Kenya. Multilateral donor organizations included global and African organizations. With regard to the sample group of NGOs, both one NGO based in Kenya as well as international NGO were requested. Hubs here are similarly classified as a non-governmental entity (and also not specifically private sector) which are likewise contacted and subsumed under NGOs. National and international tech companies as essential actors in digital transformation as a sample group mainly included international and national tech companies. Startups were classified as an individual subgroup. Academia as a sample group refers to universities. Last but not least, the public sector actors mainly include Kenyan governmental organizations. However, this actor group was impossible to recruit for an interview. However, this sample is nonetheless covered by references of the other interviewee stakeholder groups about the role of the Kenyan government.

3.2.2 Preparation of Interview Guide

An additional preparatory step to fieldwork is the preparation of the interview guide. In this dissertation, the interview guide preparation followed two steps: Firstly, topics and themes from the literature review (Chap. 2) were collected (macro planning). Here, all relevant themes were selected regardless of scope. In a second step, (micro planning) possible questions were formulated and assigned to topics. Then, the order of questions in a topic was adjusted according to the line of questioning and in order to not only cover all themes, but also to avoid repetitions. As a last step, the amount of questions was adapted to the given time frame of interviews (around 60 minutes) and an introductory passage was formulated.

Only one interview guide was prepared for all groups of participants to a) establish comparability, b) generalizability, and c) because all participants were tested for both digitalization and development cooperation. (The result was that that skills were either in one area or another and interface topics were sometimes difficult to answer, see also Sect. 4.4.4 and Sect. 6.2). As a consequence, the questions were only slightly adapted to the interviewees background and group belonging.

3.2.3 Interview Pretest

To limit challenges during the interview and data gathering, this dissertation has conducted pretests prior to the actual data collection. In a first phase, the actual development of the guideline-based questionnaire (Sect. 3.3.1), the formulation of questions and answer categories was tested, for instance in order to decide on alternative questions and the structure of the questionnaire (Weichbold 2014). This was especially challenging since the researched impact of the digital transformation is a cross-cutting issue, so that sometimes overlaps occurred. Besides more standardized reviews of pretests, the informal evaluation of colleagues has also proven its worth at this stage of pretesting and is accepted in the empirical science literature (Campanelli 2008; Weichbold 2014). In accordance, these first pretests were conducted with academic colleagues. Moreover, the results i.e., the questions of these pretests, were adapted (e.g., through rewording, addition of new questions or removal of questions, or rearrangements). Thus, these first pretests were also used as an instrument for quality control and improvement in the qualitative research.

Secondly, after the first pretest, the adapted questionnaire was tested again in line with Weichbold (2014). Usually this is conducted in the actual field. However, pretest procedures in the field were deliberately not conducted in this research project for two reasons: On the one hand, as emphasized in Sect. 3.2.1, the sample size was already limited due to the restrictions by the COVID19 pandemic. Hence, an additional 'waste' of the already small sample size was intentionally avoided. On the other hand, as also discussed in Sect. 3.2.1 and Sect. 3.3.1, the research design of expert interviews exposed specific conditions applying to the expert status of the interviewees. Hence, this second pretest was instead conducted with participants from academia (in October 2020). Consequently, the pretest in such a laboratory setting aimed at testing whether the respondents understood the questions, if the questions were asked in a comprehensible manner and also for the researcher, whether the guide is clearly and in sufficient font size, so that the interviewer does not lose track (Döring and Bortz 2016). For this pretest, the method 'Think Aloud' from cognitive research was applied. Here, respondents were asked to express their thoughts in the course of answering the questions. In this way, the question understanding and the answer process are to be disclosed in order to obtain indications of misunderstandings, misleading associations or difficulties in selecting the answer categories (Faulbaum, Prüfer and Rexroth 2009; Weichbold 2014).

Thirdly, an additional phase of pretesting was chosen. After conducting and transcribing the first three interviews, the interview transcripts were discussed in a expert group from academia. This method was chosen, because it was noticed that the interviewees mainly belong to either the field of digitalization or the field of development cooperation. Therefore, this third pretest was used to ensure that both heterogeneous groups can answer the questions. As Sect. 3.2.1 demonstrated and Chap. 6 will further emphasize, it has already become obvious in the pretest and field access, that experts at the intersection between digitalization and development cooperation are rare. Thereby some degree of pretesting in the field was achieved as well.

In sum, the pretest was conducted to check the survey tool (formulation of the individual questions and the questionnaire as a whole), but also its technical implementation (e.g., for digital conduction of the interviews in field work, font size of the questionnaire for the researcher), sampling and the conduct of interviews (Döring and Bortz 2016; Weichbold 2014). Hereby, assurance procedures for the survey instrument and the entire survey designs was gathered (Weichbold 2014:299).

3.2.4 Field Access

A final step in fieldwork preparation is the actual field access. Gaining access to the field under study is more essential in qualitative research than in quantitative research, since the contact between researcher and people (here interviewees) is closer and/or more intense (Flick 2014). The term 'field' here refers to institutions categorized as stakeholder groups operating in Kenya (Sect. 3.2.1). Moreover, the communicative and interactive process between researcher and participants is essential for (verbal) data collection. Hence, a preoccupation with field access "[...] serves not only methodological or research-pragmatic purposes, it also yields insights into structures and sequences in the research as a social event, and into the field of action that is under investigation" (Wolff 2004:196).

To begin with, it shall be noted that, although the sample is composed of representatives of organizations, field work was not intended for organizational observations in detail. Instead, the approach here is more based on a macro institutional level, since the comparison of different organizational groups is key to this research and thereby implications for political deciders will indirectly be further derived. Moreover, given that institutions are the present field of study and also operate in the field of a country (Kenya), the special role of organizations as field will be discussed. Generally in the qualitative research literature, access to institutions is considered as more complicated (Flick 2014). The reasons for that and how this research project has dealt with these challenges will now be expressed:

Firstly, in organizations different levels are involved in the research (access) process. Since the sample of this dissertation is, among others, comprised of bilateral and multilateral donor organizations that are highly governmental, highly official and highly sensitive to diplomacy issues, the authorization process needed to be considered for this research. After initial challenges and further internal consultancies of some of these institutions, agreement on degree of anonymity and data handling was achieved with the researcher ("Work of agreement", Flick 2014:160). Hence, all of the interviewees' data was handled without provision of information on the participant.

Secondly, in research of institutions the researcher is considered as a "client" (Flick 2014:159), so that procedural routes are also labeled as "research preventers" (Wolff 2004:197). Hence, the researcher has to request interviews in formal ways. In this research project, each potential interviewee was requested formally and individually via email by attaching both participant information sheet and a declaration of consent for research ethical reasons as previously discussed. The declaration of consent was mostly filled out by the participant prior to the interview. In some cases the participants needed to be reminded so that they were filled out after the interview.

Thirdly, research "[...] represents an intrusion into the life of the institutions to be studied" (Flick 2014:160). In order to minimize such disturbances, the interview scope provided in the interview request to potential interviewees was limited to 60 minutes, which the researcher estimated as a minimum scope to sufficiently conduct the interview and as a minimal scope of disturbance (further implications see Sect. 6.2). At this point it is worth mentioning that some participants voluntarily

(!) chose to conduct longer interviews with up to 120 minutes (e.g., I5 and I6), while others were only available for 30 minutes (e.g., I24).

Fourthly, after access to the institution was established, access to the actual employees i.e., representatives of the selected institutions was the next step in conducting field access. This step was the most challenging for this research project. In the field work and interview preparation phase institutions could be identified, suitable employees however were not easily identifiable. Here the researcher used the snowballing method (Sect. 3.2.1) as well as the communication channel of sending requests to press offices or similar bodies within the organizations for recommendations (although the latter in most cases turned out to be unsuccessful).

Fifthly, this research project has started with interview requests to experts in development cooperation due to existing contacts of the researcher and due to the fact that this subgroup of the bilateral stakeholders offered further interview possibilities in the intersection of digitalization and development cooperation, which was not necessarily given in the other groups except for multilateral institutions. Hence, this research project chose to start with organizations engaged at the intersection of digitalization and development cooperation and then to focus on the more specialized stakeholder groups (either with experience in digitalization or development cooperation).

For full disclosure it is to be noted that the field was exited, when the final interview was completed. Due to the mentioned circumstances of this research project and the fact that thereby no field trip in the classical sense could be conducted, field exit likewise is of limited relevance here.

3.3 Collection of Verbal Data: Guideline-based Expert Interviews

Interviews are one of the most common types of qualitative data collection. One advantage of qualitative interviews lies in their potential to add complexity to phenomena and in providing the opportunity to make marginalized positions heard. The use of interviews as a survey method, in general but especially for this dissertation, can be justified as follows:

Firstly, the appropriateness of the method is always based on the research question (Silverman 2006). In this dissertation, the research question asks how digital ecosystems in developing countries can be coordinated for sustainable digital development (Chap. 1). Secondly, "[...] qualitative interviewing is particularly useful as a research method for accessing individuals attitudes and values" and provides "better access to interviewees' views, interpretation of events, understandings, experiences and opinions" (Byrne 2012:209). In this research project, the research question is directed towards the experts' experiences from working in the field of digital transformation in Kenya. Hence their views, understandings and experiences are vital for this research so that interviews are a suitable choice. Moreover, due to the fact that the digital transformation is not only a recent but also ongoing phenomenon, this dissertation follows a more constructivist approach regarding the interview situation as a bilateral construction of the interpretation of reality on the part of the respondents and the interviewer. Accordingly, the data collected can only reflect the discourse within which the knowledge and experience of the respondents are embedded (Silverman 2006). Thirdly, in addition to the here required specific expert knowledge, but also due to the exploratory nature of this research, the interview method is chosen. Moreover, as opposed to quantitative approach such as standardized questionnaires, the interview setting gives the researcher the possibility to intervene during the interview for instance if questions emerge (e.g., due to unclear terms or comprehension questions). The semi-structured guideline questionnaire likewise enables the researcher to directly intervene in the event of queries for providing support. This will be further discussed in the following.

3.3.1 Interview Type: Guideline-based Expert Interviews

Expanding on these introductory remarks around guideline-based expert interviews, the setting of different types of scientific interviews need to be discussed in more detail. Firstly, interviews are mainly classified according to the degree of their standardization. Accordingly, a distinction is made between open (or unstructured/non-standardized), semi-standardized and standardized interviews (Döring and Bortz 2016:357). This dissertation has chosen a semi-standardized interview in form of a guideline-based interview. Semi-standardized interviews are conducted with the help of guidelines, so that relevant aspects of the research topic are recorded in the guidelines before the survey. While respondents' statements can be compared systematically on the basis of the guideline-oriented interview, the researcher should ensure that there is room for certain flexibility in the conversation situation. Accordingly, Schreier (2013) for instance classifies the following question types in a guideline interview: introductory questions, to get accustomed to the interview situation, guideline questions and ad-hoc questions that aim at deepening unexpected aspects. In this dissertation, all three question types are phrased.

Secondly, the most common forms of partially standardized interview types include the focused interview, the problem-centered interview and the expert interview (Kruse 2014). Per definitionem of this dissertation, the *expert interview* has been chosen, so that only this form of semi-standardized interview types will be elaborated further. Expert interviews are considered to be a special application of guideline interviews (e.g., Flick 2009; Helfferich 2014). As a specific characteristic, respondents are identified as representatives of a particular group whose expertise is to be collected on a particular issue, whereas in the other types of interviews the respondents are more likely to be understood as affected or lay people whose biographical and personal relation to a particular problem is to be determined. In addition, an expert interview assumes that respondents have knowledge that is not accessible to other actors in the field of action (Flick 2014:227–228), which is particularly relevant to this research project regarding the fields of both development cooperation *and* digitalization. This is also relevant given that expert knowledge is attributed to greater objectivity than lay knowledge, but it should also be pointed out that expert knowledge is both situational and personal so that a subjective interpretation must be assumed in the transfer of knowledge (Helfferich 2014:570).

An initial difficulty around expert interviews is first of all the composition of the sample (Sect. 3.2.1). Prior to conducting the interview, it must be clarified in a comprehensible manner, who can be considered as an expert in the particular field of action. Meuser and Nagel (2009) for instance define expert knowledge especially in contrast to lay knowledge as follows:

> "A definition of the term expert can best be arrived at by reviewing the features distinguishing expert knowledge from other forms of knowledge like everyday knowledge and common-sense knowledge. In scientific research an individual is addressed as an expert because the researcher assumes—for whatever reason—that she or he has knowledge, which she or he may not necessarily possess alone, but which is not accessible to anybody in the field of action under study. It is this advantage of knowledge which the expert interview is designed to discover, and it is an exclusive realm of knowledge which is highly potential because and in as far as it is linked with the power of defining the situation." (Meuser and Nagel 2009:18)

According to the authors, interviewees are considered experts if they acquire special knowledge through their activities, thereby giving them privileged access to information, which is linked to a specific functional context and is socially institutionalised (Meuser and Nagel 2009:20). For Bogner and Menz (2009) the selection of experts has a fundamentally relational character: "This means that the question of who is to count as an expert for purposes of method always has to be answered in relation to the concrete field of operation in which the expert acts, and with reference to the investigative spectrum of the empirical study being carried out." (Bogner and Menz 2009:54). In this dissertation, interviewees are considered experts according to their activity or employment in the digital development field, and are either experts in development cooperation, digitalization and/or its intersection. They were interviewed in their specific institutionalized role, in this case employability in this field in either academia, donor organization, Kenyan government, NGO or (local) private tech sector.

After elaborating on this dissertation's understanding of experts, it is further to be established how and why expert interviews can be used in a research design. Expert interviews in social science can be used according to the following three objectives: exploration, systematization or theory extraction (Bogner and Menz 2009:46). Exploration is here achieved through inductive approach as well as systematization through coding and theory extraction through interpretation. For the latter it is to be further emphasized that this dissertation has developed a theoretical framework for network governance in digital ecosystems, which can be transferred to various other empirical use cases such as different countries.

3.3.2 Style and Settings of Interviews

After discussing the chosen method of guideline-based expert interviews, this section will elaborate on the actual style and settings of the conducted interviews. Usually, interviews as a qualitative design are conducted in a face-to-face setting between the researcher and interviewee(s). As already mentioned, face-to-face interviews were impossible due to the global COVID19 pandemic, which inhibited the planned field work. Instead, for these practical reasons, the interviews were conducted online via digital conference tools (e.g., Microsoft Teams, Zoom, Google Meet, further discussed in Flick 2014:233–236). This research project organized its online interviews in a synchronous form resembling a face-to-face meeting. Interviewees were initially greeted in video and audio format. For the actual interview however most participants requested to not record the video, only the audio. Hence, the cameras were turned off. This has implications for the interview setting as it more resembled an interview via telephone than an actual face-to-face interview (see further Sect. 6.2). Here differences between digital and non-digital interviewing were thus observed. With regard to observation during the collection of verbal data through interviews, field notes were not taken since the researcher was not physically present in the field and no possibility of observation was provided (also due to cameras turned-off in digital interviews).

Interviews were conducted in the period from October 2020 to December 2021. During this time 23 interviews were conducted, which are distributed as illustrated in Table 3.2.

Table 3.2 Conducted interviews (own illustration)

Stakeholder group	Interview Number
Bilateral Donor Organizations	I3, I4, I5, I6, I13, I18
Multilateral Donor Organizations	I12, I14, I17, I19, I20
NGOs	I1, I7, I11, I15, I2
Private Tech Sector	I16, I21, I9, I10
Startups	I23, I24
Academia	I22
Kenyan Government	0

3.4 Analytical Process: Transcription and Interpretation of Verbal Data

After discussing the process of collecting verbal data, this last section will discuss this dissertation's analytical approach to transcription and interpretation of the verbal data gathered in the guideline-based expert interviews. Generally, interviews can be analyzed by coding and categorization (Flick 2014:236). For this purpose, the analytical steps of transcription and data management as well as of qualitative data/content and network analysis will be elaborated.

3.4.1 Transcription and Data Management

As a first step of the analytical process, the recorded interviews were transcribed. While in qualitative research various transcription systems are proposed (Kowal and O'Connell 2004), a standard regarding the degree of exactness and transcription strategies has not been established and is debated (e.g. Flick 2014; Kowal and O'Connell 2004). The interviews in this dissertation were transcribed directly after the collection of the verbal data (as proposed by Kuckartz and Rädiker 2019:41). Through this approach, it was possible to directly and timely reflect on the conducted field research already after the first interviews (Flick 2014). Although it is suggested in the methodology literature that it is not necessarily mandatory to transcribe whole recordings but only relevant passages (Kuckartz and Rädiker 2019:42), due to the complexity of the studied matter in this dissertation's expert setting where possibly every utterance is of value, and also for transparency reasons, this dissertation uses the strategy of detailed word-for-word transcription. More specifically with regard to transcription rules, the transcription system of Kuckartz and Rädiker (2019) was followed in this dissertation. Here, the rules proposed by the authors are elaborated in the following with regard to their specific usage and adaption in this dissertation (Kuckartz and Rädiker 2019:42):

1. Each speech contribution is transcribed as a separate paragraph. To increase readability, after the paragraph, a blank line is added.
2. Paragraphs for the interviewer are introduced as 'R' (for researcher) while the interviewees are introduced by 'I' (for interviewee). Moreover, numbers for the interviewees are added to the abbreviations (here: I1, I2 etc.) in order to distinguish between interview partners. Since the interviews in this dissertation were only conducted by the researcher herself, there is only one 'R' throughout all interviews.
3. Speech is transcribed verbatim, hence not phonetically or in summary form. Dialects are not transcribed but translated as accurately as possible into the standard form, here standard English.
4. Language and punctuation are standardized where necessary. The word order, definite and indefinite articles, etc. are retained even if they contain errors.
5. Clear, longer pauses are indicated by ellipsis points in brackets, here in form of '(...)'. Depending on the length of the pause in seconds, one, two, or three points are followed (as proposed by Kuckartz and Rädiker 2019:42). In this dissertation, mostly three digits are used since the exact duration could not always be obtained due to technical disturbances during the interview (e.g., due to disrupted internet connection). Moreover, in passages with pauses longer than 10 seconds '[long pause]' is used.

10. External interruptions or interferences are noted in double brackets stating the cause, such as '[cell phone rings]'.
11. Vocal utterances made by both the interviewee and the interviewer are noted in simple brackets, such as '[laughs].

13. Incomprehensible words and sections are identified by '[unk]' meaning 'unclear'.
14. All information that would allow an interviewee to be identified is rendered as anonymous.

In contrast, the following rules proposed by Kuckartz and Rädiker (2019) have not been followed due to limited applicability to this dissertation (ibid.:42):

6. Intentionally stressed words are underlined.
7. Very loud speech is indicated by writing in capital letters.

8. In contrast to Kuckartz and Rädiker (2019), affirmative or agreeing utterances made by the interviewer and/or interviewees such as 'mhm', 'aha', are transcribed for transparency purposes (also in order to follow the design of exact transcription).
9. Short interjections made by the other person are further indicated by '\\'. In contrast to the authors, in this dissertation a new paragraph is then started.
12. For videos: nonverbal actions are placed in simple brackets, such as (opens the window), (turns away), or similar. This is not applicable to this dissertation as not all interviews were conducted in a video setting.

In contrast to the transcripts, interviewees' utterances such as 'mhm', 'eh', 'ah' as well as ellipses for pauses '(...)' are omitted in the text body of the empirical analysis to improve readability.

As usual in the field, transcription in this dissertation was performed digitally with the help of a software program. Due to several concerns that made established systems such as MAXQDA or f4 unsuitable for the purposes of this dissertation, a custom software tailored specifically to the requirements at hand was employed. While this software is not publicly available yet, it offers the same set of features as the well-known existing systems. That is, it combines both the functionality for initial transcription from audio to a textual representation, as well as annotation and analysis capabilities for the later coding phase. More specifically, the interview audio files are imported into the software and associated with essential metadata, such as the interview number and a brief description of the interview setting. Subsequently, an initial rough transcript of the audio file is generated by means of a speech-to-text machine learning model and then refined by the researcher in order to eliminate remaining errors. This is facilitated by an accurate association between the transcribed text and the original audio file, which allows the researcher to navigate within the audio file by pointing to a location within the text and vice-versa. Once transcription is complete, a hierarchical codebook is iteratively created by annotating relevant text passages with the appropriate codes. Finally, the annotated interviews are thoroughly analyzed by grouping, filtering and searching the annotated text passages.

As mentioned above, this software offers several advantages over alternative systems. Firstly, all data is stored exclusively on the local computer running the software, ensuring that no potentially sensitive data is inadvertently leaked through cloud services frequently found in other existing programs. Secondly, the software employs a data model that allows the transcripts, the codebook and most importantly the annotations to be exported in a format suitable for integration with the LaTeX typesetting system employed for the original manuscript. This integration enables the researcher to specify a semantic reference to a text passage, which is automatically replaced by LATEX with the correct page and line numbers as they appear in the final document. In summary, this custom software both increases data security and reduces the number of inaccuracies in the transcripts and in the analysis process.

As a last step, the transcribed interviews were further adapted to the interviewees requested form of anonymity as indicated in the signed declaration of consent, so that for instance names or passages where the interviewee requested anonymization were anonymized. Statement of the exact name has been chosen to avoid for reasons of uniformity, data protection and since the interaction of stakeholder groups is the goal of research with the interviewees serving as representatives for each group. Moreover, were specific organizations are referred to in the empirical analysis, these are general comments that cannot be attributed to the interviewees' organizational affiliation. Simultaneously, this serves the goal of generalization. Furthermore, it is important to note that Interview 16 was transcribed, but only used as background information with only indirect quotation given the circumstances around the declaration document. Moreover, as per their request, I8 was not transcribed or used. Hence, it is not listed in the interview list, but numerated for full disclosure. As a consequence of the sequential numeration, the 23rd interview is referenced as I24.

3.4.2 Qualitative Content Analysis and Network Analysis

After the transcription of the interviews, the gathered data was analyzed using qualitative data analysis. According to Flick (2014), qualitative data analysis can be defined as "[...] the interpretation and classification of linguistic (or visual) material" (ibid.:370) with the aim of deriving statements that can be generalized. A further aim of qualitative data analysis is to firstly describe the phenomenon to be studied (Flick 2014:369). In this dissertation this phenomenon refers to the experiences and insights into the interviewees' roles in the digital ecosystem of Kenya. The research focuses on the case (here actor groups in the case study of the Kenyan digital ecosystem) as well as its features and links between actors. Moreover, the focus is on comparing these stakeholder groups according to similarities and differences, which is necessary for addressing the research question as to how (and by whom) digital networks can be governed.

Similar to the passage on research design (Sect. 3.1), coding was not conducted by the method of Grounded Theory, because this dissertation used open coding (emerging from the verbal data) but also thematic coding, where initially a few codes were applied after (a prioiri) emergence from the theoretical background (also since the questions in the semi-standardized expert interview setting are likewise linked to the theoretical background). For the analysis, the above stated software was used.

Qualitative Content Analysis

For systematically describing the meaning of qualitative data, in this dissertation the qualitative content analysis was chosen. Qualitative content analysis is an evaluation method that processes materials such as texts, which occur in the data collection context of social science research projects such as transcripts from interviews (see e.g., Helfferich 2014; Mayring and Fenzl 2014; Schreier 2014). In comparison to other text analytical methods, qualitative content analysis is used the most frequently (Mayring and Fenzl 2014:543). The advantage of qualitative content analysis as an evaluation method is that it can be used to deal with relatively large amounts of material, but it remains qualitative and interpretive, meaning that latent structures of meaning can also be captured. Due to the rule-based nature of the procedure, the qualitative content analysis is easy to understand from an intersubjective perspective (Mayring and Fenzl 2014:543).

Moreover, qualitative content analysis is defined by its category guidance. The category system is the analysis instrument by which only relevant text passages are marked that relate to the categories (Mayring and Fenzl 2014:544). In accordance to scientific practice, these deductive categories of this dissertation are supplemented by other categories that are derived inductively from the material in order to address all aspects of meaning of the data material, which cannot be fully explored ex ante (Schreier 2014). Hence, it is ensured that the categories reflect the data material. This flexibility lies within qualitative content analysis as it "[...] combines varying portions of concept-driven and data-driven categories within any one coding frame" (Schreier 2014:171). Qualitative content analysis is highly systematic, since every part of the material is examined, also concerning its relevance to the research question (Schreier 2014:171).

This dissertation follows the inductive category building in qualitative content analysis according to Mayring (2014). Mayring (2014) distinguishes three techniques of qualitative content analysis, namely summary (text reduction), explication and structuring. Here, the techniques of structuring was applied where particular aspects of the material were filtered out in reference to theoretical background but still open to emerging material. Hence the categories were only partly formulated in advance in a mix of deductive and inductive category assignment with the majority of inductive category building (Mayring 2015:373). In sum, this dissertation followed the proposed steps of Mayring (2014) in inductive category building, similar to the schematic description in Fig. 3.1.

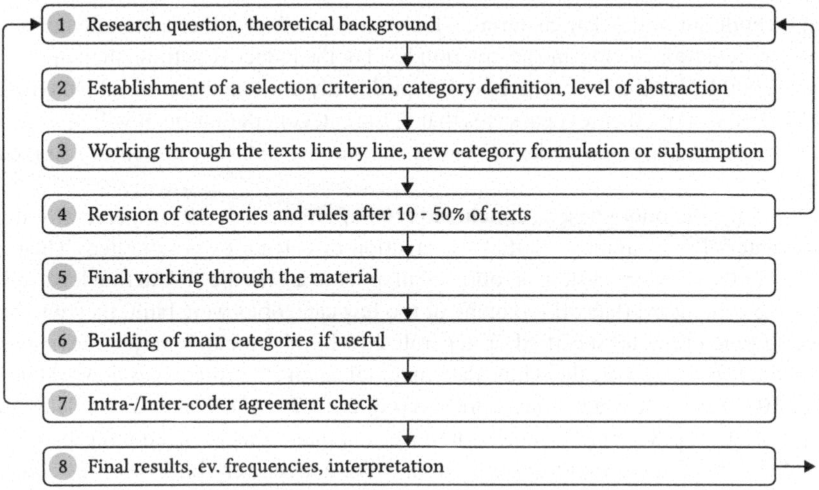

Fig. 3.1 Process of inductive category building in qualitative content analysis (own illustration based on Mayring 2014:80)

As a first step, the research question and the theoretical background were formulated (Chap. 1, Chap. 2). Secondly, "[w]ithin the logic of content analysis, the level or theme of categories to be developed must be defined previously. There has to be a criterion for the selection process in category formation. This is a deductive element and is established within theoretical considerations about the subject matter and the aims of analysis." (Mayring 2014:80–81) In this dissertation the research question already specifies the topic according to which the categories are formed and only those parts of the data material that are relevant to answering the research question are evaluated. The coding unit is defined as a clear meaningful element in the text, the context unit is defined as the entire interview to be evaluated and the evaluation unit is the entire data material, i.e., all conducted and transcribed interviews (Mayring 2014:84). The level of abstraction of the categories is predetermined by the research question (Chap. 1). Analogous to the author, in this dissertation concrete factors that promote or hinder coordination would be determined as an abstraction level of the categories (Mayring 2014:84).

Thirdly, the material is examined line by line. Thereby the danger of analyzing the material only according to assumptions and biases is limited. In accordance to qualitative research, this is done in an iterative process by going through the material repeatedly and by modifying the coding frame (Schreier 2014:171). If a category can be built out of the data material, which corresponds to either a already defined category regarding theme and abstraction level or the inductive setting, then the first code is build. Naming the category develops directly through the material (Mayring 2014). In cases of other text passages that fit the category definition, it will be coded similarly. Otherwise new codes are built or subsumed under the previously formed category.

Fourthly, the pilot phase is concluded after approximately 10–15 percent of the data material is examined, so that the creation of categories is saturated. Then a review of the category system is followed in order to assess the abstraction level and possible content overlaps. If too many or too few categories were built, they will be revised (and either be specified or generalized). In this research project, revisions were made by reviewing the whole data material again after three interviews (hence after 10–15 percent in accordance to Mayring 2014:79–81). This is also related to step 5 (Fig. 3.1), where the whole material (each interview) is coded after the final development of the category system. In this dissertation every interview was coded initially and then, after all interviews were coded, each one of them was coded again according to the then finalized coding frame (double coding). Hereby the quality of the category definitions is examined (Schreier 2014:171). As a sixth step, main categories valuable for the research question were built (Mayring 2014:79–81).

As a seventh step, the *intracoder and intercoder reliability* check is conducted, both of which are important quality criteria for content analysis (Mayring 2014; Mayring and Fenzl 2014). *Intracoder* reliability is ensured through an additional review after the completion of the analysis regardless of the categories and thereby functions as an indicator for the stability of the approach and for measuring reliability. In this dissertation, intracoder reliability was done by the researcher herself through encoding the material again after a certain period of time and then comparing the correspondences of the categories. For the *intercoder reliability*, matches between two different coders ensure objectivity of the approach (Mayring and Fenzl 2014:546–547). In academic research, the intercoder reliability is measured mostly with Krippendorf's Alpha or Cohen's Kappa. However, it is a matter of discussion in the literature as to which threshold values should be used for the intercoder agreement in qualitative content analysis, since these methods are mainly used in the quantitative content analysis (Mayring 2014:114). Moreover, "[i]n qualitative research however, a perfect agreement between different analysts can hardly be reached, because interpretative elements (even if extreme rule guided like in Qualitative Content Analysis) always bear a subjective element. So we must be a little bit more modest." (Mayring 2014:114) Hence Mayring (2014) delivers three different alternative approaches to intercoder consensus: a) giving only the texts to be analyzed and the research question(s) to a second person, b) giving a second coder the texts to be analyzed together with all content-analytical rules (procedure, units, category definition and level of abstraction for inductive category formation, coding agenda for deductive category assignment), and c) the second coder has insight into the whole material, definitions, and codings of the first coder and acts as a supervisor. All of these procedures however can be very time consuming so that the author proposes that "sometimes only a partial inter-coder agreement test is carried out. Only parts of the textual material (random samples, exemplary text portion, difficult text portions) are selected" (Mayring 2014:114). Given the innovative and interdisciplinary nature of this dissertation as well as the required expert knowledge (especially in development cooperation and digital transformation), the latter approach proposed by Mayring (2014) is pursued in this dissertation. More specifically, parts of the material were discussed in an expert round-table where the second coder acted as supervisor of the content analysis of for instance the codebook (option c).

Finally, as per step 8, the final results are gathered from the category frame, where possible frequencies, similarities and differences between the interviews are interpreted (Mayring 2014). Here it shall be emphasized that, for the purpose of this research, the interviews where then assigned to actor groups. Although some interviewees have declared full usage of their data, for the purpose of transparency, coherence and comparability as well as generalizability, only the expert role in the specific institution was of research interest. Hence, the interviewees were assigned to groups as proposed in Table 3.2 (government, bilateral and multilateral donor organizations, private tech companies, NGOs, academia). This grouping is derived from the declaration of consent which was given to the interviewees prior to the actual interview as mentioned beforehand int his chapter. For the purpose of a more detailed analysis, the private sector group here was further divided to multinational, national and start-up private companies. This is based on their location of operation/registered headquarters and company age, more specifically on their self-declaration. Generally, experts classify themselves as mostly in senior positions of well-known institutions in the field (but is anonymized here as per their declarations). Most importantly, the codebook itself as the major source for analysis and interpretation is comprised of codes, definitions and examples.

(Qualitative) Network Analysis

In addition to the before mentioned qualitative content analysis, an additional method was applied in this dissertation. Since the analysis of digital ecosystems and the interaction of the multi-stakeholders in the ecosystem is at the center of the research question, an analysis of this particular network type of the Kenyan case study is additionally methodologically operationalized through the qualitative network analysis.

The classic network analysis is based on mathematical graph theory, which makes it possible to translate the sociomatrixes into sociograms. Sociomatrixes represent the relationship between people in '1' for the existence of a relationship, and '0' for the absence of an acquaintance. The graphic representation of the sociomatrix is the sociogram in which the actors as nodes and the actor relationships as lines between the nodes are represented (Carrington and Scott 2012:2, Fuhse 2018:39–41). Qualitative methods only played a subordinate role in network research for a long time (Fuhse 2018:135). Specifically since it is still disputed whether a purely qualitative network analysis even exists as standardized procedures are used here, especially for the visualization of the network structure (Diaz-Bone 2007; Herz et al. 2015). Stegbauer and Häußling (2010) on the other hand argue that network research is closely linked to qualitative content analysis. Thereby qualitative network studies can indirectly serve the purpose of triangulation, as "mechanisms of structure building" (Weyer 2000:254, see also Trezzini 2010).

Firstly, it needs to be distinguished between networks and network context. According to Jansen (2006), network limits can be defined through a) the identification of relevant actors as well as b) through assessing which kind of network practice is to be analyzed (e.g., if the network relation is determined by exchange of information or resources, if a concrete interaction is given or if membership of a specific institution is decisive) (Jansen 2006:71–75). Since these can be interwoven and because the subjective views of the actors have to be considered, the boundaries of the network cannot be determined completely before investigation in qualitative research. Moreover, as in this dissertation, the definite boundaries are the result of investigation and conducted research (Löblich and Pfaff-Rüdiger 2011:639–640). Hence, the authors propose that the decisive factor for defining boundaries of networks is the research question and both steps can be conducted simultaneously: "Both steps of defining the network boundaries—choosing the actors belonging to it and deciding about the type of relations to be investigated within the selected set of actors—have to be appropriate for the research question." (Löblich and Pfaff-Rüdiger 2011:639–640). In this dissertation, the network (here the digital ecosystem) is defined by the present actors as well as by their interaction (Chap. 2). Similarly, in this dissertation the argument of Fuhse (2018) is followed, since qualitative methods in network analysis are not predominantly used for identifying network structures, but especially for exploration of network contexts or of subjective interpretation patterns of network relations (Fuhse 2018:16–17). In this dissertation the exploration and interpretation of ecosystem (as the network) relations is of research interest.

Secondly, qualitative methods in network research need to be further elaborated, especially and how (and why) they are used in addition to qualitative content analysis in this dissertation. Qualitative methods in network analysis are suitable for investigating the influence of the network structure on actions and perceptions of the network actors (Hollstein 2012:407). Hence this is used in this dissertation to analyze interaction and cooperation patterns of the network actors as well as their network interpretations, from which action orientations and attributions of meaning in relation to the network can be derived (Hollstein 2012:407). This dissertations aims at explaining network dynamics, which addresses the conditions of constitution of networks, and at analyzing factors that lead to an alteration of network structures (Hollstein 2012:408). Network dynamics are understood as a combination of the actors' interpretations and interactions, thereby addressing both network practices and interpretations. Regarding analyzing influencing factors, qualitative methods such as interviews are used for network analysis, also when focussing on network effects. Network effects can be analyzed both on the actor level and the structural level. This dissertation analyses network effects on the structural level for explaining the 'success' or 'failure' of networks by asking actors about their perceptions of the network, contextual factors, possible problems and recommendations for action in the network context (Hollstein 2012:408). This is followed in this dissertation, which delivers insights into answering the research question how ecosystems can be governed for a sustainable digital development. For this purpose a qualitative design is used for analyzing the networks, because the research field of digital ecosystems (in developing countries) is not yet sufficiently explored. Hence the qualitative network analysis serves the purpose of an exploratory study, fitting this particular research field of this dissertation, for identifying the important themes, actors and cooperation forms within the network. The preferred method for data gathering in this exploratory qualitative network analysis are expert interviews in combination with an analysis of secondary sources (Hollstein 2010:460). Hence it is possible to gather these results through using a qualitative approach, even though mixed-method designs are also often used for analyzing the network interaction (Hollstein 2012:408).

Interviews as a method within network analysis can also serve the purpose of analyzing ego-centric networks through so called *network maps* (Fuhse 2018:142–145). In the interviews, network maps are employed through identifying the subjective perception patterns and relation patterns of each interviewee, which provides an arrangement of concentric circles around the ego depending on proximity and distance. However, this method is not only time consuming (which can be simplified through software programs such as UCINET or Pajek, see Fuhse 2018:142–145), but also network cards are suitable only to a limited extent for the purpose of developing the whole overall network integrating statements and putting them in relation (Straus 2010:533–535). This dissertation however aims at demonstrating the digital ecosystem of Kenya as a whole based on the aggregated interview perspectives and there identified patterns. Hence, this dissertation has chosen to use an alternative strategy to individual name maps for structuring data as a basis for the overall network in form of *name generators*, in line with Wolf (2010).

Name generators are a standardized method, where the interviewees are specifically asked to name actors with whom they are related. These name generators can be modified for exploring specific forms of network interaction (Wolf 2010:471–473). Regarding the specific research question of this dissertation, this method serves the purpose of identifying concrete personal interactions of the multi-stakeholders in the digital ecosystem of Kenya. This approach was followed for every interview, similar to the approach proposed by Pfaff-Rüdiger and Löblich (2018), who determine a network of relationships between the interviewed actors based on the ego-centered network maps and the interview data by typologizing the relationships and mapping them in a binary matrix. On this basis, the relationship network is then visualized. Standardized methods are therefore the main survey method for determining full networks, although it must be possible to differentiate the network from its context. This is done in this dissertation through querying the list of the relevant network actors (as for instance proposed by Jansen 2006:76–79). More specifically, the interviewees were asked about the actor groups of governmental sector, donor organizations, private sector companies and NGOs. Thereby, the network can be evaluated according to the centrality or power and prestige of the actors, according to density, or inductively according to local groups and cliques within the network (Fuhse 2018; Hanneman and Riddle 2012).

More specifically, this dissertation analyses the interview results according to activity and influence. The results of the qualitative content analysis will be graphically demonstrated so that it can be established who are active actors and less active actors in the digital ecosystem of Kenya (activity here measured by participants' responses on question of activity), while simultaneously assessing their influence (measured by degree of inclusion in digital ecosystem e.g., through power and voice assessed by interview participants). This delivers a (relation) matrix, which is demonstrated in Table 3.3. This matrix will be completed with regard to the Kenyan digital ecosystem in Sect. 4.4.3 and in Table 4.1. Firstly, in Table 3.3a the matrix of levels of activity and influence of respective multi-stakeholders are illustrated with lines intentionally left blank since the values of each actor are to be determined in the empirical analysis (see Table 4.1). Secondly, Table 3.3b lists possible activity and influence values ranging from 'active' (activity level) and 'high' (influence level), indicated by '▲', to 'latent' (activity level) and 'low' (influence level), indicated by '▼'. Classifications here are conducted through the qualitative content analysis and hence depend on interviewees' responses. In instances where no clear assessment by interviewees is possible (e.g., in cases of no agreements among experts or lack

Table 3.3a Activity and influence matrix of multi-stakeholders in digital ecosystems indicated by values (own illustration)

(a) Matrix template

Actor Group	Activity	Influence
Academia		
Donor Organizations		
Hubs		
Kenyan Government		
Kenyan Startups		
NGOs		
Private Tech Sector		

Table 3.3b Possible values for the activity and influence matrix in digital ecosystems (own illustration)

(b) Possible activity and influence values

Activity	Influence
▲ Active	▲ High
▲ Mostly Active	▲ Moderately High
~ Debated	~ Debated
▼ Mostly Latent	▼ Moderately Low
▼ Latent	▼ Low
? Unknown	? Unknown

of responses), the value of '**?**' is used. The value '**~**' is used if activity or influence levels are highly debated among expert interviewees so that no clear result is provided. Against this background it is important to emphasize that, given the qualitative (and not quantitative) research design, these classifications are prone to subjectivity.

Conclusively, network analysis in this dissertation is used in addition to qualitative content analysis as it adds an additional method for analyzing the interview data and provides a tool for visually demonstrating the results. Both qualitative content analysis and network analysis will be synthesized and analyzed with regard to the case study of the Kenyan digital ecosystem in the following chapter.

Empirical Evidence: Digital Ecosystem in Kenya

4

In this chapter, the digital ecosystem in Kenya will be analyzed following the design introduced in Chap. 2. The goal is to gather empirical insights into the implications of the digital transformation for governance structures and government action as well as for sustainable development in Kenya. As discussed in Sect. 2.2.5, and as highlighted by expert interviewees such as I13, the development of digital ecosystems is not only of theoretical, but also of empirical or practical importance.

Firstly, similarly to the theoretical literature review on digital ecosystems in Sect. 2.2.5, the difficulty of defining a (digital) ecosystem is shared by some of the expert interviewees such as I5. Moreover, various definitions are used among the interviewees. According to interviewed experts, an ecosystem can be defined as the format, form or organization of connections and interactions that build a network (I18), for a specific purpose (I18), in a specific field with common topics (I13) and on different levels (I5). The actors in the ecosystem are in interaction with each other so that an ecosystem can also be described as an "interaction mechanism" (I5). However, multi-stakeholder activities in the ecosystem not necessarily entail working together but rather independently (I13). Moreover, (digital) ecosystems are described in reference to their environment, not in the sense of ecological nature, but in the sense of context (I15). Actors in the Kenyan digital ecosystem are not only in interaction with each other, but also interact with digital technologies or ICTs as "everything interacts with everything" (I15). Additionally, the conducted interviews suggest using the term 'ecosystem' interchangeably and synonymously with 'networks' (e.g., I10; I5; I18). I18 specifically associates both ecosystems and networks with connections between actors where the ecosystem has the role of serving as a format or form of those connections. The ecosystem can hence further be distinguished into subgroups or sub-ecosystems (I13). In sum, this dissertation provides further empirical insights into an understanding of digital ecosystems.

C. Stötzel, *The Digital Ecosystem of Kenya*,
https://doi.org/10.1007/978-3-658-45553-8_4

Secondly, the digital ecosystem in Kenya will be further explored in reference to Chap. 2 by analyzing the contextual factors surrounding the ecosystem (e.g., the economic context, social context, political and infrastructural context) in order to address the digital implications for the country's development (the D in ICT4D). Here Sect. 4.1 will deliver insights into the influence of contextual factors on the Kenyan digital ecosystem (macro level), which this dissertation argues are important conditions that need to be addressed in governing the digital transformation and more precisely the digital ecosystem for a sustainable development. Similar to the logic proposed in Sect. 2.2, after analyzing the contextual factors, the role of digital technologies and services as a force (meso level) influencing both the macro level (ICT in ICT4D) and the micro level, will be discussed in Sect. 4.2. These potentials of the digital transformation as a driving force for development will be balanced in Sect. 4.3 discussing challenges in terms of digital development barriers, i.e., factors that hinder sustainable development of the digital ecosystem in Kenya.

Thirdly, the developed framework will be transferred to the Kenyan digital ecosystem (Sect. 4.4) in analyzing the inner workings of the ecosystem (micro level) through the four derived determinants. Finally, Sect. 4.5 will introduce the 'outcome' of the analysis in proposing two coordination 'solutions' for sustainable digital development in and of the Kenyan digital ecosystem.

4.1 Network Context (Macro Level): Environment of the Kenyan Digital Ecosystem

This dissertation argues that it is initially important to understand the contextual factors of digital development in ecosystems for governing for a sustainable digital development. In line with this dissertation's design and theoretical background (Sect. 2.1), the contextual factors of digital development in Kenya need to be discussed when analyzing the digital ecosystem. The contextual factors spin around the ecosystem and have influence on the inner workings as well as outcomes of the digital ecosystem as illustrated in Fig. 1.1. Hence, they serve as influencing factors and can also be assessed as preconditions for a sustainable digital development in emphasizing requirements that need to be considered. For this purpose, this section identifies contextual factors through data collection via expert interviewees in Kenya and through their respective analysis based on qualitative content analysis. More specifically, the contextual factors include the political, economic, social and infrastructure context as well as the development cooperation setting and as well as the crucial role of coordination for sustainable digital development in the Kenyan digital ecosystem.

As a brief introduction into the context of the Kenyan digital ecosystem, Kenya is a presidential representative democratic republic in Sub-Saharan Africa. With a GDP of USD 95 billion, it is classified as a lower-middle income country (USAID 2021). With regard to digital development, Kenya is one of the fastest-moving digital economies in the Sub-Saharan Africa region along with the Ivory Coast, Nigeria, Ghana, and South Africa (so-called KINGS). This narrative of "Africa Tech Rising" is mainly due to the mobile revolution as Africa leapfrogged from a few landlines to massive mobile phone use in just a decade (Osiakwan 2017:60, see also Sect. 2.2.2). Moreover, Kenya has been a longstanding leader in technology innovation across Sub-Saharan Africa (USAID 2020). The country positions itself as the 'Silicon Savannah' (Graham and Mann 2013; Akamanzi et al. 2016). The narrative around Silicon Savannah is also part of the current digital development discourse as discussed in Sect. 2.2.1. In addition to the literature review around the ICT4D discourse provided in the referenced section, the interviewed experts in this dissertation likewise observe the mentioned Kenyan self-perception and digital ambitions as a regional pioneer (I13). Comparisons to the namesake 'Silicon Valley' are expressed with regard to the US Silicon Valley as both a rolemodel, but also a source of potential benefits for the Kenyan digital ecosystem given the increased interest (I14). According to interviewed experts such as I24 and I13, this narrative of the Silicon Savannah is also mainly driven by foreign multinationals or expats in Kenya.

Against this background, the highlighted importance of contextual factors for studying digital ecosystems (Sect. 2.1), is further emphasized by this dissertation's findings of the qualitative data analysis. I18 emphasizes that "[...] digital needs to be specified in the context and put in the context of a discussion", so that the country's different contexts are crucial factors for analyzing digital development (see also I3; I12). One of these specific contexts is for instance seen in the infrastructural setting in the country or also in the region. I2 here emphasizes that developing countries in general, but specifically Kenya had different infrastructure equipment than developed countries that needed to be addressed before engaging in digital development. This has hence put them at an disadvantage in terms of digital development, but also at an advantage since "[...] that has given these countries the time to and to learn a little bit more from what's going what's going on around them and learning from the mistakes that might have been done in the developed world" (I2).

Another way to approach contextual factors and ecosystems, is the notion of 'environment'. In this research project, environment is not understood in environmental terms, but more in the sense of surroundings and context (as similarly discussed in Sect. 2.1 as the macro level. Since both context and environment describe a macro level perspective on digital ecosystems, they are also often used interchangeably and synonymously in the conducted interviews. Further underlining the importance of assessing contextual factors, I22 for instance highlights that "the thing is one has to understand the environment". The data analysis of this research project further leads to the following results in determining and understanding the importance of contextual factors for the development of digital ecosystems:

Firstly, the digital transformation affects the economy, society and politics which together create an environment with support structures such as framework conditions or legal conditions (I6). Hence, legislation and policies likewise function as environmental or contextual factors for enhancing digital transformation, also in the business environment (I9). For innovative entrepreneurs in Kenya, the degree of a facilitative environment is also determined by "how easy or difficult is it for [unk] for a Kenyan to start a business" (I14), as similarly expressed in the *Ease of Doing Business Index* (see also Sect. 2.3.5). In the case of Kenya, expert interviewees assert that the former Permanent Secretary in the (then) Ministry of Information and Communication, Bitange Ndemo (2005 to 2013), laid the foundations for digital infrastructure and "a regulatory environment that enabled MPESA to come up" (I7). In sum, the digital transformation itself demonstrates a "rapidly changing context" (I17).

Secondly, it is important for digital development (and respective actors) to create "an environment that's more supportive of of innovations" (I9), which in the case of Africa is assessed as an environment where a lot of innovations happen (e.g., I19). Here the goal is specifically to curate an environment that facilitates and leverages digital transformation (I14) while also mitigating its risks (I13). In sum, an enabling environment is required for digital development and sustainable development of digital ecosystems emphasizing the need for an holistic approach that recognizes the importance of the environment (I20). Thirdly, the extent and quality of environmental factors can be further used as distinguishing features between the digital development of regions, such as the Global North and the Global South (I13), which is also mirrored in concepts such as the digital divide (see e.g., Sect. 2.3.4) and leapfrogging (Sect. 2.2.2).

Fourthly, and of special relevance for answering this dissertation's research goal of understanding interactions in the Kenyan digital ecosystem, environment is understood synonymously with ecosystem due to actors' and their (inter-)actions (e.g., I15). Hence 'ecosystem' is used to describe the environment i.e., context in which actors operate. The multi-stakeholder interactions in the (Kenyan) ecosystem likewise determine contextual factors, which is for instance highlighted in the description of "[...] an extremely dynamic extremely competitive environment" (I7).

Multi-stakeholders in the ecosystem hence operate in an environment that is characterized by competitive forces. Competition is also a further indicator when assessing power relations which will be further discussed in Sect. 4.3.4. An overview of key multi-stakeholders in the Kenyan digital ecosystem is provided in Sect. 4.4.2. The specific roles the multi-stakeholders' take with regard to shaping the here discussed environment, will be further elaborated in Sect. 4.4.5.

Lastly, in line with this dissertation's understanding of environment in terms of context, the environment is comprised of factors such as infrastructure, regulation, policy, education that determine growth and development, as summarized for instance by I14 and in Sect. 2.1.3 and as likewise mirrored in the OECD's 'Digital Economy Framework' (OECD 2018a). Those factors will be further discussed in more detail in the following subsections.

4.1.1 Economic Context: Kenyan Digital Economy

The economic context of countries is one essential part of their development status, while it is also an important area of influence of the digital transformation. As argued in the theoretical background (Sect. 2.2.5), economic value and contributions as well as their interrelatedness with digital ecosystems is crucial for the analysis of the Kenyan digital ecosystem. Moreover, the economic context is one important determinant in which an ecosystems operates and by which it is influenced. To begin with a general economic development perspective, Kenya's economic growth averaged 5.7 percent between 2015 and 2019, making it one of the fastest growing economies in Sub-Saharan Africa (The World Bank n. d. r). Around 35 percent of Kenya's GDP is generated by the agriculture sector (The World Bank n. d. c) with an employment share of 54 percent of total employment (The World Bank n. d. e). Economic growth however is mainly based on services with 54 percent of GDP as of 2020 (The World Bank n. d. q). More specifically, the ICT sector is set to account for up to 8 percent of the country's GDP through IT-enabled services and create more than quarter of a million jobs by the end of 2021 according to International Trade Administration (2021).

With regard to the composition of the Kenyan economy, it is noteworthy that Micro, Small and Medium Enterprises (MSME) are the backbone of the Kenyan economy. As elaborated in the most recently available Kenya National Bureau of Statistics MSME Survey of 2016, MSME were estimated to have contributed approximately 33.8 percent to national GDP in 2015 with approximately 81.1 percent of employment (Kenya National Bureau of Statistics 2016).[1] However, informal enterprises represent the majority of these MSME by 96 percent of microenterprises, 92 percent of medium enterprises and 89 percent of small enterprise (Federation of Kenyan Employer 2021:2). With regard to this dissertation's research interest, one person enterprises (or micro enterprises) play a crucial role in inclusive and sustainable digital development in that they are argued to be a main target group for digital technology implementation: "[...] ninety percent of our business is a small medium enterprise so again if we need a sustainable digital development we need to think of the small enterprises how do we how does technology help them?" (I19). At this point, the question is how can MSME reach the scale that is economically asked for and foster a sustainable digital economy. However, this economic perspective is not of specific interest for this dissertation and hence leaves room for further research.

To complete the contextual factors of the Kenyan (digital) economy, the informal economy accounted for 83 percent of total employment in 2019 while also providing 90.7 percent of new jobs in 2019 (Republic of Kenya 2020a). With regard to economic strength and performance, there are significant regional differences in Kenya. Of the 47 counties in Kenya, the urban counties of Nairobi (21.7 percent), Nakuru (6.1 percent), and Mombasa (4.7 percent) made the largest contributions to the national GDP. The economic center is Nairobi as the main contributor to the national GDP (Wankuru 2019). These counties also account for the highest population rates. It is also noteworthy for this dissertation's focus on sustainable and inclusive digital development that 60 percent of the Kenyan economy is centralized in Nairobi (I22). This poses a specific challenge for the coordination of the Kenyan digital ecosystem and for sustainable digital development as this dissertation will further elaborate in Sect. 4.4.3.

According to the *Ease of Doing Business Index*, the environment for setting up a company or enter business is Kenya is assessed as relatively easy ranking 56th of 190 economies as of 2020 (The World Bank n. d. d). This however is discussed controversially among the interviewees, especially by the interview group of startups, which will be discussed in more detail with regard to entering the Kenyan digital ecosystem in Sect. 4.4.3.

---

[1] According to Kenya's Micro and Small Enterprises Act 2012, micro enterprises are defined as a firm, trade, service, industry or a business activity whose annual turnover does not exceed Kenya Shillings (Kshs). 500,000 and whose total employees are less than 10 people. A small enterprise has an annual turnover of between Kshs 0.5 to 5 million, with the number of employees in the range from 10 to 49 people and medium sized enterprises have between 50 and 99 employees (Kenya National Bureau of Statistics 2016).

Digital Economy

More specifically with regard to multi-stakeholders, especially private sector actors in the Kenyan digital economy and ecosystem, the Kenyan telecommunications sector is dominated by three mobile network operators and three operational mobile virtual networks operators (ITU 2018b:90):

- SAFARICOM, 40 percent owned by the UK mobile group Vodafone and 35 percent by the Government of Kenya, has the largest share
- AIRTEL, a subsidiary of the Indian mobile group
- TELKOM KENYA, taken over by Helios Investment Partners in June 2017
- EINSERVE, a virtual mobile network operator owned by a Kenyan financial group
- SEMA MOBILE
- MOBILE PA

These are not only important multi-stakeholders in the Kenyan digital ecosystem, but also build one sample group of this dissertation's qualitative interview research.

After this brief overview of the Kenyan economy, findings concerning the digital economy will be analyzed to deliver empirical insights into the implications of the digital transformation (with the economic digital development being one context). This is a fine balance since digital transformation is not only a sector itself, but has implications for all other sectors being a *cross-cutting issue*, as emphasized for instance in I13. Moreover, it is necessary for understanding the context of the Kenyan digital ecosystem to note that the terminology of 'digital economy' and 'digital ecosystem' are not clear-cut. This dissertation's findings contribute to this as both are often used interchangeably (see e.g., I13 and I14). Additionally, digital economy is also used interchangeably with (digital) 'market'. Moreover, digital economy is also referred to as "ICT economy", both in the conducted expert interviews (e.g., I21), but also in ICT4D discourse as discussed in Sect. 2.2.5. Additionally, the interview statement "the digital economy or in terms of digitalization" (I6) might suggest that even digitalization or digital transformation is first and foremost associated with economic development. This is especially noteworthy as it poses some challenges for interpretation of this dissertation's findings (see further Chap. 6). As similarly discussed in Sect. 2.2.2, the potential for leapfrogging in Kenya will be analyzed in more detail from the expert interviewees' perspectives in Sect. 4.2.1.

For further analyzing the economic digital development of Kenya, expert representatives such as I11 argue that transformative (digital) innovations have shaped the Kenyan economy. In that, the digital transformation is claimed to be different from previous mega-trends or revolutions as digital is not only omnipresent, but also an enabler for every organization to conduct their activities (I20). This is further supported by I2 asserting a stimulating effect of innovations for businesses or industries (including traditional ones). Secondly, digital transformation impacts economic sectors themselves, so that it has the potential to foster an internal sectoral development. This not only applies to the ICT sector, but also to more "traditional sectors" such as manufacturing or agriculture as examples of "the so-called old economy" (I6). Especially in agriculture there is seen "a huge potential to increase" (I4). Generally, the perception is shared that "the digitalization is everywhere, in every sector, in every industry" (I17). For businesses operating more traditionally in Kenya however, I11 shares the experience that it becomes increasingly harder to conduct business as there is a lot of (digital) transformation. This emphasizes the need for (traditional) industries and businesses to adapt to the digital transformation in order to secure a sustainable future growth (and not be left behind).

The most important determinant in the digital development context of the Kenyan digital ecosystem however is *entrepreneurship* (see e.g., Sect. 2.1.4). As mentioned above, entrepreneurship is a key component of the Kenyan digital economy, especially with regard to MSME. More specifically, President Uhuru Kenyatta declared that Kenya is a "start-up nation" (Republic of Kenya 2015), reminiscent of the previously discussed notion of 'Silicon Savannah'. Not at least since then is Kenya getting "the attention of the world when it comes to entrepreneurship and high-tech business opportunities" (Osiakwan 2017:67). According to the annual African Tech Startups Funding Report 2020, Kenyan tech startups raised a record amount of funding in 2020, securing a combined total 19.1 billion Kshs (Jackson 2021), which makes Kenya Africa's funding leader. This is supported by the Foreign Direct Investment (FDI) net inflows, which in Kenya were 1.4 percent (of the GDP) in 2019 (The World Bank n. d. f) noting an increase in FDI flows by 27 percent to USD 1.6 billion with investment in diverse sectors (UNCTAD 2019). This narrative around Kenya's digital entrepreneurship is likewise confirmed by the interviewed experts who describe Kenya as a "[...] very entrepreneurial country and entrepreneurs constantly young people coming up with different business ideas and a lot of them are innovative [...]" (I14). This is supported by I17. Due to this narrative, the focus on Kenyan entrepreneurship is also visible in the ICT4D projects focussing on entrepreneurship, especially donor funded ones, as similarly discussed in Sect. 2.1.4. Despite the vast amount of digital startups in Kenya and even in Africa, only a few reached scale on international level with an African grown so-called 'unicorn' still missing (I20). The role of startups in the Kenyan digital ecosystem from a digital development perspective as well as their interactions will be further elaborated in the following (e.g., Sect. 4.4.5).

4.1.2 Political Context: (Digital) Governance

Another contextual dimension of digital ecosystems is the political context (see Sect. 2.1.2). Kenya was a former British colony until its independence in 1963. From 1969 until 1992, Kenya was governed by a single-party system with regular elections and a National Assembly. The (re)-introduction of multi-party elections in the 1990s however led to significant election-related conflicts in 1992, 1997 and 2007 (Cheeseman, Collord and Reyntjens 2018). Since its independence, two ethnic groups (the Kikuyu and Kalenjin) have dominated the presidency, so that ethnicity remains a very important organizing principle in Kenyan politics, although the 2010 constitution intended to reduce the role of ethnicity in elections. President Uhuru Kenyatta was sworn in for a second and final five-year term on 28th November 2017 (The World Bank n. d. r) followed by William Ruto in the elections of August 2022. Although Kenya holds regular multiparty elections, the *Economist Democracy Index* assesses Kenya as a hybrid regime (democratic and authoritarian elements) with the country ranking 95th of 167 in its report of 2020 (The Economist 2021). According to the *Freedom House Index* 2020, Kenya is assessed as "partly free" (score 48 of 100), with political rights only reaching 19 of 100 points (Freedom House 2021a). Similarly, internet in Kenya is likewise assessed as "partly free" with data privacy remaining a concern (Freedom House 2021b). Free internet is not only a precondition for everybody to participate in the digital transformation and hence for inclusive human development within their rights (I3), but also affects the work of the multi-stakeholders, such as development cooperation (I2). Moreover, it demonstrates an interdependence of actors (I3) in the provision and maintenance of free internet.

The state of the Kenyan governance institutions can be further exemplified by the *World Bank's Worldwide Governance Indicator* (as of 2020), where Kenya demonstrates weaknesses in all six categories as discussed in Sect. 2.1.2, namely: voice and accountability, political stability and absence of violence, government effectiveness, regulatory quality, rule of law, control of corruption. Government effectiveness in Kenya demonstrates a relative strength with 39.42 (The World Bank n. d. t). According to the source, Kenya's greatest governance weaknesses prevail in 'Political Stability and Absence of Violence' (14.5 out of 100) and 'Control of Corruption' (21.15). Corruption thereby is one major obstacle to Kenya's (digital) development. This is likewise supported by Kenya's score of only 31 (out of 100) and ranking 124th of 180 countries in the 2020 *Corruption Perception Index* (Transparency International n. d.). This obstacle along with further obstacles will be discussed in more detail in Sect. 4.4.4.

On a multilateral level, Kenya is a member of institutions such as the African Union (AU) and Smart Africa. A further characteristic of the Kenyan political system is its multi-stakeholder approach provided in the constitution article number ten placing a constitutional requirement on anyone making public policy to consult stakeholders (e.g., I15).

Political Stability

With regard to this dissertation's research question, political stability is one important factor explaining how coordination of digital ecosystems can contribute to a sustainable digital development (especially by whom). As discussed in Sect. 2.1.2, state authority, capacity and legitimacy are crucial factors for the functioning of governmental institutions. Since (state) capacity and authority in the digital transformation are of specific relevance (Sect. 2.5.2), this will be discussed in more detail in Sect. 4.1.2 and Sect. 4.4.5. Moreover, this dissertation argues that the digital transformation has a profound influence on state authority, capacity and legitimacy. With regard to legitimacy (Sect. 2.5.2), I4 states that it is difficult for multi-stakeholders, including the Kenyan government, to find and establish their legitimacy due to the amount of actors present in the digital ecosystem. Moreover, this is linked with power asymmetries and capacity (Sect. 2.5.2, Sect. 4.3.4), which this dissertation argues are essential factors in governing digital ecosystems for sustainable digital development. This will be further discussed in more detail regarding the Kenyan digital ecosystem in Sect. 4.4.4.

To further underline the importance of political stability for a sustainable digital development in Kenya, both political stability and good governance build a key area of the SDGs, especially SDG 16 with the goal of strong institutions, for which governance is essential (Monkelbaan 2019:v). Expanding on Kenya's performance in terms of governance and political stability, empirical findings of the qualitative data analysis deliver further insights into this dissertation's research goals into the political context of digital ecosystems in developing countries such as Kenya:

Firstly, political stability as an important factor of sustainable digital development (in Kenya and in general) is further confirmed by I3. Promoting and ensuring political stability, particularly in fragile contexts, is one area where development cooperation is active, e.g., through ICT projects (I5). The role of development cooperation will be further discussed in Sect. 4.4.5, but it here emphasizes a challenge for countries such as Kenya an an additional layer of actors (in contrast to countries of Global North). Regulations of the digital transformation's implications, such as data protection, are also issues that are governed differently in different regime types, whether it is democratic, autocratic or hybrid. I3 asserts that this is even more problematic for countries without a functioning or stable democracy. Political stability also serves the function of ensuring trust of multi-stakeholders in the political or governmental actors and actions, which is further discussed in Sect. 4.4.3.

Secondly, and in line with the above mentioned *World Bank Governance Indicator*, political stability as a crucial factor for sustainable (digital) development is further expressed in corruption. As argued for instance in Sect. 2.1.2, corruption is a major barrier for digital development and the development of the Kenyan digital ecosystem, since good governance is a precondition for sustainable development. This argument is shared by expert interviewees, for instance arguing: "You can have a thriving ecosystem here but I think there's a lot about equity and social inclusion that needs to happen in a society is a very unequal society and of course corruption has got a lot to do that." (I17) The interlinkage between corruption and sustainable digital development in Kenya is further emphasized by expert interviewees referring to an all encompassing systemic corruption (I5; I10) that hinders digital development. More specifically, corruption not only on the public sector level but also on the private sector level, including the foreign tech companies active in Kenya, is experienced as impeding the development of the Kenyan digital ecosystem (I7). Here the Kenyan government is seen in the governance role of a legislator to address this problem through regulation and legislation (I10, see further Sect. 4.4.5).

Thirdly, the correlation between the digital transformation and corruption serves as an additional factor for explaining constraints in the development of the Kenyan digital ecosystem. While the potential of digital technologies to address corruption is at least discussed by some interviews such as I10 and I18, the interview findings suggest that the interest in using digital technologies to tackle corruption is not necessarily pursued by public sector organizations in Kenya. More specifically, corruption could be one explanatory factor for preferring analogue processes instead of digital ones (I10). This perception is shared by I21, which will be discussed in more detail in Sect. 4.4.4. Due to the small amount of data of this dissertation however, the aspect of corruption cannot be fully tested and thereby leaves room for further research.

Digital Policies

A more specific component of the political context of the Kenyan digital ecosystem are the before mentioned regulations and laws that likewise constitute the context in which the ecosystem operates and by which it is influenced (Fig. 1.1). As previously mentioned at the beginning of this section, the legal conditions of a country around digital transformation are crucial. Laws and policies are important elements of governance, also in the context of the digital transformation, which is likewise highlighted by I11 and I18 with the latter arguing that policies, strategies and legislation are "[...] crucial as a baseline and foundation for all for all the the other digital develop activities taking place". With regard to roles in this governance context, the role as a regulator and legislator is attributed to the government setting the legal framework (I22), which will be further discussed in Sect. 4.4.5. In Kenya, the Ministry of Information, Communications and The Digital Economy has the core responsibility for ICT policies in Kenya.[2] The Information, Communications and Technology Authority (ICTA) under the Ministry of ICT is further tasked with streamlining the management of all government ICT functions. A further overview of governmental stakeholders in the digital transformation in Kenya is provided in Fig. 4.1. The Kenyan government has passed a number of digital strategies, policies and bills, with the following selection:

- Vision 2030
- The National Broadband Strategy 2023
- Digital Economy Blueprint 2019 (Republic of Kenya 2019)
- The Kenya National Digital Master Plan 2022–2032
- Startup Bill 2020 (Republic of Kenya 2020b)
- Electronic transactions: Kenya Information and Communication Act, Rev. 2009 (UNCTAD n. d.)
- Consumer protection: The Kenya Information and Communication (Consumer Protection) Regulations, 2010 (ibid.)
- Privacy and Data Protection: The Data Protection Act, 2019 (ibid.)
- Cybercrime: Kenya Information and Communication Act, 1998 and The Computer Misuse and cybercrimes Act, 2018 (ibid.)

According to DLA Piper (n. d.), the digital policy landscape in Kenya can be classified as 'limited'. However, the country has legislations on electronic transactions, consumer protection, privacy and data protection and on cybercrime, which is quite advanced in comparison to the state of other African countries (UNCTAD n. d.). With specific regard to digital policies, Kenya's position as a leader in the digital transformation in Africa was reinforced by the *Digital Economy Blueprint* in 2019. This blueprint addresses the development of the digital economy, improving digitalization in trade, governance, infrastructure and entrepreneurship (Republic of Kenya 2019). The blueprint's importance is further emphasized by this dissertation's interview data. However, it is important to highlight that legislation, regulation and regulative or legislative frameworks are often used synonymously by expert interviewees. Against this background, it is further noteworthy that policies in this

[2] For further information see https://ict.go.ke/. Formerly known as Ministry of ICT, Innovation and Youth Affairs. Thereafter it will be referred to as Ministry of ICT.

dissertation include specific acts or bills, blueprints and legislation, and also legal frameworks. For example, the blueprint is often mixed up or used interchangeably with 'framework' in expert interviews, such as I19. In that, the Digital Economy Blueprint is assessed as "very ambitious" (I17), incorporating the goal to leapfrog the digital economy (I17), and is also attributed the potential to serve as a rolemodel or good practice for other countries in the region (I6). I6 further asserts that the European Union (EU)'s 'Team Europe Initiative for Digitalization' is also aligned to the Digital Economy Blueprint in supporting the Kenyan government. This is further confirmed and specified by I17 as there are overlaps with at least three pillars of the blueprint (here overlap is not understood as negative overlaps, such as doubling, but rather emphasizing the alignment). A critical perspective on the blueprint is shared by I6's claim that it is only a rough agenda with no operational plan. Since April 2022 the blueprint has been replaced by 'The Kenya National Digital Master Plan 2022–2032'.[3]

In addition to the before mentioned digital economy blueprint, the startup act (e.g., I24; I19) and the data protection laws were specifically highlighted and discussed controversially among expert interviewees. Taking the example of data protection and privacy laws in Kenya, these are discussed among experts as follows: On the one hand, data protection laws in Kenya are perceived as "in place" (I19) with I6 even assessing Kenya's data protection as "[...] maybe the most comprehensive data protection act in the African continent". Here parallels to the European General Data Protection Regulation (GDPR) as a rolemodel are further noted (e.g., I13). On the other hand, the Kenyan legal system is assessed as rather weak by I10, especially in comparison to the ones of Europe, and also in terms of a "lack of adequate framework conditions" (I6). Due to these asserted flaws around data privacy, I10 for instance shares the possible danger of exploitation and selling of data (e.g., in form of private information) due to flaws in legislation. This is similarity expressed by I13 referring to the value of data extraction, where data does not remain and therefore benefit the local economy. Hence, not only the flow of value from the local economy abroad, but also the need for revision of regulations are highlighted. The latter especially refers to foreign tech companies for instance regarding their ease of entry into the Kenyan digital ecosystem since a lack of regulation makes it easy for them to enter (I2). This will be further discussed in Sect. 4.4.3.

The consequences of 'weak' laws, in Kenya but also more generally in the Global South, are further demonstrated in I2: "[...] in the Global South these laws and regulations are not really very developed yet and then also that means they're not

[3] Thus, the new Digital Master Plan was introduced after this dissertation's data collection phase and hence is not part of the empirical interview analysis.

necessarily supporting their own local industry which opens up the doors to to foreign companies that come in way more advanced". Hence, weak legislation around digital transformation topics, also with regard to the private sector, poses a weak spot for governments and can potentially endanger their local economy and ecosystem. Hence, further policies for supporting the growth of local innovation and industries are highlighted (e.g., by I2). A further challenge is emphasized with regard to the enforcement of data privacy laws due to limited capacities (I6), which will be further discussed in Sect. 4.4.4.

4.1.3 Infrastructure Context: Access to Digital Technologies for Development

As elaborated in Sect. 2.1 and in line with OECD (2018a), infrastructure is an important contextual factor for (digital) development. This applies to 'basic' infrastructure such as health care, access to electricity and water, but also to access to digital infrastructure. For instance, the before mentioned urban-rural gap is also emphasized by many households not having access to piped drinking water (only 34.2 percent) (Gitogo 2020). Even more crucially with regard to digital development, only 70 percent of the Kenyan population have access to electricity (The World Bank n. d. a). In comparison to other developing countries this is rather high, but not sufficient for an inclusive (digital) development.

Mobile cellular penetration in Kenya is 114 per 100 inhabitants as of 2020 (The World Bank n. d. l). The penetration level of feature phones is at 67.9 percent and the penetration level of smartphones is 53.4 percent (Communications Authority of Kenya 2021a). With regard to fiber optic coverage, fixed internet technologies offered in Kenya include fixed wireless, Asymmetric Digital Subscriber Line (ADSL), cable modem and optical fiber (ITU 2018b:90). In 2016, the National Optic Fiber Backbone network had been deployed to all 47 counties in the country, in addition to several other fiber backbones such as The East African Marine System, Eastern Africa Submarine System, Lion 2 and SEACOM (ITU 2018b:90). These are numbers with regard to physical digital infrastructure access, which need to be further distinguished from quality of access. The international internet bandwidth in Kenya is 10,217.46 Gigabit per second (Gbps), which sees an increase in comparison to June 2020 of 38.2 percent (Communications Authority of Kenya 2021b).

For a sustainable digital development and participation in the digital transformation, access to infrastructure such as connectivity and electricity, but also to hardware and skills are crucial. These are further emphasized in this dissertation's interviews as basic requirements for digital development of the ecosystem (e.g., I7; I18; I13).

More specifically, infrastructure and access are barriers to the sustainable digital development in Kenya, as it is "[...] still struggling with some basic things we needs to be met for example infrastructural things [...]" (I23). Without digital infrastructure and connectivity, digital transformation cannot evolve as it is the "enabler for all the activities" (I13). Hence, connectivity serving as the basis for digital development is highlighted by interviewed experts. Whether this need is addressed and whether the importance of infrastructure is seen by multi-stakeholders in the Kenyan digital ecosystem is further discussed in the following passages around electricity, connectivity and Third-generation (3G) as well as in Sect. 4.3.2.

Access to Electricity

One crucial part of development, already prior to the digital transformation but increasingly important, is electricity. Before digital development via adoption and use of digital technologies can be achieved, the respective infrastructure is required and needs to be implemented. Infrastructure for digital development, also phrased as "digital infrastructure" (I12) for instance includes electricity (I10). Especially for connectivity purposes (see below) electricity is highlighted as the foundation for digital development (I9). Electricity hence also influences access to digital technologies and services that are available via the internet, since without electricity there is no internet as it would hamper their use, also for innovation purposes (I14).

With regard to Kenya, electricity is lagging as only 70 percent of the Kenyan population have access to electricity (The World Bank n. d. a). This is especially relevant for rural areas (digital divide) where according to I10 "[t]here's no internet, there's no electricity, nothing." Not only the existence of electricity, but also the stability or quality of electricity is an important factor for sustaining connectivity (I20; I2). This emphasizes a dependency on electricity for (sustainable) digital development that poses a challenge to sustainable and inclusive digital development in Kenya. Examples where a lack of electricity influences digital development in Kenya are for instance education and digital skills (e.g., I10), as schools are assessed as not equipped for digital development with regard to enhancing digital skills, due to a lack of foundational elements such as electricity. This is further confirmed by I20, highlighting a vicious circle of having no electricity or devices in schools standing in contradiction to plans of bringing connectivity to such school settings.[4] This will be further discussed in Sect. 4.4.4.

Hence, a lack of electricity further prevents the use of digital devices such as mobile phones due to lack of battery charging (I20). As another factor impeding

[4] Recent statistical data on the percentage of schools with internet access could not be obtained with regard to Kenya, see e.g., UNDP 2020:371.

digital development in terms of electricity is the cost thereof (I17), again to an increased extent in rural Kenya (I17). Given the expressed importance and need for electricity, the significance of electricity in ICT4D projects will be discussed in more detail in Sect. 4.3.2.

Access to Connectivity

Additionally, electricity likewise enables connectivity. Connectivity here mainly refers to access to the internet via broadband or mobile, which will be discussed in more detail below (Sect. 4.1.4). Among expert interviewees, access to the internet is assessed as a foundation of digital development (e.g., I18), as the basis (I13) and a resource (I4). This is especially relevant with regard to equality and the digital divide, since connectivity is mentioned as lacking in many areas (I20). While some expert interviewees share the experience that the Kenyan government is aware of the importance of connectivity (I13), multi-stakeholder activities in connectivity projects are perceived as less attractive and hence not so much focused on despite its importance (further discussion in Sect. 4.4.3):

> "And and then doing the basics like connectivity is obviously not so attractive but in the end is the underlying an enabler for all the activities and all the technologies were talking about then when it comes to blockchain and AI." (I13)

This dissertation's analysis demonstrates that connectivity provision or expansion is one area of digital development where ICT projects are not necessarily suitable or adapted to the Kenyan context, similar to discussions around a mislead focus of projects in the infrastructure context provided in Sect. 2.3.1. This can be further derived from I20. Synthesizing these arguments, connectivity and electricity are areas where shortcomings are observed by expert representatives (see further e.g., I20). Another similarity between the impact of connectivity and of electricity on sustainable digital development in Kenya, is a common challenge identified in the data analysis: affordability (for access and usage). Not only is connectivity perceived as "extremely expensive" (I4), but also affordability of devices for accessing connectivity is a challenge (I20).

In order to counteract these mentioned challenges of access to connectivity, the need for new technological methods beyond fiber optics in providing affordable connectivity, especially with regard to rural areas in Kenya, is expressed (I17). One of these solutions is for instance seen in community networks to expand connectivity to unconnected areas (I20). As the internet in Kenya is mostly accessed via mobile technology, mobile connectivity such as Second-generation (2G), 3G, Fourth-generation (4G) or 5G is crucial, especially since these are seen as an enabler. Hence, access to 3G (and the other standards) is discussed in more detail in the following.

Access to 3G

3G (along with 2G, 4G, 5G) is one part of connectivity and access to digital technologies and services and serves as an enabler for digital development. Thereby access to 3G likewise functions as a precondition for participation in digital development and the Kenyan digital ecosystem (Sect. 4.4.3). Access to and the availability of 3G is one component, the other is the quality of 3G. Firstly, the interviews acknowledge the common availability of 2G, 3G and 4G in Kenya (e.g., I21; I19; I17). Nonetheless, they assert only limited availability and speed (I19). In Kenya, 95 percent of the population are covered by 3G as of 2021 (ITU n. d.), which is high in comparison to the Sub-Saharan African average with a 3G network coverage of 81 percent in 2020 (GSMA 2021b). This emphasizes the role of mobile devices and technologies (see further Sect. 4.2.1, Sect. 4.2.2). In line with this dissertation's argument that infrastructural preconditions are highly important for sustainable digital development and the digital ecosystem, the governance challenge is to ensure and being aware of the role of 3G, especially with regard to inclusive development e.g., of rural areas (I19). With regard to 5G provision in Kenya, China's dominance is highlighted by I17, which will be further discussed in Sect. 4.3.4.

4.1.4 Social Context: Societal Development in the Kenyan Digital Ecosystem

After discussing the economic, political and infrastructural context in which the Kenyan digital ecosystem operates and by which it is influenced, the societal context will be elaborated in this section as a final contextual determinant. From a general development perspective, Kenya is rated with medium human development as it ranks 143rd (of 189 countries) in the *Human Development Index (HDI)* 2020. Despite the relative improvement in Kenya's human development and economic growth (The World Bank n. d. r), a large proportion of the population experiences chronic poverty, as 36.1 of the population live below the national income poverty line (UNDP 2020). The GDP per capita, purchasing power parity in Kenya is relatively low with USD 4,452.2 (2020) in comparison for instance to USD 12,095.9 in South Africa and USD 54,263.6 in Germany (The World Bank n. d. g). Moreover, the unemployment rate in Kenya is estimated to be at 3 percent (2019), while the Kenyan labor force participation rate was 74.6 percent (The World Bank n. d. s,n). These factors have implications for the sustainable and inclusive digital development of Kenya in terms of affordability and participation in the digital transformation, which will be discussed in more detail in Sect. 4.4.4. Fixed-broadband subscriptions per 100 inhabitants on the other hand are 0.6 in Kenya in contrast to 13.6 of the World's average (ITU 2018a:90), which emphasizes the leapfrogging effect and the dominant role of mobile technologies (as discussed above and in Sect. 2.2.2, see also Sect. 4.2.1). Nonetheless, only 30 percent of the population (approximately 55 million inhabitants) use the internet as of 2020 (The World Bank n. d. h,n), with costs or affordability being one of the impeding factors (further discussion in Sect. 4.4.4). It is noteworthy that Kenya has a very youthful population with 35.7 million Kenyans (75.1 percent) being below 35 years. However, Kenya struggles with a high youth unemployment rate at 39 percent (Gitogo 2020). The majority of the Kenyan population (72 percent) lives in rural areas (The World Bank n. d. p), which is important to consider with regard to the digital divide (Sect. 2.3.4, Sect. 4.3.6).

As of 2018, the literacy rate of Kenya was 82 percent meaning that the majority of Kenyans are literate (The World Bank n. d. k). According to the *World Intellectual Property Organisation's Global Innovation Index 2021*, it is the second most innovative country in Sub-Saharan Africa after South Africa (World Intellectual Property Organization 2021). More specifically, Kenya was one of only three Sub-Saharan African countries as of 2018 that were close to the African Union's target of 1 percent of GDP invested in R&D with around 0.8 percent (UNESCO 2018). Despite the relatively high literacy rate and government expenditure in education and research, digital literacy and skills that are needed to participate in the digital transformation and the digital ecosystem are challenges in Kenya. This will be discussed in more detail in Sect. 4.4.4.

These stated challenges for human development and their implications for sustainable digital digital development in Kenya are further discussed among expert interviewees. Generally, the impact of the digital transformation on the Kenyan society is noted (e.g., I11). More specifically, the perception is shared that there are various developmental hurdles in Kenya that are needed to be overcome (I24). As mentioned, poverty is one major challenge for the Kenyan society. With regard to its implications also for digital development, it is important to highlight the role of local differences in global conversations (for instance in comparison to the ones in the Global North). For instance, it is stated that Kenyans struck by poverty, especially in rural areas, prioritize discussions around food before discussions around digital technologies (I15).

These developmental issues further have an impact also on local innovators developing digital solutions for solving local problems given that the local conditions are different (especially in comparison to other development contexts). Hence, with regard to sustainable digital development, it can be argued that there is a gap between the 'fit' of local innovations and foreign innovations for local contexts (which also has to do with culture, socialization etc.). I24 as one representative of a Kenyan tech actor argues that the focus in finding innovations in Kenya differs from the one in a developed country such as the US. This is similarly perceived by I4 arguing that "[...] there's so many innovative solutions being created which would not have been would not have been created if if the innovators in the Global South were thinking inside the box of the Global North". The arguments demonstrate that it is crucial for sustainable digital development in Kenya to take these contextual factors into consideration, especially when designing ICT projects of multi-stakeholders (see further Sect. 4.3.2). This directly leads over to the next subsection where the role of development cooperation in Kenya's digital ecosystem will be discussed.

4.1.5 Development Cooperation and ICT4D Projects in Kenya

In Sect. 2.1.6, collaboration, cooperation and coordination have been discussed as based on a continuum. Since coordination is an essential part of this dissertation's research question, the role of coordination in the Kenyan digital ecosystem for sustainable digital development will be discussed in a separate section (Sect. 4.1.6), to which all the other following chapters refer back to. Emphasizing the role of cooperation for a development cooperation angle on ICT4D in Kenya, the three concepts of cooperation, collaboration and coordination are often used interchangeably by interviewed expert representatives (e.g., I1). This not only alludes to the importance of common understanding and awareness (Sect. 2.5.1, Sect. 4.4.3, Sect. 4.4.4), but also complicates its analysis given the lack of terminological distinctness and possible biases.

Expanding on the role of cooperation between multi-stakeholders in the Kenyan digital ecosystem, expert representatives such as I5 and I1 highlight the importance of cooperation. I11 even calls for closer cooperation among multi-stakeholders in the Kenyan digital ecosystem: "[...] if we all came together and just did one thing for all that would make it better." Cooperation in the ecosystem is further perceived from an interaction perspective in terms of conversations and exchange between multi-stakeholders (I3). Similarly, the data analysis suggests that cooperation is used synonymously with alliances (I2). This finding further supports this dissertation's focus on governance formats for sustainable digital development, with alliances being one of them (Sect. 4.4.6).

Besides approaching cooperation from an interaction perspective, a second angle especially relevant to this research project refers to development cooperation efforts. The goal of some interviewed multi-stakeholders, especially donor organizations, is to promote and foster cooperation between (here mostly) European members and cooperation with (global) partner countries and also with the private sector (I12). Cooperation between donor organizations and the private sector however is not without challenges, for instance "[...] as a public sector organization we have limitations in how we work with the private sector" (I17). An additional challenge to cooperation is competition (e.g., I18), which is assessed as one factor impeding coordination and sustainable digital development that will be discussed in a separate section (Sect. 4.3.4). Further challenges of multi-stakeholder cooperation in the Kenyan digital ecosystem, especially with regard to development cooperation, will be discussed in more detail in the next subsection.

Aid Fragmentation in Development Cooperation

Before analyzing fragmentation in Kenya, the development cooperation environment is noteworthy. Kenya is the third largest ODA recipient of the OECD-DAC in Africa as of 2019 OECD (2019c). Among the top 10 OECD-DAC donors in Kenya are for instance the United States, United Kingdom, EU institutions and Germany as per ODA 2019-2020 average (OECD n. d.). Moreover, China as a non-DAC and SSC donor, is the third largest official donor to Africa after the United States and the European Union as of 2018 (Overseas Development Institute 2018). Kenya for instance is an official BRI country (Lau 2020) and part of the *China Africa Tech Initiative*, which aims at bringing startups and Small and Medium-sized Enterprises (SME)s together for collaboration and (technology) transfer (China Africa Tech Initiative n. d.).

Against this background and as discussed in the theoretical background (Sect. 2.1.5), aid fragmentation is a challenge for development of countries such as Kenya. Some expert interviewees likewise highlight the existence of aid fragmentation in Kenya (I4; I12) and assess it as an obstacle for development (I6). Some donor organizations themselves are criticized by expert interviewees due to following a fragmented approach in their development efforts such as the UN with all its organizations (I2). Furthermore this is an example for the number of actors and fragmentation (Sect. 4.4.3). A further instance of fragmentation in the Kenyan digital ecosystem is seen in the different speed of donor activities that contribute to fragmentation: "So we have some that have been forerunners, some some are little bit lagging behind and only coming on board now and that has led to a very fragmented indeed approach in terms of of donor support within within Kenya" (I13). This has to do with donor's digital capabilities among other factors, which will be further discussed in Sect. 4.4.4. In line with I2, this dissertation argues that the different interests of multi-stakeholders including donors, not only contribute to fragmentation in the Kenyan digital ecosystem, but also impedes coordination for sustainable digital development. This will be discussed in more detail in Sect. 4.3.1 and Sect. 4.4.3.

More specifically with regard to the number of development actors in the Kenyan digital ecosystem, Kenya for instance had 39 aid relations in 2014 (OECD 2014). This not only emphasizes the vast amount of donors active in Kenya, but would classify Kenya as an 'aid darling' (Davies and Klasen 2019). This dissertation argues that the large number of actors in the Kenyan digital ecosystem, not only but including donor organizations, is one factor impeding sustainable digital development and demonstrates the need for coordination (Sect. 4.1.6). The data analysis of this research project discusses experts' experiences of an increased aid fragmentation (e.g., I2) in the Kenyan digital ecosystem.[5] Further expanding on the implications of the donor landscape for the functioning of the Kenyan digital ecosystem, the impact on Kenya's (digital) development of being a 'donor darling' is discussed controversially among interviewees. While being a 'donor darling' comes with increased attention, also in terms of investment (further details in Sect. 2.1.5, Sect. 4.4.4) as illustrated by I3, it likewise impedes digital development. More specifically, donor darlings in the Kenyan digital ecosystem are seen in hubs and also some startups. The consequences for these stakeholder groups are that "[...] maybe the ones who know how to write proposals best and the other ones who don't know how to write a proposal just never pass the first stage and to to be considered for a partnership" (I4), i.e., favoring already known ones. This perception is shared by I13, arguing that in the area of of digital innovation and startup support there are a lot of donor darlings with the same organizations receiving support. This leads to a concentration of (donor) support systems on a few specific startups or NGOs instead of broadening target groups and "coordinating better who gets which type of support from from which actor or organization" I3. This favoritism is not only observed in terms of financial support, but also in form of support programs such as innovation challenges (I14). Thereby donors potentially create a lock-ins in that donor darlings have advantage due to having knowledge of donor proceedings to be considered for support, which amplifies asymmetries and unsustainable development. This uncoordinated and fragmented setting of (donor) support in the Kenyan digital ecosystem increases transaction costs for instance due to dependencies on donor requirements (I17), which will be discussed in more detail in Sect. 4.4.4.

This fragmented and uncoordinated approach of donors (and private sector) in the Kenyan digital ecosystem is likewise mirrored by I17 experiencing a donor satu-

[5] The interviewees did not refer to a specific timeframe to which their statements or experiences of an increase apply. Assumingly, their statements stem from their time as part of their role as organization's representatives in Kenya.

ration and venture capital saturation. The data analysis of this dissertation identifies digital development and ICT4D projects as one area of donor saturation, which is further discussed in Sect. 4.3.1. In sum, this fragmented and unequal approach of donors towards recipient countries such as Kenya is one area where coordination as a countermeasure can achieve sustainable development. This is further confirmed by an expert representative of a bilateral donor organization: "I think the aid fragmentation that you've mentioned is is quite a big issue from my perspective as well because ja at some point this this needs to be coordinated in order to to have it make it more efficient" (I3). This is especially relevant for this dissertation's focus on digital development as these raised issues likewise apply to the digital transformation context. The digital field in Kenya is characterized as "[...] quite fragmented in terms of different sectors and sectors the digital economy" (I6), emphasizing the need for coordination as "there's a big need of building capacity here to streamline it, to coordinate it more [...]" (ibid.).

The implications of fragmentation in development cooperation for the sustainable development in the Kenyan ecosystem will be further explained in the following using the case of ICT4D projects in Kenya. This dissertation will provide insights into possible solutions as elaborated in the theoretical framework, with regard to a discussion of governance roles, governance structures as well as possible formats of a meta organization and digital platform for coordination, which all are analyzed with insights from the expert interviews (Sect. 4.4.5, Sect. 4.4.6, Sect. 4.5).

4.1.6 Coordination for Sustainable Digital Development

Synthesizing the findings of the previous section, the importance of coordination of multi-stakeholders' activities in the Kenyan digital ecosystem becomes obvious. Similarly to Sect. 2.1.6, coordination is the basis of this dissertation's research question so that all following sections relate back to this one. More specifically, coordination of digital ecosystems will be elaborated in this section in providing the argumentation frame for the following passages. Here, both coordination enabled through digital technologies (e.g., I1) and coordination through network governance will be discussed in more detail in Sect. 4.4.1 and Sect. 4.5. This dissertation's interview data analysis further identifies coordination in the Kenyan digital ecosystem in terms of a) the current state of coordination in the Kenyan digital ecosystem especially regarding coordination of multi-stakeholder activities, b) intergovernmental coordination in the Kenyan (digital) public sector, as well as c) possible impacts of coordination with regard to transaction costs and efficiency. These will be discussed individually as follows.

Current State of Coordination in the Kenyan Digital Ecosystem

Beginning with a), the current state of coordination in the Kenyan digital ecosystem, these findings expand on the discussion around ICT projects and fragmentation (Sect. 4.3.2). More specifically, I12 highlights that currently "there's so many players and as I in the there is a confusion [...]". More importantly for this dissertation, overall coordination is stated as lacking (I6). While I17 argues that there have been previous attempts of (European) donors for coordination, I5 sees potential for coordination of (all) multi-stakeholders and I3 further argues that more coordination is needed. The need for coordination in the Kenyan digital ecosystem however is discussed controversially among the interviewed expert representatives, especially with regard to *what* exactly needs coordination.

More specifically, I5 and I11 highlight the importance of coordination for sustainable digital development. Coordination in this dissertation refers to coordination of multi-stakeholders' activities in the Kenyan digital ecosystem. I15 highlights: "Yes there must be there needs to be more you know there has to be more coordination there has to be more coordination of the actions". This is supported by I5 with regard to the very complex set of different stakeholders. The coordination of multi-stakeholder engagements is claimed to further enable "[...] collaboration potentials that are currently not not really recognizable" (I5). This calls for (even more) visible collaboration enabled through coordination, which could for instance address such a lack of collaboration (I20), as further discussed in Sect. 4.4.3. Summarizing the experts' calls for coordination of multi-stakeholders activities that support this dissertation's argument, coordination among actors is needed for more holistic and sustainable solutions, long term impact and for addressing obstacles (I6). Notwithstanding this enthusiasm, I18 on the contrary argues that coordination is not needed with regard to the activities of all actors, but only of key actors.

Coordination of Multi-stakeholders' Activities

With regard to these key actor groups (b), the actor groups of *donor organizations* are key actors in this research project for sustainable digital development in Kenya. Expanding on the argumentation of Sect. 2.1.5 and Sect. 4.1.5, coordination of donors for decreasing fragmentation is further identified in the data analysis of this dissertation. For instance, I6 argues that "there is a tremendous need of coordination, streamlining and harmonization of donor activities and development activities", highlighting the importance of coordination in development cooperation. Donor coordination would further contribute to sustainable and inclusive digital development in Kenya through counteracting notions of favoritism or donor darlings (I3). Coordination of donor organizations and their engagements is further claimed to enable a more visible collaboration and potential synergies (I5). This feeds into the developed framework precondition of Sect. 2.5.1, which will be further discussed with regard to visibility in the Kenyan digital ecosystem in Sect. 4.4.3.

A further component of coordinating multi-stakeholders activities identified in the data analysis refers to the actor group of the *Kenyan government*. More specifically, coordination here refers to inter-governmental activities for achieving a sustainable digital development in the context of the digital transformation as a cross-sectoral challenge (see e.g., Sect. 2.1.3, Sect. 2.2.1). With regard to the Kenyan digital ecosystem, inter-governmental coordination is assessed as essential by I19. One specific example where inter-governmental coordination would contribute to a sustainable digital development is seen in "[...] one thing government is doing is unfortunately as it's building roads ends up destroying ICT infrastructure that has already been built." (I21) In this instance, various Kenyan ministries are claimed to work parallel to each other (Sect. 4.4.3). According to I6, this also demonstrates a need for capacity building of the Kenyan government to coordinate the digital economy being quite fragmented in terms of different sectors. However, as argued in Sect. 4.1.5, development cooperation efforts in Kenya are likewise prone to duplication of activates due to the discussed lack of donor coordination. Hence, not only donor coordination would further counteract this fragmentation and instead foster sustainable digital development, but coordination would also allow an alignment of development cooperation activities to the agenda and priorities of the Kenyan government for avoiding duplications as "experienced a lot in the past when it comes to donor activities." (I6) The role of coordination in terms of alignment will be further discussed in Sect. 4.4.3.

In contrast to the discussed multi-stakeholder groups of donor organizations and the Kenyan government, coordination is discussed as not necessary by expert interviewees with regard to digital entrepreneurship and the *innovator scene*. An expert representative of the private sector argues that coordination would stifle innovation for which "[...] you need a certain level of chaos" (I21) and "you can't really innovate with coordination" (I21). This argument however needs to be put into the context of actors' (self-)interests, here business interests, so that certain biases cannot be ruled out completely. Nonetheless, this perception is further shared by the expert interviewee of a bilateral donor organization (I5) and also of a Kenyan NGO, with the latter arguing that coordination "doesn't fit well the digital spirits and the they don't want to be coordinated doesn't work really" (I1). A reason for the unique role of the innovator scene in coordination of the Kenyan digital ecosystem is seen in their different approaches, such as in terms of ideology and agility: "[...] there doesn't need to be so much coordination really because it is you know very liberal approach and just very agile approach things pop up and tried out and things work and things scale or not" (I1). Hence, on the one hand, coordination of the innovator scene is discussed as counterproductive to innovation by the expert representatives. On the other hand, the findings suggest that the degree of coordination is an important determinant for a sustainable and inclusive digital development. Regardless, it underlines the challenge of balancing the actors' interests and coordination as well as coordination challenges for a sustainable digital development.

Coordination for Minimizing Transaction Costs and Increasing Efficiency
Synthesizing these findings on the current state of coordinating multi-stakeholder activities and expanding on coordination potentials, the following benefits of coordination for sustainable digital development are identified in the data analysis (c):

Firstly, as discussed in Sect. 2.1.5, coordination of multi-stakeholders' activities in the Kenyan digital ecosystem can contribute to *minimizing (transaction) costs*. More concretely, I20 argues that it's very costly not to coordinate, emphasizing the correlation between coordination and sustainable digital development in Kenya. The potential for minimizing costs however is contested by I1 arguing that coordination itself is costly: "[...] is simply too much overhead too much work and friction and cost and everything". A use case where coordination can minimize costs however is seen with regard to minimizing costs through finding and speaking with one common voice (Sect. 2.5.1, Sect. 4.4.3), especially with regard to the interaction of multi-stakeholders with the Kenyan government: "So by virtue of working together in a coordinated manner and with a single voice we hope not to overwhelm them in terms of this dialogue and and conversations." (I17) This is one example for the role of both voice (Sect. 4.4.3) and digital capacities of the multi-stakeholders (Sect. 4.4.4). Moreover, this applies specifically to donor organizations but also to all other multi-stakeholders as argued by the expert representative of a Kenyan hub emphasizing the importance of coordination for (generating) a common understanding: "[...] just understanding if we all get to understand what coming together and coordinating might do for the for digital transformation then we we could move this much faster." (I11) This is closely related to costs for the Kenyan government that emerge from power asymmetries and dependencies between multi-stakeholders in the digital transformation, which will be discussed in more detail in Sect. 4.3.4 and Sect. 4.4.4. Another dimension of coordination for minimizing transaction costs is transparency, in line with transparency as a precondition for coordination and governance of the developed framework (Sect. 2.5.1). Transparency in the Kenyan digital ecosystem for a sustainable digital development will be further discussed in Sect. 4.4.3.

Secondly, and closely linked to minimized costs, coordinating the activities of multi-stakeholders in the Kenyan digital ecosystem potentially increases efficiency (efficiency of multi-stakeholder activities and thereby network efficiency, see further Sect. 2.5.1). With regard to the empirical interview data, coordination of actors and projects could potentially increase efficiency and impact, as underlined by I7 and I15. This specifically is discussed with regard to coordination needed for decreasing aid fragmentation and thereby increasing aid effectiveness (I3). More concretely, coordination for increased efficiency is further perceived with regard to mobilizing more resources as coordination enables having an overview over (equal) resource distribution (I14; see Sect. 4.5.1). However, the potential for increasing efficiency through coordination is potentially lower if the motivation for coordination is just to tick a box (I20). This raises the question of how, and through whom, coordination can be achieved which will be discussed in Sect. 4.4.4, Sect. 4.4.5, Sect. 4.4.6 and Sect. 4.5. Moreover, the role of digital technologies for coordination in the Kenyan digital ecosystem in the context of sustainable development will be discussed in the next section, moving from the macro to the meso level of analysis.

4.2 Digital Technologies for Development in the Kenyan Ecosystem (Meso Level)

As discussed in Sect. 2.2, the digital transformation influences digital ecosystems in offering both potentials and challenges to development. It has also been established in the theoretical background that a common language and understanding of the digital transformation is crucial for its governance (Sect. 2.5.1). This section will deliver empirical insights into the 'ICT' part of 'ICT4D' in the Kenyan digital ecosystem through the qualitative data analysis. Similarly to the design proposed in Fig. 1.1, digital technologies here are analyzed as a meso level operating between and influencing both the macro level (context) and the micro level (inner workings). Whether ICTs further overtake governance roles in the Kenyan digital ecosystem will be discussed in more detail in Sect. 4.4.5. Whether they provide a governance form or tool will be discussed in Sect. 4.4.6. This present section will follow the respective theoretical background (Sect. 2.2) in that the potential of ICTs for leapfrogging will be critically discussed by experiences from the Kenyan case. Furthermore, it will be discussed which technologies influence the Kenyan digital ecosystem. Here, the focus lies on the technological level, which poses specific challenges for sustainable digital development in the Kenyan digital ecosystem, which will discussed in more detail in the next section (Sect. 4.3).

Before analyzing digital technologies in the Kenyan digital ecosystem, it is important to first highlight that the understandings and definitions of the digital transformation differ between interviewed experts, similarly to differing understanding in academic literature (Sect. 2.2). Digital transformation is used synonymously with other terms such as 'digitization', 'digitalization' or 'digital space' in the expert interviews (e.g., I12; I6). This represents a terminological confusion as highlighted for instance in I18: "[...] the danger of the term and then the process and discussion been very vague because digital digital is very broad term and the broad statement". More importantly, a lack of a common understanding of digital transformation is emphasized for instance by: "And I think we're falling very often in the trap of considering an end in itself without understanding what is it meant to lead us to." (I17) The importance of a common language and understanding for the Kenyan digital ecosystem and for a sustainable digital development will be discussed in more detail in Sect. 4.4.4. Hence, if digital technologies (and their functioning or use) are not understood properly by involved multi-stakeholders (I22), it hinders digital development. The is observed especially with regard to digital technologies such as AI with I18 sharing the experience that artificial intelligence is often not fully understood. AI as one exemplified digital technology will be explored in more detail in the next subsection (Sect. 4.2.2).

Secondly, digital transformation is ascribed the potential to foster development such as economic, political, social development in the sense of a driving force (as discussed in Sect. 2.2.2). In this section, it is not (yet) discussed as to whether ICTs are an actor that can take an active governance role such as the role of a driver. The role of driving the digital transformation in the Kenyan digital ecosystem will instead be further discussed in Sect. 4.4.5. Most importantly for this dissertation's analysis of the Kenyan digital ecosystem, I5 for instance highlights that "innovation or digital ecosystem is becoming increasingly important". Noteworthy here is the reference to the digital ecosystem being used interchangeably with innovation. The digital transformation and digital ecosystems (generally but especially in the digital development context) are assessed as

> "[...] still quite new I would say and therefore all the engagements with regard to helping digital ecosystems are rather new, meaning that before of course they were engagement in and setting up networks and and coordinating the mechanisms but also engaging with these these new set of of stakeholders maybe that's actually a trend" (I5).

This further underlines this dissertation's research focus on the correlation between coordination, sustainable digital development and the Kenyan digital ecosystem from an actor interaction perspective (Sect. 2.1.4, Sect. 2.5.1) as well as from a governance perspective (Sect. 2.4.3, Sect. 2.5.4). In sum, digital technologies in the digital ecosystem of Kenya are identified "as a key space transformational space" (I14) and "in the case of Kenya it's definitely a a driver like a momentum" (I6). The difference and uniqueness of the digital transformation, in comparison to other megatrends (Sect. 2.2), is assessed as allowing "[...] opportunities for development that were not there before" (I4) and "[...] that digitalization can make a difference and can help in the development" (I12). Such opportunities for development are for instance seen in the potential for leapfrogging as discussed in Sect. 2.2.2, which will be discussed with regard to Kenya in the following subsection.

4.2.1 Leapfrogging

In line with the theoretical background (Sect. 2.2.2), leapfrogging is one crucial dimension of ICT as drivers for development (ICT4D) of countries such as Kenya. The leapfrogging potential in Kenya is discussed controversially among the interviewed expert representatives in this research project. On the one hand, experts such as I10; I9; I4 share their perception of the digital transformation as a potential for development in Kenya. In this context, ICTs are ascribed a unique role for development as "through whatever aspect of life you think in Africa, it is only the technology that can leapfrog it." (I19) For I5 on the other hand, the leapfrogging potential only exists in theory and the assumption that integrating digital technologies in processes will immediately change the whole process outcome is overestimated. Instead, for seizing the leapfrogging potential, other factors such as (legal) framework conditions are necessary, which however are currently perceived as lacking (I6). More importantly for this dissertation's research question around coordination for sustainable digital development, the lack of coordination of multi-stakeholders is one factor that is "[...] hard to keep this leapfrogging process going" (I5). After discussing the general observations of experts on the leapfrogging potential in Kenya, the different angles of leapfrogging (economic stages, societal development and structural development) will be discussed in more detail in the following, with the latter two specifically emerging from the interview data analysis.

Leapfrogging: Economy
Similar to the theoretical section on leapfrogging (Sect. 2.2.2), the value of the digital transformation especially for economic development is further discussed by expert interviewees with regard to Kenya. Here digital technologies are seen as "playing an important role in Kenya for for the creation of economic growth, for the creation of jobs and for employment promotion" (I6) and for "catching up with other countries not only in Subsaharan Africa but worldwide" (I6). Moreover, digital technologies and innovations (I11) are perceived by expert interviewees as enabling the "whole digital you know marketplace really" (I21). The greatest leapfrogging potential is seen by expert interviews, in line with theoretical discussions in Sect. 2.2.2,

with regard to the banking sector or financial payments. The leapfrogging effect here is seen in "[...] to not go from bank account to online banking but really straight into mobile banking" (I2), which is different to the development of banking systems in countries of the Global North which are argued to be dominated by "[...] a very traditional banking system" (I5). For instance, "[...] especially in the African context there are a lot of very smart solutions out there to leapfrog" (I5). MPESA from Kenya is one of the most prominent examples (e.g., Nan and Markus 2019; Lee 2019; Graham 2019). With regard to sustainable and inclusive digital development, MPESA is one example of including "former excluded users are able to use e-money transfer services in a convenient, secure, and affordable way, circumventing traditional costs and risks associated with physical cash." (Nan and Markus 2019:468) Moreover, leapfrogging of technological stages is especially enabled through mobile phones (e.g., via which MPESA is operated) as discussed in Sect. 2.2.2 and Sect. 4.2.2. While these findings are not necessarily new, they deliver further empirical confirmations from the Kenyan experience.

Leapfrogging: Society

Besides the leapfrogging potential for the Kenyan economy, expert representatives likewise highlight the transformative influence of the digital transformation for the Kenyan society. Here experts assess the leapfrogging potential of the digital transformation as a societal change (I2; I12; I11; I5). More specifically, digital technologies as driving forces are associated with "'[...] changes in the society and with changes in policies and with changes in like potentially politics as well and and how societies are affected by it" (I2). More importantly for this dissertation, ICTs are argued to enable changes in the relationship between citizens and governments with "[...] potential to improve good governance based on digital technologies be it civic participation, be it e-government service deliveries from governments to citizens" (I4). This is similarly discussed in Sect. 2.1.2 and Sect. 2.4.1.

Leapfrogging: Structures

As a third area for leapfrogging, the data analysis highlighted the macro level factors (Sect. 4.1). Here the data suggests that leapfrogging through digital technologies is possible due to lack of (pre-)existing structures in developing countries, running contrary to the discourse of developing countries being at a disadvantage, specifically in the digital transformation (Sect. 2.1.1). The lack of existing structures and systems is furthermore seen as a potential in comparison to industrialized countries (I19). More specifically, due to the lack of structures such as policies and a start "from zero" (I18; I24) with a "cleaner slate" (I18), Kenya directly jumped to newest or most locally fitting technologies in "almost build anything themselves" (I24) without the need to update outdated systems. This likewise applies to using best practice polices already tried and tested e.g., by countries of the Global North, so that the structural leapfrogging potential likewise exists in terms of adopting latest policies (I18), which underlines the importance of learning for sustainable digital development (Sect. 4.3.5). Moreover, this structural leapfrogging potential due to a lack of legal structures is hence seen as a source for innovation (I19). The lack of regulations as offering room for digital development and structural leapfrogging is further seen in gaps between the speed of the digital transformation and regulations: "I don't think policies have necessarily caught up we see gaps in regulations and all of that and I think that's really where the potential for leapfrogging lies [...]" (I2).

These findings lead to the concluding hypothesis that leapfrogging structural digital development is 'easier' if a) fixed structures are not already preexisting and b) policies or regulations are flawed or non-existent. Most certainly, these possible potentials need to be assessed critically in relation to their downside effects, especially from a sustainable digital development perspective. This for instance includes legislation around data protection Sect. 4.1.2, which are also further discussed in Sect. 4.4.4. This likewise leaves room for further research.

4.2.2 Digital Technologies and Services Influencing the Kenyan Digital Ecosystem

After discussing ICTs as a driving force for development, this section analyzes digital technologies themselves in more detail. Moreover, as also discussed in the theoretical background, there is a global hype around certain technologies (Sect. 2.2.3). Hence a selection of such 'hyped' technologies in Kenya will be explored. For a further analysis of digital technologies influencing the Kenyan digital ecosystem, this dissertation distinguishes between local and foreign technologies (e.g., I17) and between 'basic' vs. 'advanced', meaning more innovative types of technologies without going into too much technical detail.

'Basic' Technologies

Firstly, one dimension of the digital transformation in Kenya refers to technologies enabled by the internet, but also to using "[...] very very low technologies inc like like SMS technology so that was one dimension of digitalization there was quite prominent in Kenya" (I17). Despite advances in internet access and increased availability of services through the internet (Sect. 4.1.3), one interviewee from private sector circles (I16) argues that the internet may not be the main mode of collaboration in Kenya, but instead 'normal' Global System for Mobile (GSM) messaging. More specifically, MPESA is mentioned here as the most prominent example of using USSD based systems (I21), a technology that "[...] has been bypassed never even been a feature that has been used in in UK in Germany but because it's kind of free and and using you know below the line communications channels on the mobile networks" (I21). This is a further example for leapfrogging technologies or technological stages, as discussed in the previous subsection (Sect. 4.2.1). Hence, one crucial factor for the accessibility of technology is their availability and affordability (see more in Sect. 4.4.4), which underlines the importance of such 'basic' technologies in the Kenyan digital ecosystem.

'Innovative' Technologies: Artificial Intelligence and Blockchain

Secondly, more 'advanced' technologies or more innovative types of technology are a further essential influencing factor in the Kenyan digital ecosystem. I13 for instance highlights the leapfrogging potential of innovative technologies in that "[...] there is this kind of leapfrogging indeed happening on on on certain areas when it comes to emerging tech". Of these types of technology, AI and blockchain were the technologies that have been mentioned the most by the interviewed experts, followed by references to IoT, but to a lesser extent.

However, with regard to the Kenyan digital ecosystem and sustainable digital development, this dissertation's data analysis identified challenges with regard to AI. These range from AI not being properly understood by multi-stakeholders (Sect. 4.4.4) as well as AI as a 'hyped' technology that is characterized by actor proliferation (Sect. 2.2.3, Sect. 4.3.1). Due to the incremental nature of these (rather) recent innovations, they pose a challenge for governments in terms of regulation (e.g., I13, Sect. 4.1.2) – certainly not only for the Kenyan government, but worldwide. More specifically with regard to this dissertation's analysis of sustainable digital development, the current focus on more innovative types of tech in line with the current 'trend' or 'hype' leads to the perception that using these innovative technologies is not always aligned with the specific context or needs (e.g., I7), which will be further discussed in Sect. 4.3.1. The gap between the degree of innovativeness of the used digital technology and the needs on the ground are further questioned for instance with regard to agriculture, especially in rural Kenya: "[...] if you're in agriculture which many people are many people employed and working within the field of agriculture you know the technology that is going to transform your life is is actually gonna be more mechanical" (I21). Such an underlying leapfrogging effect (Sect. 4.2.1) for rural areas in Kenya is likewise questioned by I12.

Another characteristic of innovative types of tech is that they require a vast amount of data (e.g., I18). Hence the role of data will be further analyzed in the following as it is a crucial element of digital transformation (also beyond a mere technological level).

Data

Taking up the previous argument of data as a requirement for the functioning of technologies, analyzing te role of data for sustainable digital development in the Kenyan digital ecosystem is further important. Data is not only the basis for learning of technologies such as AI, but also of multi-stakeholders that collect data (or rather information) in their activities in the Kenyan digital ecosystem. This is for instance described by I14: "[...] how do we quickly learn what is working and what is not working by testing you know hypotheses by collecting data by bringing in different partners into the form." More concretely, data is also a basis for decision making for actors. Hence it is crucial for decision makers, here especially the Kenyan government, to have the right data and information for their decision making and action (I20).

Furthermore, the handling of data, especially of personal data has direct implications for people of Kenya (I6). One implication here derives from the use of digital technologies (worldwide) in providing personal data for using digital technologies. Here one expert representative refers to the example of MPESA: "[...] here in Kenya they request for personal data every time I do you know I move MPESA or something somebody can just get all the way my ID details and my data back and then you know withdraw money you know." (I10) This is one example emphasizing the value of data and the importance of data protection as discussed for instance in Sect. 4.1.2.

Expanding on the value of data, the interviewed expert representatives emphasize the value of data as currency (I6) and also of data markets. Data markets are not only an instance for collecting and sourcing data, but are perceived as a "quite hot sensitive topic" (I6), which thereby demonstrates certain sensitivities of multi-stakeholders (see Sect. 4.4.4). More specifically, one role of donor organizations here is seen in defining data markets for the Kenyan government (I6) and in awareness raising of the Kenyan government around data as a currency (I6). With regard to data markets from a physical presence or localization perspective, data as a currency is a further determinant in geo-strategical decisions and positioning (worldwide). As one prominent example, the BRI and its data component is highlighted by I2 (see also Sect. 2.3.3, Sect. 2.3.6).

Moreover, the associated physical positioning around data centers or data markets is further discussed by expert interviewees in the context of possible *exploitation* of data. This stems from the questions of who owns which data and how is data processed. This circles back to the role of digital or telecommunication infrastructure enabling data flows. I3 questions the role of infrastructure providers and their ownership of data flowing through their cables, possibly providing them "with information others will not have". This is a further example of power and knowledge asymmetries, which will be further discussed in Sect. 4.3.4. Similarly, this is supported by I10 highlighting possible data fraud due to legislation flaws (see also Sect. 4.1.2). This likewise puts actors under pressure, along with a necessity to have and share data, even of governments for being able to use certain technologies. However, ethical concerns are raised here with regard to data and AI:

> "And also what I would bring in here is that artificial I mean one of the key requirements for artificial the intelligence to work it's actually it's a it needs a tremendous amount of data and very often the challenge is that the countries don't have that amount of the that amount of data available and it they they shouldn't in many cases" (I18)

The sensitivity of data and the competition among multi-stakeholders around data (see Sect. 4.3.4) against the background of data as a currency, raises the perception of parallels to the exploitation in the colonial past in terms of digital colonialism or *data colonialism*, which will be further discussed in more detail in Sect. 4.3.7.

Nonetheless, digital transformation is perceived as enabling information flows and transparency (I4). Transparency, especially in digital ecosystems, are one determinant for this dissertation's framework and precondition for ecosystem governance (Sect. 2.5.1), which will be further discussed in more detail with regard to the Kenyan digital ecosystem in Sect. 4.4.3.

Software

Expanding on the role of digital technologies and data, digital software is a further essential part of the digital transformation that influences multi-stakeholders' interactions. The role of software in sustainable digital development is multifaceted as for instance discussed in Sect. 2.1.4. Similarly to the previously discussed access to electricity (Sect. 4.1.3), access to software is a further important determinant for sustainable digital development (e.g., I14). Moreover, software is closely linked to the before mentioned role of data for its functionality. In this dissertation, software is not analyzed in terms of their functionality, but as a further area where sustainable digital development is challenged and as an area for coordination.

Firstly, the interviewed expert representatives mention software as one example where new dependencies in the Kenyan digital development are experienced (e.g., I5) – 'new' here referring to dependencies that arose with the digital transformation. This dependency is further expressed towards software providers, as "[...] can be very dependent on on certain components or aspects of a certain service provider" (I5).

This monopolistic power creates dependencies for software users in terms of lock-in effects, posing a challenge to sustainable (and long-term) digital development (I5), which will be discussed in more detail in Sect. 4.3.4 and Sect. 4.4.4. Moreover, a specific challenge for countries such as Kenya is mentioned by experts with regard to dependence on dominant software actors from the Global North, especially if local knowledge is non-existent (I1). Hence limited skills and capacities to develop software locally fosters this dependency (I21). The role of digital skills, literacy and capability will be further discussed in Sect. 4.4.4.

One technical challenge that emerges from this discussion around software dependency on the Global North is further highlighted by one expert representative emphasizing the issue of pirating due to software prices (I19). Costs for software in the Kenyan digital ecosystem is a crucial challenge that will be discussed in a separate section due to its importance (Sect. 4.4.4). From this, issues for sustainable digital development emerge with regard to the role of the Kenyan government and digital policies (Sect. 4.1.2), since loopholes in regulation offer the opportunity for (local and foreign) businesses to potentially exploit the use of software or digital platforms such as Instagram. Expert I19 shares the following experience of bypassing taxation systems:

> "So like for example I've seen like for for now on our of the Instagram a lot of small businesses have started using Instagram to to sell their waires[?] these are companies they're not registered they're not paying taxes but they are using Instagram all the time they flood pages with their products and they're selling, they're selling good money there because they're bypassing the system."

This challenge for sustainable digital development is closely linked to the cost of software and also to transparency, which will be discussed in more detail below with regard to open source software as a further software area.

Open Source Software A further component of analyzing the importance of data and software for sustainable digital development in Kenya is linked to open source. Open source is discussed as a possible technical means for decreasing dependencies of multi-stakeholders on multinational tech providers (Sect. 2.2.4, Sect. 2.3.6, Sect. 2.5.1) to achieve coordination and sustainable digital development. With regard to the Kenyan digital ecosystem, I1 associates open source with decreasing dependencies. Open source is further associated with transparency and open standards such as 'https' that initially enabled the beginning of the internet (I7).

Collaboration and cooperation, here in reference to open source technology, is further highlighted by I1 as an area for learning and networking for "[...] using each other's approaches and technologies and and so it can be like a a big toolbox where where digital development players just yeah take take whatever they need from it at their specific specifics and make them work". This is not only an example of a claimed "meritocratic" (I7) innovator scene, but also underlines associated low costs for open source software (I7; I22).[6] The difference in cost between open source and corporate software is be further explained as: "[...] whereas the other organizations that are not open source are very compact in terms of the trainings, licensees versions coming up changing [unk] continuing different versions every so often, security systems and software whereas the open source can't really" (I22). Hence, the difference between corporate and open source software is seen in services provided beyond to the software delivery, which is one explanatory factor for the costs and dependence of actors on corporate software. This will be discussed in more detail in Sect. 4.3.4.

Another closely linked challenge for sustainable digital development in Kenya is the needed hardware, especially mobile phones. Before hardware and the special role of mobile phones will be discussed in more detail, the requirement to provide 'mobile first' software in Kenya is highlighted by one expert interviewee of a Kenyan startup (I24). After analyzing the role of software as an determinant that is influenced by the macro level and influences the micro level, the role of hardware will be discussed in the next subsection.

Hardware

Hardware as part of digital technologies and services is another influencing factor in the Kenyan digital ecosystem. Firstly, hardware is discussed by expert interviewees as a foundation layer (I18), a resource (I10, Sect. 4.4.4) and a precondition (I20) for participation in the digital transformation. The importance is further highlighted by I14 arguing that access to hardware is a determinant for Kenya's performance in the digital transformation.

Secondly, hardware in this dissertation is not (directly) analyzed from a technological perspective, but with regard to hardware from a meso perspective as interacting between the micro and the macro level. Hence, hardware here is analyzed with regard to respective actor interactions. Most importantly, similarly to software providers, one key actor group in the Kenyan digital ecosystem are providers of hardware, such as of smartphones or computers. These providers are not only private sector (I7), but

[6] The cost of open source software is to be further debated, but exceeds the scope of this research project

international or foreign ones with manufacturing done abroad (I7). While software was discussed among expert representatives as mostly dominated by providers from the Global North and especially Western ones, the provision of hardware in Kenya is perceived as "more mixed multi-polar" (I1) between Western and Chinese providers. Among interviewed experts, China is perceived as the dominant hardware provider in Kenya (I2). The role of Chinese hardware is especially relevant in the Kenyan digital development, as already mentioned in the context of data (Sect. 4.2.2), which will be further discussed in Sect. 4.4.4. Nonetheless, the hurdle of accessing hardware in terms of costs especially for marginalized parts of the Kenyan society is highlighted by I9.

Thirdly, hardware is another instance (in addition to connectivity, see Sect. 4.1.3) where suitability of multi-stakeholders' actions is questioned, such as in the case of laptops for schools in rural areas with limited infrastructure, electricity or internet (I10). Additionally, it is a further example for actor proliferation and hype (Sect. 2.2.3, Sect. 4.3.1) and illustrates a further example for misdirected and therefore unsustainable digital development. This is further underlined by the example of the Kenyan taxation of hardware since "[...] in Kenya it's quite expensive to get a laptop because of the additional taxes that you pay" (I9) which "[...] already that's kind of a hindrance in terms of digital transformation right?" (I9). Hence, this is another example emphasizing the importance of digital policies for sustainable digital development (Sect. 4.1.2).

Mobile Devices

One type of hardware that is specifically important in Kenya are mobile phones. As previously discussed in the theoretical background (Sect. 2.2.2), mobile phones are crucial in the digital development context since they are the prominent example of leapfrogging technological stages, such as from broadband to mobile or from laptops to mobile phones. The expansion of mobile services in Africa serves as a popular example of how the continent can leapfrog technologically and thereby also economically (e.g., Graham 2019). Especially countries such as Kenya, Ivory Coast, Nigeria, Ghana, and South Africa (so-called KINGS) are often listed among the fastest-moving digital economies. This narrative of 'Africa Tech Rising' is mainly due to the mobile revolution as Africa leapfrogged (Sect. 4.2.1) the world from a few landlines to massive mobile phone use in just a decade (Ndemo and Weiss 2017). The disruptive effect of mobile phones for leapfrogging, is further seen in leapfrogging to the "first communication layer would be mobile" (I18). Hence, the importance of the primary mode of accessing the internet via mobile hardware is emphasized (I21).

In comparison to other hardware such as computers or laptops, most Kenyans use their mobile phones for digital services since "[...] quite a few people having access to other types of devices" (I21). The high mobile penetration rate in Kenya has already been discussed in Sect. 4.1.4, nonetheless it is important to highlight the importance of this hardware type. Interviewed expert representatives note that mobile phones are "[...] very key when it comes to digital penetration in Kenya" (I24). Moreover, Africa is referred to as the "[...] mobile mobile first continent" (I19) where "mobile technologies has overtaken virtually everything." (I19) Kenya leads the continent in terms of mobile subscriptions, which contributes to access of mobile related technologies or services such as apps (I14). The special leapfrogging effect of financial services with MPESA as the most prominent example is likewise discussed as a mega-shift enabled through mobile devices (I16; I2, see also Sect. 2.2.2, Sect. 4.2.1). This innovation potential in Kenya specifically is claimed to stem from the attitude of 'making the best of what you have'. One illustrative example for this is provided by I19 stating that only a phone is needed for conducting business:

> "So it means now a twenty five year old or twenty four year old has access to a phone, once they have access to a phone and they just start selling stuff they start doing their get creative talents online and they do not have to to go through the you know the bureaucratic process of setting up a company and now this is the the the sweet spot you just pay via mobile money."

However, the ability to conduct business solely via mobile phones is questioned by a (private sector) expert representative by highlighting its limits: "You cannot sit you can't do a spreadsheet on a mobile phone you can't write a program and see on a mobile phone right?" (I21).

Although mobile phones are a tool for further innovation, this hardware type likewise poses challenges for the Kenyan digital ecosystem from a sustainable development perspective. Most importantly, here again costs and affordability are essential, which is one of the major challenges stated by interviewed experts (e.g., I19; I20). Here again the role and the dominance of Chinese hardware, especially Huawei, is highlighted. While the Chinese dominance in the hardware sector has been addressed in the previous subsection, their role in the mobile phone sector is underlined since a lot of Huawei phones are perceived to be used in Kenya (I23). The expert interviewee argues here that one reason for this lies in lower costs of Huawei phones, especially in comparison to phones offered by the US multinational Apple (I23). This is likewise perceived by I19 where affordability of a mobile phone poses challenges to conducting business. Besides the affordability of the actual hardware device, affordability further includes affordability of electricity and of connectivity which likewise impacts mobile phones usage due to charging the devices (I20, see Sect. 4.1.3). This will be further discussed in more detail in Sect. 4.4.4.

Additionally, a specific challenge identified in the data analysis combines software and hardware due to the "fast changing specifications or of devices" (I19). From this the challenge of interoperability emerges, which has not been identified as a major challenge in the reviewed ICT4D literature. Here, interoperability is mentioned for instance by Navarrete et al. (2010); Heeks (2016); Aulkemeier et al. (2019) and only briefly discussed in Sect. 2.2.4 and Sect. 2.3.5. The empirical data analysis likewise delivered limited insights into interoperability, which will be briefly discussed in the following.

Interoperability

On the one hand, expert interviewees assess interoperability as a foundational element of a sustainable digital transformation (I18). This is further supported by I18 with regard to the interoperability layers of how different systems communicate to each other and enable user access, which is "[...] kind of the truly technology part of it." Moreover, the lack of interoperability efforts and their fostering is experienced as hindering (sustainable) digital development (I18). Thereby, interoperability is one example for a technological challenge of governing the digital transformation. More specifically, this is discussed as a challenge for the Kenyan government, for instance with regard to e-citizen efforts as "[...] those processes are not clear, they're using small products that are not integrated and and there's no integration between systems and all that" (I10).

One technical challenge to interoperability is for instance a transfer of technology (Sect. 2.4.3). Technology transfer between technologies developed in the Global North and their adjustment to the local context in Kenya is assessed critically by experts such as I3. I1 on the contrary sees potential for technology transfer in Kenya through adapting "some light weight version" of technology transfer beyond outsourcing. While the (physical) presence of multinational tech companies in Kenya is seen as an advantage for technology transfer (I16), this will be further discussed in Sect. 4.4.3.

After elaborating the ICT level of analyzing the Kenyan digital ecosystem and their impact on sustainable digital development, the next section will deliver further insights into challenges for sustainable digital development.

4.3 Challenge: Barriers to Sustainable Digital Development of the Kenyan Digital Ecosystem (All levels)

In line with the theoretical background of this dissertation (Sect. 2.3), the before mentioned digital transformation and digital technologies pose specific challenges to sustainable development of developing countries such as Kenya. The macro (Sect. 4.1) and meso level (Sect. 4.2) are synthesized in this section and operationalized as a challenge (determinant 3, Fig. 2.6) while also drawing upon challenges arising from the micro level. Similarly to the theoretical section, this section refers to *all* three levels. As specific challenges, hype around digital technologies, ICT project failure, digital colonialism, digital divide, differing interests of multi-stakeholders as well as power and knowledge asymmetries are identified. These will be discussed in terms of their impact on sustainable digital development of Kenya in the following.

4.3.1 Hype around Digital Transformation

In addition to the previously discussed role of 'basic' and 'innovative' technologies and services, this section adds the actor perspective (micro level) focussing on multi-stakeholders' activities in the context of these technologies. Here specifically, factors such as actors' interests and actor proliferation with notions of 'hype' are argued to impede sustainable development. This section is further closely linked to Sect. 2.2.3 as well as Sect. 2.5.1, Sect. 4.3.3 and especially Sect. 4.4.4.

To begin with, the digital transformation in the hype context is intensely debated by expert interviewees. The majority of expert representatives, for instance I18, I13 and I11 agree on digital transformation being a hype. I18 further argues that the digital transformation is not only a hype, but perceives "[...] the risk of staying at the level of hype". If this were the case, this is would not be sustainable given that a hype is among others characterized by being short-lived (I10). Only a minority of interviewees (e.g., I24; I16) perceive digital transformation as not being a hype. More specifically, I16 asserts that digital transformation is not a hype, but a reality and even a necessity. Nonetheless, I21 argues that it can be a hype depending "[...] on who you are". This leads to the following experts' discussion with the majority of respondents assessing digital transformation as a hype, with special reference to the Kenyan digital ecosystem.

More specifically, the influence of the digital transformation on sustainable digital development in Kenya delivers further insights into the experience of 'hype' in the Kenyan digital ecosystem. Drawing on the findings discussed in the beginning of Sect. 4.2, digital transformation is often not understood by all multi-stakeholders. Moreover, the multi-stakeholders' assessment of its impact on development is likewise challenging as it is often *overestimated.* For instance, the impact of digital technologies or ICTs on the development (of Kenya) is criticized by I10 asserting an overestimation of ICTs in that "probably ICTs are expected to do all that". Here overestimation and hype are further criticized by I15 since "[...] when people talk about dig digital technologies and you know they sound as if they're the next you know they're the best thing that happened to humanity." This is further supported by I17 representing a multilateral organization admitting that multi-stakeholders such as "[...] we might be overexcited with the term digitalization or digital" (see further Sect. 2.2.1). This overexcitement is further confirmed by I15 claiming that "I think you know even as we hype the usefulness of digital technologies we we sort of have not also been very open in in in all it's aspects". Openness in this instance is not further specified by the interviewee, but from the context it is being understood in terms of transparency, which is a hurdle to sustainable digital development (Sect. 2.5.1, Sect. 4.4.3). Again this might be due to a lack of understanding or knowhow or also actors' interests (Sect. 4.4.4, Sect. 4.4.4, Sect. 4.3.3). From I12 the conclusion can be drawn that digital transformation is "exciting in the first place but it it it has so much more to come along with that you need to think think". Hence openness about every component and implication of the digital transformation is required. More concretely, I15 delivers an example for this neglected openness that leads to overestimation and hype of ICTs for development:

> "For example you know it doesn't tell you that you need say a gadget, you need a gadget that is wired in a way that it can receive wifi. It also doesn't tell you that access comes with costs and within those costs it doesn't talk about affordability so in that sense yes I would say sometimes it is hype".

Hence the actual impact of digital technologies on the development of Kenya is to be discussed critically beyond the level of hype if this is not handled sustainably and inclusively (inclusive here of all challenges and potentials).

Actor Proliferation: Shiny Topic

Synthesizing the argument of a lack in common understanding of ICTs and an overestimation of their impact on development, this is closely linked to the further challenge of actor proliferation. Actor proliferation in the discussion around digital transformation as a hype endangering sustainable digital development is for instance emphasized by I12: "[...] certainly is a hype and everyone wants to do something on in the digital sector and and people don't naturally talk to each other because everyone drives is driving their own agenda and I see that with everyone is driving a program or a support program on digital skills and entrepreneurship". Engagements of multi-stakeholders in the digital ecosystem of Kenya hence are criticized with regard to actor proliferation that fosters the level of hype. Moreover, activities of multi-stakeholders in the digital sector are driven by their interest to "show that they're involved in these nice and shiny projects programs" (I3). Hence the motivation (Sect. 2.5.1, Sect. 4.4.3) here is seen as rather related to recognition and self-portrayal. Furthermore, ICT projects are not only perceived by expert representatives as "nice and shiny" (I3), "exciting" (I12), or "attractive" (I13), but even stronger as "sexy topic" (I4). This is especially expressed with regard to donor organizations:

> "However I also have the feeling that there's almost like a little bit of competition amongst donors to be involved in the sexiest topics of digital transformation and and everyone wants to get on these certain topics which ultimately means they're there is doubling of activities maybe not so much alignment, more fragmentation" (I4).

This statement emphasizes that competition among donors (Sect. 2.3.6, Sect. 4.3.4) in the context of actor proliferation increases fragmentation and doubling of activities that need to be coordinated (as argued in Sect. 2.5.2 Sect. 4.1.5, Sect. 4.4.3).

Further consequences of this donor proliferation in digital development as an appealing are for development cooperation (I5), are seen in leading to "[...] an increasing number of of ICT project or projects that would like to integrate ICT components" (ibid.). However, this demonstrates a possible fragmentation of digital development (Sect. 2.5.1, Sect. 2.1.5) and hence emphasizes the need for coordination in the digital ecosystem of Kenya. This will be further discussed in the following (Sect. 4.3.2). Actors' interests (Sect. 4.3.3), here with a special focus on donor organizations engaging in the 'hype' around ICTs, can impede sustainable digital development. More specifically, this is further confirmed by I17 asserting that digital transformation has a lot to do with rhetoric, which can be different to the reality on the ground. Similarly, I10 highlights actors "[...] implement solutions that are not a hundred percent suitable for their processes just to show that they are digital but in the essence that technology cannot even support their processes so you'll find that it's about quick fixes and not the big picture." Multi-stakeholders' ICT projects in Kenya here are criticized, which will be further discussed in Sect. 4.3.2.

Hype around Innovative Tech such as Artificial Intelligence

The discussed overestimation of digital technologies in combination with the argued differences in understanding are further reflected in the current trending innovative tech such as AI (Sect. 2.2.3, Sect. 4.2.2). While I5 asserts "there are certain specific hypes for certain topics I would say technological topics [...]", the hyped topics identified in the data analysis of this dissertation revolve around innovative tech where the focus (and investment) of actors lies (I22). This is for instance reflected in the perception of innovative tech such as AI as being more attractive than doing basics such as connectivity (I13, Sect. 4.1.3). More concretely, I18 assesses this hype of AI in the context of development very critically:

> "[...] hype and that's artificial intelligence. And very often people tend to view this as some kind of technology that by itself solves all the problems and and gives all the answers. And the and at the same time very often grossly overestimating the capabilities of this artificial intelligence and also under underestimating that requirements for artificial intelligence to work properly. And very often I think that people this talk about the artificial intelligence they actually mean statistics and statistics that can be achieved through you know what are different reporting options or or or predictive statistics and many other component of the statistic that that can enables somebody to make better better information from vast vast amount of data."

This further emphasizes that not only is AI often not understood properly, its use-cases are likewise not always suitable, which is an example of unsustainable digital development that will be discussed in more detail in the following (Sect. 4.3.2). This underlying misled focus of multi-stakeholders on innovative type of tech (as opposed to more basic ones as discussed in Sect. 4.2) is here argued to be not holistic and hence not sustainable as "[...] it doesn't matter what can a great app you can build if people don't have the hardware and the connectivity to use it and the knowhow to use it." (I7)

This is an example where the impact on sustainable digital development and on the actual need is questioned by interviewed experts. Here, I13 argues that the focus of multi-stakeholder digital engagements is misled insofar as activities involving AI technology are more attractive (Sect. 4.3.1) as opposed to 'basics' such as connectivity (connectivity has already been identified by this dissertation as an underlying enabler for digital transformation Sect. 4.1.3). The focus on these 'shiny' technologies opposed to the crucial basics such as digital infrastructure is further highlighted by I11 sharing the perception that the focal point of multi-stakeholders' activities is often misled here.

According to I18, this is also reflected in the fact that much can be achieved in ICT or governance programs in terms of sustainable digital development without using AI. Two further reasons behind this are identified in the data analysis: Firstly, the macro context of Kenya in using AI might be different from other contexts (such as countries from the Global North) with regard to the demand for AI as discussed in Sect. 4.1. This is further underlined by I23 asserting that the demand for (digital) infrastructure in Kenya might be higher than the demand for AI. Secondly, not only is a vast amount of data needed for AI (with the asserted value of data Sect. 4.2.2), using AI requires skills which is "[...] very expensive sometimes to train you know somebody in this emerging technologies" (I19). Moreover, digital skills themselves are a further challenge to sustainable digital development in Kenya and to coordination in the digital ecosystem, which will be discussed in more detail in Sect. 4.4.4.

In sum, these are insights into challenges of innovative technologies such as AI for sustainable development. On a project level, ICT projects using AI may hence be further factors to be assessed critically when analyzing their performance (Sect. 4.3.2). This dissertation shows that the current hype around digital technologies, especially innovative technologies such as AI, is a major challenge to sustainable digital development in Kenya as it not only impedes coordination of actions due to actor proliferation and not all multi-stakeholders being on the same page, but also in terms of knowledge and impact when conducting ICT projects.

4.3.2 ICT(4)D Projects and Fragmentation

As a closely related challenge for sustainable digital development in the Kenyan digital ecosystem, this dissertation further elaborates on fragmentation and possible failures of ICT projects. Here, projects focusing on digital skills and digital entrepreneurship were highlighted distinctly in the data analysis.

Firstly, the field of ICT differs from other development cooperation areas such as education or health, since it is not as "settled" as more "traditional" sectors. I13 further emphasizes that digital development currently is characterized by "[...] a lot indeed there's a lot of fragmentation" as opposed to the sector of Technical and Vocational Education and Training (TVET) for instance where the "dust has settled". The significance of digital transformation as a development cooperation field is further underlined by the increasing number of ICT projects (I5), similarly to increasing number of actors (Sect. 2.5.1, Sect. 4.4.3). This is especially relevant with regard to the role of donor organizations in ICT4D in Kenya, since most ICT projects are initially donor funded (King'ori Ndegwa et al. 2016:3). Similarly to consequences of fragmentation for sustainable development argued in the section above, donor proliferation in digital development and ICT projects is likewise mentioned as a major barrier that contributes to fragmentation (I4), as discussed in Sect. 4.3.1.

Secondly, the digital transformation influences the way development cooperation efforts are conducted, especially with regard to fragmentation of ICT projects. More specifically, I1 ascribes the digital transformation with the potential to decrease (aid) fragmentation due to simplified information flows and actor connections. I20 further highlights the impact of digital technologies on development cooperation efforts since it has become a tool to develop many implementation areas. Hence ICT4D projects are one example where the development of Kenya is influenced through the digital transformation. Moreover, it expresses how digital technologies shape multi-stakeholders' (development) activities in that ICTs are applied to a variety of aid sectors and hence enable cross-cutting efforts. This is further confirmed from a bilateral donor perspective as one interviewed expert representative provides an insight into how their organization realizes ICT4D projects ranging from digital governance, ICTs in fragile development contexts to digital entrepreneurship (I5). Hence a variety of topics and use cases are addressed in ICT4D projects in Kenya. While there is a multitude of ICT4D projects and topics from a variety of multi-stakeholders, the analysis of this dissertation identifies a focus on ICT projects on digital skills and on digital entrepreneurship (primarily by donor organizations), which are discussed in the following.

ICT4D Projects: Digital Skills

Beginning with ICT projects focusing on digital skills (see further Sect. 4.4.4), various multi-stakeholders offer projects in this field. Two of the most known digital development projects are assumingly the 'One Laptop per Child' project or the Kenyan 'Laptops for Schools' project (Friederici et al. 2017). Furthermore, Smart Africa for instance has implemented the 'Smart Africa Digital Academy (SADA)' which offers a platform aimed at providing digital skills to policymakers (I19). Its target group for enhancing digital skills here includes the governmental level with member governments and the private sector (multi-stakeholder approach). In addition to this multilateral organization, projects in the area of digital skills are also a major focus of the interviewed private sector organizations. One multinational tech company (their philanthropies department) asserts that their "[...] goal is to really enhance digital skills so just enabling everyone to be able to be comfortable with working with digital tools and to be able to use digital tools to improve their lives." (I9). For pursuing this goal, one initiative is 'training of the trainers' which aims at fostering and enhancing digital skills of government staff more long-term as this initiative enables the government to pursue this even after the projects end (I9). Another expert representative of a multinational tech company refers to their work with local partners to enhance their capabilities and skills to remain in the region (I10) and are also engaged in upskilling, especially of young Kenyans (I10). Notwithstanding this enthusiasm, a misled focus in ICT projects focussing on digital skills is experienced by expert interviewees. I1 for instance experiences a focus in interventions on output rather than on outcome and impact. This provides insights into ICT project failure due to misled focus (Sect. 4.3.2), which is similarly discussed in the following with regard to the second example of ICT projects focusing on digital entrepreneurship. Moreover, rationales of digital skills projects can be further analyzed for instance considering the discussion around private tech interests (Sect. 2.3.3, Sect. 4.3.3) and capability of donors in their ICT projects (Sect. 4.4.4).

ICT4D Projects: Digital Entrepreneurship

In addition to ICT programs focussing on digital skills, a second area that is characterized by a lot of multi-stakeholder activity and fragmentation in ICT projects in the Kenyan digital ecosystem is identified by this dissertation's data analysis in terms of ICT projects focusing on entrepreneurship. ICT projects for fostering digital entrepreneurship in Kenya are mostly related to and conducted through acceleration programs. Acceleration programs originally stem from business and are aimed at supporting and enhancing startups.

Firstly, I12 highlights the importance but also challenge of such acceleration programs for digital entrepreneurship given their high amount and associated fragmentation of entrepreneurship programs. The high amount of offered accelerators or acceleration programs by multi-stakeholders is likewise supported by I24 and further emphasizes both fragmentation and actor proliferation (Sect. 4.3.1) that are argued to pose a challenge to sustainable digital development. A further insight into the multitude of acceleration programs in the Kenyan digital ecosystem is provided by some interviewed expert representatives and their acceleration programs. One multinational tech company highlights their project of a Vocational Training Institute (VTI) located in various parts of Kenya (I9) in addition to acceleration programs offered by a hub (I11). With regard to multilateral donor organizations, I14 for instance refers to the 'Accelerator Lab Network' within UNDP with one lab focusing on Kenya. Another mentioned accelerator is the SDG accelerator (I14). These are just a few examples for the multitude of acceleration programs offered by multi-stakeholders in the Kenyan digital ecosystem.

Secondly, ICT projects using acceleration programs for digital entrepreneurship are discussed controversially among expert interviewees. On the one hand, the emergence of acceleration programs for digital entrepreneurship is aligned with the digital transformation as "[...] this hubs and accelerators are needed I think the fact that they're coming up as they are, it shows the need" (I14). Moreover, one expert representative of a Kenyan startup highlights the importance and benefit of acceleration programs for local entrepreneurs, which are claimed to be located "in almost every every big city" (I23). This statement leads to the assumption that accelerators or acceleration programs seem to follow a more decentralized approach and hence can be seen in contrast to the generally attested centralization of the Kenyan digital ecosystem on Nairobi (Sect. 4.4.3). As a deeper study here exceeds the research goal and gathered data of this dissertation, it offers potentials for further research. Moreover the interview data of this Kenyan startup representative did not specifically reveal whether their startup had been part of any acceleration program, although I23 suggests that it did not participate in any of such acceleration programs. Potential insights into the reasons behind this is further analyzed in the following passage.

On the other hand, interviewed experts question the impact of ICT acceleration programs, especially with regard to their fragmentation. In addition to the before mentioned high number of acceleration programs in the Kenyan digital ecosystem (Sect. 4.1.5), expert representatives further see additional threats to sustainable digital development in their experience of startups being stuck in acceleration programs or handed from one acceleration program to another. This included both acceleration programs offered by the private sector but also offered by donor organizations. I5 (bilateral donor) for instance shares the experience of Kenyan startups being stuck in acceleration programs offered by bilateral donor organizations and being handed from one donor acceleration/incubation program to another and "[...] when you would like to measure the impact you can't really see any any improvement because the the startup actually sticks in this incubation phases still". This supports this dissertation's research question in so far as that donor coordination (here of acceleration programs) could counteract this fragmentation and unsustainable impact on digital startups for a sustainable digital development (see Sect. 2.1.6 and Sect. 4.1.6). This stagnation of tech startups in acceleration programs is similarity perceived by one interviewed Kenyan startup (I23) arguing that further scaling efforts are missing in these programs. This is one instance where the limits of acceleration programs in terms of impact on sustainable development of Kenyan startups are highlighted. This is further confirmed by one representative of a hub (I11): "[...] the programs are structured in many ways we might there is only so much the accelerator program can do for you right?".

For the interviewed Kenyan startups, scale and funding (Sect. 4.4.4) are mentioned as major components that are lacking in these acceleration programs (I24). Hence, processes for the phase after completion of the program are detrimental for the further development and potential success of startups that have participated in such a program. This is why there seems to be a competition (Sect. 4.3.4) between acceleration programs and venture capital as one interviewed startup shared their experience that they avoided acceleration programs in favor of acquiring venture capital while acknowledging awareness of "quite a number" (I24) of donor acceleration programs.

Thirdly, and further expanding the analysis of limitations to the acceleration programs in the Kenyan digital ecosystem, I11 offers an additional criticism towards ICT entrepreneurship programs in arguing that these are not aligned with the specific environment. This emphasizes this dissertation's argument of the important correlation between the environment e.g., legal environment, and success of startups or local digital economic growth (see Sect. 4.1). This gap (context and impact/outcome) is further confirmed by I2 criticizing acceleration programs as being unsustainable as successful Kenyan tech startups do not remain in the Kenyan digital ecosystem and/or do not transfer knowledge to other Kenyan startups.

Ultimately, this dissertation argues that a coordination of multi-stakeholders, here with regard to ICT projects and specifically acceleration programs for digital entrepreneurship, would contribute to a sustainable digital development as a countermeasure to this fragmented setting. This is further underlined by the next subsection discussing the failure of ICT4D projects with regard to the Kenyan digital ecosystem.

ICT4D Project Failures

As discussed in the beginning of this section, various factors contribute to the performance (success or failure) of digital ecosystems, with overestimation of ICT and discussed hypes (Sect. 2.2.3, Sect. 4.3.1) as well as fragmentation of projects being two of the factors contributing to failures. Further factors are related to the other challenges discussed in this section such as actors' interests (Sect. 2.3.3, Sect. 4.3.3) that are likewise mirrored in determinant 1 (Sect. 2.5.1) of this dissertation's framework. This subsection however will analyze where ICT projects of multi-stakeholders in the Kenyan digital ecosystem are assessed as failing by expert interviewees and hence lead to unsustainable developments. Moreover, this subsection is a further contribution to this dissertation's argument that (more) coordination between multi-stakeholders regarding their activities in the Kenyan digital ecosystem can possibly counteract a misleading focus and also decrease unsustainable outcomes of ICT projects (see also Sect. 2.3.1).

Failed Suitability of ICT4D Projects As previously indicated in Sect. 4.3.2, a misled focus of multi-stakeholders' ICT projects is one major determinant for failed suitability of measures and processes and thus contributes to unsustainable digital development in Kenya. Firstly, it has been indicated in Sect. 4.2.2 and Sect. 4.1.3 that there are infrastructure gaps in Kenya. Additionally, gaps between ICT projects focussing on infrastructure and the ones focussing on e.g., digital skills are identified as instances of misled focus in ICT projects (e.g., I10). This closely ties in with the highlighted argument of infrastructure as the basis for digital development, hence the focus of ICT projects should (in the first place) focus on or at least consider infrastructural settings and needs. More specifically, infrastructure is seen as detrimental for the performance of each ICT project, as I20 argues that initiatives will not work if the infrastructure is not in place. Such misled directions in ICT projects, including the ones of the private sector, are experienced for instance by I10 (private sector) criticizing ICT projects for starting at the 'wrong' end, i.e., focussing on digital skills projects while electricity, hardware (and digitally skilled) workforce are missing.

One reason for this misdirection identified in the data analysis of this dissertation is seen in actor proliferation and hype (Sect. 4.3.1, Sect. 4.3.1, Sect. 2.2.3). This is further underlined by the hype around certain innovative tech such as AI which is assessed as a more "attractive" field to conduct an ICT project than "doing the basics like connectivity" (I13) as discussed in Sect. 4.2.2 and Sect. 2.2.3. A further reason here are differing interests of the diverse set of multi-stakeholders in the Kenyan digital ecosystem (Sect. 2.3.3). While the impact on these diverging interests will be discussed in more detail in Sect. 4.3.3, one example here is provided by I19 with regard to private sector companies' engagement in IoT that stands in contrast to the 3G availability and quality in Kenya (Sect. 4.1.3). This gap between actors' interests (here business interests) and needs for a sustainable digital development in Kenya are further confirmed by I14, sharing the experience of a "difference between what solutions get to scale up what solutions get to do more marketable as opposed to the need based solutions".

Failed Sustainability of ICT4D Projects Secondly, it emerged from the data analysis that fragmented or failed ICT projects can lead to flawed sustainability, i.e., unsustainable digital development. According to Okeyo and Kioko (2016), the sustainability of the ICT projects in Kenya is considered as not very established (yet). This is further highlighted by Wamoto (2015) arguing that stakeholder engagement in a majority of government projects is not adequate enough to ensure that sustainability of these projects is guaranteed. Possible reasons for this are seen in the management of various stakeholders, their expectations and the development of a seamless working synergy for the benefit of the organization (Wamoto 2015; Ominde et al. 2020). This is in line with this dissertation's argument that coordination of stakeholder activities in the Kenyan digital ecosystem is crucial for sustainable digital development. Moreover, the addressed 'management' and expectation management will be further analyzed with regard to ecosystem governance in Sect. 4.4, as opposed to stakeholder management on the project level such as by Wamoto 2015; Ominde et al. 2020.

The empirical data analysis of this dissertation further identified failed sustainability of multi-stakeholder activities in ICT projects especially by development cooperation actors in Kenya, and also with regard to the general ICT projects level in terms of longevity and durability as part of sustainable digital development (Sect. 2.1.3). Firstly, with regard to development cooperation (projects) in Kenya, I4 for instance questions its sustainability (noteworthy here is the self-assessment of their own group). Unsustainable digital development is further experienced in the discussed increasing number of actors and of ICT projects in Kenya (Sect. 4.1.5, Sect. 4.4.3). One identified reason is seen in the discussed hype and actor proliferation (Sect. 4.3.1), illustrated using the example of a multilateral organization that even wants "to increase the number of digital development projects and initiatives" (I12) despite the discussed donor saturation (Sect. 4.1.5). Another reason mentioned in the context of failed sustainability of ICT projects is the parallel i.e., duplicative working of multi-stakeholders (Sect. 4.4.3). Here, I2 highlights one example of an open data government platform in Kenya supported by a multilateral organization, which was similarly developed in parallel by a Kenyan ministry, so that eventually one project was canceled which "left them run into into the the the nothingness if you want I mean it was a huge effort it was a lot of money spent for nothing" (I2). This is argued as one example for failed efficiency and effectiveness of activities or projects in the Kenyan digital ecosystem (see further Sect. 4.4.3). While these two examples refer to multilateral organizations, it is to be emphasized that this not only applies to this actor group but similarly to others.

More specifically, further examples for failed sustainability of multi-stakeholders' ICT projects refer to the missing follow-up after the project ends. I5 for instance shares the experience that

> "[...] what I mean with sustainability is what can happen is when the project ends and there was a certain service provider involved and the service provider was quite expensive or or the staff that you needed had to be very specialized that means that there's a high risk in the end of the project that this kind of service or this kind of software or this kind of infrastructure cannot be sustained or or managed or handled any longer".

This is further shared by I10 illustrating the example of a missing followup of hardware distribution and use. Besides missing followups, duration is identified in the data analysis as a further constraining factor that impedes sustainable development of ICT projects. Here, the short-term design of projects is questioned with regard to long-term impact as the essence of sustainable digital development (Sect. 2.1.3). This is highlighted by I20 arguing that governments need to ensure sustainability through long-term thinking for instance in that sustainable PPPs are put in place (Sect. 4.4.6). Moreover, "quick fixes" (I10) are mentioned as an ICT project approach and also political cycles (Sect. 4.4.3, Sect. 2.5.2, Sect. 4.4.4). More specifically, more "short term" (I3) engagements of governmental institutions are named as impeding the projects "from being implemented or carried out successfully" (I3). Additional sustainability issues are addressed in the following Sect. 4.3.3, which are also closely related to ecosystem efficiency and effectiveness (Sect. 2.5.1, Sect. 4.4.3).

4.3.3 Interests of Actors

Another challenge that impedes sustainable digital development identified by this dissertation are differing interests of multi-stakeholders (Sect. 2.3.3) in the Kenyan digital ecosystem. Moreover, this dissertation argues that these differing interests are not only a challenge to sustainable digital development, but also to its coordination in line with network governance (Sect. 2.4.3). Interests of stakeholders here refers to their pursued goals or self-interests in the Kenyan digital ecosystem, which is to be differentiated from willingness or motivation (Sect. 2.5.1, Sect. 4.4.3). In the conducted expert interviews, stakeholder interest, focus and objectives are often used interchangeably by expert representatives. In the following, each of the stakeholder groups' interests (Kenyan government, bilateral and multilateral donor organizations and private sector) will be analyzed with regard to the Kenyan digital ecosystem while also highlighting the role of shared objectives for sustainable digital development and especially for coordination. The interests of NGOs are implicitly analyzed in more detail in Sect. 4.3.4.

Firstly, the general interest of multi-stakeholders in the Kenyan digital ecosystem and digital sector (such as the digital economy) is underlined by expert interviewees, for instance "so there's a lot of high level interest" (I17) and lot of investment interest and dynamism (I20).

However, and secondly, despite this high level of interest of multi-stakeholders in the Kenyan digital ecosystem, the data analysis of this research project identifies different levels of interest among the actors, especially in terms of self-interests. Interactions among multi-stakeholders in the Kenyan digital ecosystem are claimed to be based on how each one benefits (I11), i.e., in pursuing own interests also in the form of "own agendas" (I3). Moreover, the different interests of multi-stakeholders are further highlighted by I13 and "ultimately everyone has their own interests in in in the game" (I2). More specifically, "everyone" here includes "[...] almost every organization, government and company has their own interests at the end of the day" (I9). This directly emphasizes a challenge for sustainable digital development in terms of collaboration and coordination, namely a lack of shared objectives and a common ground. I4 further highlights that finding the common denominator is the core of every multi-stakeholder interaction in the ecosystem. Shared objectives (or the lack thereof) is also part of this dissertation's identified network rationale (Sect. 2.5.1) and of defining an ecosystem (I13, Sect. 2.2.5). Moreover, it is essential for achieving alignment between actors, which will be discussed in more detail in Sect. 4.4.3.

Against this background, I14 further supports this dissertation's argument that different objectives of actors not only hinder collaboration among multi-stakeholders, but also impede coordination due to an ambivalence of shared objectives. While the interviewed experts highlight the importance of shared objectives (also in Sect. 4.4.4), a lack of shared interests or objectives among the multi-stakeholders are experienced by some expert interviewees in the Kenyan digital ecosystem (I10; I4). Similarly, it is described as challenging to find joint objectives considering multi-stakeholders different interests (I12). As a consequence, it is assessed as difficult to find a shared solution which gets further complicated due to differences in understanding (I7, Sect. 4.4.4). Hence a lack of shared objectives that are essential for sustainable digital development is impeding coordination of multi-stakeholder activities in the Kenyan digital ecosystem for achieving this sustainability goal (Sect. 2.1.3).

Thirdly, common interests of multi-stakeholders in the Kenyan digital ecosystem are furthermore emphasized with regard to the before mentioned success of ICT projects or programs (Sect. 2.3.1, Sect. 4.3.2). This is emphasized by I3 sharing the experience that implemented projects and programs are those where different interests can be aligned (otherwise they are not implemented). Most importantly for this dissertation, this especially applies to the interests in the digital transformation in Kenya, which is discussed in this dissertation also with regard to notions of "hype" (Sect. 4.3.1). Here hype and interests are closely linked as "everyone wants to do something on in the digital sector" where "everyone drives is driving their own agenda" (I12). 'Agenda' in this context is also associated with actors' objectives or targets such as profit or sustainability targets (I21). Hence, differing interests and self-interests of multi-stakeholders are one cause for fragmentation of action in the Kenyan digital ecosystem (I12), which endanger sustainable digital development as further discussed in Sect. 4.1.5 and Sect. 4.3.2. The fragmentation is likewise perceived with regard to different levels of interests, with I15 for instance highlighting that "[...] every time there is an issue that needs stakeholder engagement you will always find there are some more sectors that are more interested than others or have a stake or have a higher stake than others." These mentioned stakes are likewise determinants for power dynamics between multi-stakeholders, which is discussed in more detail in Sect. 4.3.4 and is also reflected in using their voices (Sect. 4.4.3). Additionally, the data analysis of this dissertation identifies interests of multi-stakeholders as not only being crucial factors for the performance of ICT projects, but also for the success of governance formats such as MSP where "each group or or each each stakeholder has a different let's say assumptions requirements interests and also ideas so such a multi-stakeholder partnerships are quite complex to manage and also to create impact [....]" (I5). This impact of MSPs on sustainable digital development in Kenya as an organizational format for coordination of multi-stakeholders actions is analyzed in more detail in Sect. 4.4.6.

Fourthly, for analyzing impacts of differing interests of multi-stakeholders in the Kenyan digital ecosystem on sustainable digital development, this dissertation argues that it is crucial to explore where the differences in interest lie between the actor groups. The individual interests of multi-stakeholder groups will be analyzed in more detail in the following.

Interests of the Kenyan Government

Beginning with the interests of the Kenyan government in the digital transformation, manifold political interests play a role in the Kenyan digital ecosystem (I5). From the data analysis of this dissertation, three common themes among expert interviewees emerged that likewise apply to public sector actors: actor proliferation, degree of interest to foster digital transformation (or not) as well as biases in interest. With regard to the Kenyan government, they serve as two sides of the coin:

On the one hand, expert representatives highlight the awareness and interest of the Kenyan government in the digital transformation in Kenya (e.g., I3, see also Sect. 2.5.1, Sect. 4.4.3). However, I19 shares the observation that "[..] unless a government has a very specific focus on digital economy they will end up doing everything and nothing at all because their efforts are scattered." Underlying notions of hype (Sect. 4.3.1) are also perceived with regard to the Kenyan government with I21 (private sector) asserting "politicians maybe glory hunting for their own own profiles", which might serve as an example of self-interests. I10 further asserts that personal interests may be one reason for impeding the digital development. More concretely, the expert provides a further example demonstrating the difference between digital and more physical or tangible projects to gather public attention and visibility (see further Sect. 4.4.3) in that a "[...] bridge is far more demonstrateable to that to people at level than showing them you made this investment in a in a digital thing they can't necessarily see it it's intangible." (I21). This is supported by I10 sharing the experience of being a multinational tech company representative that "there is a lot of resistance because nobody wants to buy the technology, people want to do things manually [...]". Here the expert questions the actual interest of the Kenyan government in promoting the digital transformation given public sector digital knowledge and a certain passiveness (I10, see also Sect. 4.4.4). This might not only be an example for self-serving interests, but also emphasizes the role of political stability and good governance in the Kenyan digital ecosystem. Corruption here is mentioned as a further factor when analyzing passiveness towards digital development (Sect. 2.1.2, Sect. 4.1.2). This is likewise shared by I21 highlighting the correlation between disinterest in digital transformation and corruption (I10). These factors therefore are identified as a specific challenge also for coordination and governance (Sect. 2.5.2) in the Kenyan digital ecosystem and will be discussed in more detail in Sect. 4.4.4.

On the other hand, in its decision making processes the government faces not only the challenge of digital complexity (Sect. 2.5.2, Sect. 4.4.4), but also the presence of differing interests of multi-stakeholders. Addressing different interests of multi-stakeholders, e.g., in policy making processes, while balancing with their own interests, poses a challenge to the Kenyan government (I6). More specifically, private sector interests have a further impact on sustainable digital development with expressed instances of policies being skewed towards private sector interests (I19, see further Sect. 4.3.4). This is especially the case if advisors to the Kenyan government stem from the private sector (I2) and thereby influence the Kenyan government. Advisors roles in sustainable digital development is further discussed in Sect. 4.4.5. Hence, the differing interests of multi-stakeholders in policy making processes pose a specific challenge to the Kenyan government in achieving sustainable digital development since this creates a setting of power asymmetry, for instance due to (inter-)dependence (Sect. 4.3.4, Sect. 2.3.6). These insights further demonstrate one dimension of differing multi-stakeholders interests that influence government action and also hinder coordination of action. This is underlined by I20 highlighting that coordination is in the interest of (aid) beneficiaries such as the Kenyan government thereby emphasizing the role of coordination for sustainable digital development.

Interest of Bilateral and Multilateral Donor Organizations

This directly leads over to the data analysis on interests of bilateral and multilateral donor organizations that (similarly to other stakeholders) have the interest of advising the Kenyan government in topics around digital transformation in Kenya. Donor organizations in the role of advisors in the Kenyan digital ecosystem will be discussed in more detail in Sect. 4.4.5. More importantly for this section, interests of multilateral donor organizations in the Kenyan digital ecosystem are for instance expressed with regard to the single digital market between the EU and AU as an example for a shared interest between two regional organizations (I2). The data analysis did not identify further explicit interests of multilateral donor organizations in the Kenyan digital ecosystem (but will be indirectly discussed in the following).

With regard to interests of bilateral donor organizations in the Kenyan digital ecosystem, two dimensions of interests are highlighted by expert interviewees (see also Sect. 2.1.4, Sect. 2.3.3): political and economic interests (I17). Factoring in the digital transformation and sustainable digital development, I17 argues that the interests of donors in the Kenyan digital ecosystem differ. This is for instance due to donors' interests being reflected in their regulations and protocols, which is a barrier to sustainable and inclusive digital development as discussed in Sect. 4.4.4. Similarly, one challenge to sustainable digital development here are settings of aid fragmentation emerging from these different donor interests (I2, likewise discussed in Sect. 4.1.5). Another dimension of donor interest is associated with the interest of actor proliferation (Sect. 4.3.1) with an increased interest from multi-stakeholders, including donors, in Kenya (I3). Another result of donor interest emerged with regard to collaboration between donors and the private sector where opposite interests collide (I12; see also Sect. 2.3.6, Sect. 4.3.4). The role of profit here is one crucial interest that divides public and private sector actors, which leads over to an analysis of private sector interests in the Kenyan digital development.

Interest of the Private Sector

With regard to interests of private sector actors in the Kenyan digital ecosystem, the data analysis identified two main interests: profit and CSR. To begin with business interests, one expert interviewee identified profit interest, especially of foreign tech companies, to enter the Kenyan market (I5). Following the market logic, the interests of private sector organizations in the Kenyan digital ecosystem further include "to brand themselves" (I19) and "to capture market" (I19). Additionally, the interest is seen in positioning themselves, even to the extent of becoming a monopolistic power in the ecosystem (I2, see also Sect. 4.3.4). This interest is mentioned with regard to private tech companies, especially multinationals (I9). Moreover, business development is a further interest that is likewise perceived as a prerequisite for companies to collaborate as "[...] they will only collaborate if they see a business case to collaborate" (I4). One such 'business case' for tech companies is for instance argued to be "[...] using Africa or African countries as a test ground for solutions that they are aiming to implement in the Global North [...]" (I3). Crucial here are profit interests, which are one factor mentioned as incentive to engage in and support digital transformation in Kenya as private sector actors do not engage "[...] entirely for free" (I15). Other interests mentioned are "talent acquisition, affordable talent, outsourcing, corporate social responsibility" (I1). The latter will be discussed in the following.

The interest of private sector actors to collaborate with other multi-stakeholders driven by business interest is likewise highlighted. For instance, engaging in and being part of networks such as in the form of associations (Sect. 4.4.6) is a private sector interest in using synergies as it opens business opportunities with other members of the network (I21). Hence, collaborations are pursued in case tech companies see a business interest, which can be ambivalent, especially in regard to collaboration with Kenyan startups (I4). Hence, internal or self-serving interests (I9) are important factors to consider. With specific regard to sustainability or durability, the private sector is claimed to seek long-term relations with other multi-stakeholders, for instance in the context of long-term investments (I12). For this, certain preconditions need to be fulfilled such as trust (ibid.), which in this dissertation is derived as one precondition for sustainable coordination (Sect. 2.5.1, Sect. 4.4.3). In advising the government the private sector presents their own interests (I2, Sect. 4.4.5). Further consequences of these interests are discussed in more detail in Sect. 4.4.4 and Sect. 4.3.4.

As a second area of private sector interests identified by the data analysis of this research project is CSR. In addition to the business interests, CSR interests (e.g., I1) are argued to be an alternative in cases where corporates do not see a business case (e.g., I4). These are cases where the interest lies in "[...] giving back to the community so to say we are not only just advising organizations but we're also investing in the population [...]" (I10), i.e., through digital skills projects by the interviewed multinational company. This likewise potentially serves an internal purpose of talent acquisition (I1). CSR activities are mostly conducted via philanthropy departments of corporates or their "philanthropies side" (I9). CSR activities of foreign tech companies are however also criticized as "it's sugar coating" (I21), which will be discussed in more detail in Sect. 4.3.4 and Sect. 4.4.4.

4.3.4 Multi-Stakeholder Power Asymmetries

As an additional major challenge for coordination to achieve sustainable digital development, this dissertation emphasizes power between multi-stakeholders in digital ecosystems (Sect. 2.3.6, Section 2.5.2). In line with the argumentation presented in these theoretical sections, the relevance and importance of power relations for sustainable digital development in the Kenyan digital ecosystem is further highlighted in the data analysis of the conducted expert interviews. The importance of considering power relations between multi-stakeholders in the Kenyan digital ecosystem is further highlighted by I13 underlining the relevance of this dissertation: "[...] I think that's also something that is not taken into account enough, what are the power relations of of the actors between themselves when we have multi stakeholder mechanisms." Hence this section will analyze power relations between the multi-stakeholder actor groups in the Kenyan digital ecosystem. Power here mainly refers to the power position of actors obtained through resources such as financial resources and especially digital related know-how and knowledge (Sect. 2.3.6, Section 2.3.6). These are also closely related to voice as an explaining factor for dominance as to which actor groups is listened to and whose not (Sect. 4.4.3). This dissertation argues that these challenges for sustainable digital development can be addressed through coordination and governance, which is why the next sections will build on that and discuss governance roles and formats for overcoming these challenges in line with the proposed theoretical framework (Sect. 4.4, Sect. 2.5).

Before analyzing which actor group is assessed by expert interviewees as dominant or less dominant, it is to be established where dominance in the Kenyan digital ecosystem is observed. Here I14 argues that "[w]hoever controls resources then becomes dominant." As a consequence, this dominant actor then executes power and dominance over (inferior) actors, so that they "[...] sometimes might not align truly." (I11) Here the correlation between alignment, actors' interests and resources become further obvious (Sect. 4.3.3, Sect. 4.4.4). Resources are an important factor determining dominance, which can be observed to the extent of concentration of resources and thereby of power as "there is definitely a gap and the gap can be attributed to resources" (I14). Resources ranging from technological, personal to financial resources will be discussed in more detail in Sect. 4.4.4. Moreover, the interviewed expert representatives in the Kenyan digital ecosystem describe the Kenyan digital ecosystem as characterized by competitive thinking between multi-stakeholders (I4) operating in a "[...] context of an extremely dynamic extremely competitive environment" (I7). Competition endangers collaboration, which hinders a sustainable digital development as supported by I15: "[...] we need an ecosystem where you know different people work with each other based on collaboration and not competition". The data analysis identified two angles of competition in the Kenyan digital ecosystem: one, competition on the country level for achieving digital development and two, competition between actors. Expanding on the first perspective, the argument of Kenya becoming competitive in terms of digitalization with their growing digital economy in comparison to other countries (I6) and the competition of countries for development cooperation or aid (I18) are identified in the data analysis. The existence of power asymmetries between multi-stakeholders (perspective two) in the Kenyan digital ecosystem however is discussed controversially among expert interviewees:

Firstly, power asymmetries in the Kenyan digital ecosystem are identified by expert interviews I5, I22 and I23. More concretely, I10 asserts that multi-stakeholders in the Kenyan digital ecosystem are not at eye-level. According to I12, these power asymmetries are especially intense the higher 'the price to win' is. With special regard to the actor groups, I9 perceives imbalances specifically between the Kenyan government and other (mostly private sector) stakeholders in the digital transformation with the government being "[...] a little bit behind in terms of what's happening in tech and so they may not be completely uptodate in terms of what's going on and so they might push more out more somewhat outdated forms of of the digitalization". Contrary to these perceptions, I11 shares the experience of interactions being at eye-level. Hence, the available interview data provided mixed results on power asymmetries on a general level.

Secondly, power imbalances are further observed on both the micro level (interactions in the Kenyan digital ecosystem) but also on a global or regional level. A regional dominance in digital transformation is alluded to that is "split between American and and China" (I17) as both are stated as origins of digital technologies (I3). More specifically this refers to the (private sector) actor groups, since "the US is pretty much dominated by activities with regard to Google, Facebook so really tech giants" (I5). A US dominance in digital technologies is likewise highlighted by I6. As elaborated in Sect. 4.2.2, a dominance by China in the hardware sector in Kenya is highlighted by expert interviewees. Nonetheless, a difference between US and Chinese companies' dominance in the Kenyan digital ecosystem is seen in companies such as Huawei dominating infrastructure sector (e.g., Huawei providing the 5G network in Kenya), whereas the US is perceived as much stronger in the digital entrepreneur ecosystem (I17). Following the US and China are some European companies (I5). On the contrary however are experts such as I14 denying a perceived dominance by the West. I1 supports this by perceiving the dualism between the two regions as "more mixed multi-polar". A possible regional domination reminds of notions of (digital) colonialism, which will be discussed as a separate challenge in Sect. 4.3.7. The question as to where and between whom these power imbalances are experienced, will be analyzed in more detail in the following subsections.

Power Asymmetry: (International) Private Sector vs. Startups

What the data analysis demonstrates is that dominance in the Kenyan digital ecosystem is mostly seen with regard to the private tech sector, especially the international one. Generally, the power of building the digital economy is attributed to the private sector (e.g., I12), specifically the mobile money sector is seen as a dominant sector (e.g., I22; I14). Moreover, competition in the Kenyan digital ecosystem is likewise perceived between the private sector actors themselves. With regard to multinational tech companies, "[...] they have very fierce competition here" (I22). This is supported by I9 (multinational tech company) stating that they do not collaborate with competitors: "[...] I don't think we tend to work with competitors as much."

Besides the competition between multinational tech companies, further competition in the ecosystem is experienced between local (Kenyan) tech companies and the international ones (e.g., I2). More specifically, dominance of the foreign private sector is observed regarding big tech companies dominating over Kenyan startups in the ecosystem. This is seen as one explaining factor for a limited *voice* of Kenyan startups (I19, see further Sect. 4.4.3). This power imbalance between the bigger foreign tech companies and local Kenyan tech companies, especially startups, is further highlighted by I14 with regard to foreign tech companies having "more money to scale, a momentum [?] to deploy you know to do outreach". Foreign tech companies getting more momentum than startups is further perceived by I10. As one consequence, this imbalance leads to the expert perception that Kenyan startups likewise feel inferior, which is a disadvantage to sustainable digital development (I21) and supports the hypothesis of Kenyan startups and multinationals not being at eye-level (I10). Kenyan startups face an additional layer of competition, since they not only compete with bigger tech companies, but also with other startups. This for instance applies to the context of ICT funding programs or acceleration programs as argued for instance by I3 and as previously discussed in Sect. 4.3.2.

Power Asymmetry: (International) Private Sector vs. Kenyan Government

Expanding on private sector power asymmetries, and most importantly for this dissertation, a power imbalance between the private sector and the Kenyan government is emphasized by expert interviewees. Specifically, private sector dominance is experienced in terms of close relations with the Kenyan government, even to the extent of favoritism:

> "[...] the first thing the fact that some stakeholders have muscles and so they throw their weight around especially the businesses and they tend to work more with governments. So you find that their views are given more I think that's a univers[al] disadvantage you find that their views are given more weight than probably other stakeholders." (I15).

While I22 does not see the Kenyan government at a disadvantage in the digital transformation and interaction with the private sector, I19 however assesses that the Kenyan government gets left behind as the private sector "muscle in the ICT sector in Kenya is quite strong and many times they they go ahead of the government so you find the government gets left behind." The expert interview data highlights dependencies in the digital transformation as one explaining factor for these (unsustainable) power asymmetries, but also knowledge and information asymmetries (as discussed in Sect. 4.3.5). More specifically, I22 emphasizes the existence of a knowledge vacuum between the private sector and the Kenyan government. Similarly, I3 highlights information asymmetries due to digital infrastructure possession as one additional layer of power asymmetry in the relationship between the private sector and the Kenyan government. Additionally, power asymmetries are highlighted by expert interviewees with regard to resource asymmetries where the private sector again dominates, especially in terms of financial resources in the digital transformation (I19). This will be discussed in more detail in Sect. 4.4.4.

In sum, the data analysis of this dissertation identifies four dimensions of power imbalances between the two actor groups of the Kenyan government and private tech sector: a) the private sector overtaking governmental functions such as policy making, b) influencing the Kenyan government from within in their role as advisors, c) competition between the two actor groups, for instance in the labor market e.g., around digital skilled workforce, and d) competition also on technical level (possible open source usage of Kenyan government as opposed to commercial digital services). Those will be discussed as follows.

Private Sector overtaking functions of the Kenyan Government Firstly, the data analysis highlights a dominance of the private tech sector over the Kenyan government, i.e. "[...] over the government's functions" (I2). One crucial element of the government's 'functions' is policy making. Here, the expert interviewees discuss the role of the private sector as having such a strong speaking power that they influence policies (e.g., I19). The private sector is experienced by expert interviewees as deeply involved in policy making and influential in terms of "lobbying for policies and supporting government to figure out the digital transformation road plan and what they want to do." (I9) This dominating influence is even assessed to the extent of "[...] especially external digi tech companies dictate how which services are being delivered" (I4). Dominance as here expressed in form of forcing, even implies that the Kenyan government in some instances is perceived by the expert interviewees as "not necessarily free to decide themselves" (I2). One indirect effect of this discussed 'favourtisism' is that the power of foreign tech companies in Kenyan policy making likewise effects local tech companies, for instance with regard to tax policies as highlighted in Sect. 4.1.2. More specifically, foreign tech companies not paying taxes in the Kenyan digital ecosystem are highlighted by expert interviewees (e.g., I2). This is assessed as problematic for instance in comparison to local tech companies and also as a source for stifling innovations due increased telecommunication taxes (Sect. 4.1.2).

Moreover, the perceived impact of influencing decisions is likewise perceived as impacting the local Kenyan tech companies. For instance, I2 (international NGO) shares the experience that policies are not necessarily structured to support the growth of local industries but rather foreign ones, which counteracts sustainable digital development due to power imbalances and dependencies.

Synthesizing these perspectives of private sector actors influencing the Kenyan government, this power asymmetry between the legitimately elected government and the (not legitimately elected) private sector poses a danger to sustainable digital development, especially with regard to equality and inclusion. Against the background of these interview results, a possible risk of state capture (i.e., fully overtaking governmental functions and responsibilities) is assessed as "[...] state capture is not probably not so far away or is not as far away as we might think" (I3). This is further discussed in the following paragraph.

Private Sector influencing the Kenyan Government from within Secondly, this dominance of the private sector's relation with the Kenyan government in the digital transformation is further manifested in their role as advisors to the government. While the previous passage has highlighted power asymmetries between the private sector and the Kenyan government from the perspective of two individual actor groups, a further finding of the data analysis is that the private sector additionally influences the Kenyan government from *within*. Hence, the closeness (Sect. 4.4.3) of the private sector and the Kenyan government is further explained by the private sector advising the government also from within the government. This, among other reasons, roots from the fact that governments are being advised in the digital transformation, for instance by donor organizations or the private sector (also on other topics), which is discussed in more detail in Sect. 4.4.5. Due to limited digital capabilities of the Kenyan government in comparison with the private tech sector (as discussed in Sect. 4.4.4), governments such as the Kenyan government "[...] believe it would be better to bring in the people who have that apparent knowledge and then they almost bring in their on enemy" (I2).

Influencing the Kenyan government from within the government structures is further emphasized with regard to the private sector actors being "[...] very strong very strong to the extent that they sit on the board and when it comes to making the digital policies for Kenya they take they practically come in and [chuckles] take it over and to drive it." (I19) Notions of 'taking over' are further experienced in influencing of policies driven by the companies' business interests (I2). To underline this argument, I2 provides a specific example:

> "[...] that's exactly what happened they have people [...] from China and from America sitting in in high positions at the government advising the government on how to run their their their like digital transformation, their digital economy and so on and inevitably it resulted in contracts for them, so I think they have a surprisingly strong strong position within these governments and they have they've structured a lot of the the policies that we have in place today."

This demonstrates a dominance of foreign tech companies advising the Kenyan government resulting in policies influenced to suit business interests. This relation between the international private sector and the Kenyan government leaves room for further research.

Synthesizing these two perspectives, it can be stated that power of the private sector over the Kenyan government in the digital ecosystem endangers the government's capacity and the state's legitimacy given that the private sector actors are not legitimately elected (as discussed in the theoretical background Sect. 2.1.2 in reference to Grävingholt et al. 2015). These are examples of power imbalances undermining the Kenyan government in the digital transformation. Hence, this dissertation argues that power imbalances are one important factor impeding sustainable digital development. Nonetheless, while they in sum may lead to the conclusion that the government is inferior to the tech companies, some expert interviewees highlighted the crucial and unique role of the Kenyan government in the digital ecosystem and in this power relation setting. Although statements show that the private sector is influencing policy, the government is also still perceived as dominating in that "[...] the government is the owner the owner of even the private sector the way it does things." (I22). This alludes to the government's unique role in exerting control and dominance though regulation and legislation, for instance regarding private sector's actions (I4). However, limited capabilities, dependencies and the specific challenge of simultaneously balancing regulation and leaving room for innovation can put governmental actors in a position of less control (I7).

Competition between Private Sector and the Kenyan Government Thirdly, competition in the Kenyan digital ecosystem is observed between private and public actors, for example in the labor market around digitally skilled workforce. This is a further example of *brain drain*, which is a challenge in Kenya that will be discussed in further detail in Sect. 4.3.5. The presence of multinational tech companies in the Kenyan digital ecosystem has an effect on the public sector, on local companies and also on innovation and startups in terms of access to digitally skilled workforce. On the one hand, the competition among private tech companies regarding skilled workforce, is illustrated for instance by I10 (private sector). This has an impact on local tech companies since international tech companies such as Microsoft, Google, Facebook and Huawei (I21) "[...] hiring aggressively in Nairobi, there are a lot of engineers being pulled from local companies" (I7). This not only leads to a shift in employment (brain drain), but also has an effect on the (expected) salary levels. In a country with a relatively high (youth) unemployment rate (Sect. 4.1.4), this has an additional effect on innovative startups that similarly face workforce competition (I10).

In addition to the competition for digitally skilled workforce among private sector actors, the data analysis further identified competition between public and private actor groups as one barrier to sustainable digital development in the Kenyan digital ecosystem. While this can likewise be observed in other sectors as well, this competition is for instance experienced by I12 in ICT capacity building programs sending public sector staff abroad: "[...] what happens is that the public sector is losing out on staff because they receive a higher capacity and then they are more interesting for the private sector." This is especially controversial since training programs abroad are also initiated for instance by donor agencies, e.g., in some ICT projects (Sect. 4.3.2). In sum, while competition around skilled workforce among private sector actors and with public sector actors has benefits of job creation and enhancing digital skills, it likewise has the opposite effect of brain drain in terms of talent going abroad or working for a foreign company rather than for local entities. If the Kenyan public sector is also not competitive with the private sector salaries, this further strengthens the digital skills gap. Moreover, differing interests between multi-stakeholders foster competition and hence impede coordination. This limits digital capacities of the Kenyan government and their ability to foster sustainable digital development.

Competition between Multinational Software vs. Open Source Software
Fourthly, the dominance of multinational tech companies is further identified in the data analysis in their provision of digital services such as software in contrast to open source software. As discussed in the theoretical background (Sect. 2.2.4), open source software is discussed as one technological option to decrease dependencies and lock-ins (see e.g., Sect. 2.3.6). With regard to the Kenyan digital ecosystem, the interviewed experts perceive a power struggle between multinational software providers and open source in the Kenyan digital ecosystem (e.g., I22) as well as a dependence on multinational software (I23). This poses a special challenge for the Kenyan government as experts argue that the government has to "[...] to choose between Microsoft and open source software" (I23). On the one hand, the power asymmetry stems from a cost perspective and "[...] the perception that maybe open source software is a bit cheaper [...]" (I22). Notwithstanding this enthusiasm, if and

how open source is cheaper, it is to be further debated (e.g., Sect. 2.2.4). Due to costs of software licenses (Sect. 4.4.4), a lack of capabilities (Sect. 4.4.4) and the development context of Kenya (Sect. 4.1), this dissertation emphasizes the specific challenges for developing countries such as Kenya in contrast to countries from the Global North with regard to sustainable digital development. As one example, advisors to the government such as multilateral organizations are highlighted as promoting open source software to the Kenyan government (I19).

On the other hand however, the power asymmetry is likewise indirectly fostered by the government rejecting proposals towards implementing open source software as "[...] governments don't like open source software [...]", which gives an advantage to multinational software "[...] to get more business because they favour large reputable company they give good support" (I23). This dissertation provides insights into possible reasons behind this, such as lock-ins (Sect. 4.4.4), limited (internal) capabilities and also the dominating power of private sector companies over the Kenyan government, for instance in policy making (which might be applicable to not only the Kenyan government). The latter explanation is supported by I19 naming the power and "muscle" of the private sector as one potential reason why the Kenyan government rejected this cited open source proposal. However this needs to be further explored in future research as the gathered data here only provides this limited insight.

Power Asymmetry: Donor Organizations vs. Kenyan Government

Switching back to the power asymmetries between actor groups in the Kenyan digital ecosystem, further power asymmetries are identified among donor organizations themselves and also between donors and the Kenyan government (here mostly bilateral). Beginning with an analysis of the power asymmetries among donor organizations (bilateral and multilateral), they are driven by the actor's interests, actor proliferation, fragmentation and competition as previously discussed (e.g., Sect. 4.3.1, Sect. 4.3.3).

Firstly, competition among donor organizations in the Kenyan digital ecosystem are discussed by the expert interviewees. For instance, I4 shares experiences of donors in competition "to be involved in the sexiest topics of digital transformation" (as further argued in Sect. 4.3.1). More concretely, donor organizations themselves compete on a project level as they "[...] see themselves as players in the ecosystem who offer their own programs" (I4). This competition between donors for (ICT) projects follows a tender process with an "winner takes all approach" that is "[...] a barrier to having multi-stakeholder project" (I21).

Secondly, the power asymmetries between donors and the Kenyan government is already inherent in the country's development status and in development cooperation activities since aid flows from donors to beneficiaries who "[...] are at the receiving end" (I2, as discussed in more detail in Sect. 4.1.5 and Sect. 2.1). This creates a setting of dependencies where beneficiaries such as Kenya are dependent on (technical and financial) aid of international donor organizations (Sect. 4.4.4). One expert representatives further alludes to settings of donor organizations overtaking government responsibilities, similarly to the previous discussion around the private sector overtaking government functions, given the assumption "[...] that donors will do it for them if they don't do it." (I4) Since this perception is only identified by one expert interviewee, this provides potential for further research.

Power Asymmetry: Kenyan Government vs. NGOs

A final power asymmetry derived from the analysis are power asymmetries between the Kenyan government and NGOs. Voice here serves as an indicator for being recognized and included, so that the lack of recognition is one factor determining less power (see Sect. 2.5.1). With regard to NGOs in the Kenyan digital ecosystem, the expert interviewees experience less recognition of the Kenyan government towards NGOs with NGOs given less priority in policy matters (I15). This experience is further illustrated using the example of the 2016 ICT policy review (I15), which will be discussed in more detail Sect. 4.4.3. Hence, NGOs here are less included and perceived as having less power in the context of ICT policy processes in Kenya. However, it is important to highlight that these experiences are shared by an expert representative of a Kenyan NGO. Therefore, these asymmetries in the relation between the Kenyan government and NGOs will be discussed in more detail in Sect. 4.4.3.

Power Asymmetry: Conclusion and Solutions

To conclude this section, power asymmetries between various multi-stakeholder groups in the Kenyan digital ecosystem are experienced by interviewed experts. Not all of them interact at eye-level, which has been established as a crucial factor for sustainable and especially for inclusive digital development (Sect. 2.1.3). What this section further highlights is the follow-up question as to whether the dominant actors are also the most active ones. This will be discussed in more detail in Sect. 4.4.3.

Moreover it begs the question how to overcome these power imbalances in the ecosystem. Possible 'solutions' to the discussed power imbalances in the Kenyan digital ecosystem can be summarized as follows: Firstly, the data analysis suggests that the Kenyan government, as the only legitimately elected actor group, can decrease dependencies on the private sector by paying more attention to and working closer with NGOs to achieve sustainable digital development (I19). Moreover, dependency can be decreased through standards and more self-developed solutions, especially in digital infrastructure, which will be discussed in more detail in Sect. 4.4.6.

Secondly, partnership formats (at this point not further specified) to reduce power asymmetries are highlighted as possible 'solutions' (e.g., I4), which are derived as one governance determinant in the here proposed framework (Sect. 4.4.6, Sect. 2.5.4).

Thirdly, digital tools can counteract such power asymmetries as highlighted for instance in the form of open source (Sect. 2.2.4) or also digital platforms (Sect. 2.5.5), especially when putting power relations in the context of sustainable digital development achieved through governance and coordination. These solutions will be discussed in more detail in Sect. 4.4.

4.3.5 Multi-Stakeholder Knowledge Asymmetries

Closely linked to the discussed power asymmetries between various multi-stakeholder groups in the Kenyan digital ecosystem are knowledge asymmetries. As emphasized in the theoretical background (Sect. 2.3.6), knowledge is a crucial determinant in the digital transformation and the development of the Kenyan digital ecosystem. Knowledge here refers to the role of (special) knowledge in the digital transformation, knowledge degrees of multi-stakeholders, foreign and local knowledge as well as learning and experiences. Knowledge is especially essential against the background of complexity brought on by the digital transformation (Sect. 4.4.4).

Beginning with an analysis of the role of knowledge in the digital transformation, interviewed expert representatives highlight that digital technologies enable accessing information and knowledge that might not have been available before (I23). Moreover, access to (digital) knowledge is assessed as crucial for sustainable digital development (I14). More importantly, this is reflected in a expressed need for knowledgeable actors in the Kenyan digital ecosystem, especially in areas such as cybersecurity (e.g., I2) or digital technologies (e.g., I18) such as artificial intelligence (e.g., I18). Knowledgeable actors are claimed to be needed, especially in the process of starting digital businesses, "[...] across the board so that's where they need to collaborate with others and and where they need to bring in the necessary knowledge" (I2) and for instance in the "[...] formulation of policies of of political contexts and and and national national issue" (I3). The latter however depends on the degree of closeness in terms of inclusion, which is contested and one factor impeding sustainable digital development (Sect. 4.3.4, Sect. 4.4.5, Sect. 4.4.3).

Distribution of Knowledge among Multi-stakeholders

While knowledge in this dissertation is analyzed as a determinant in the Kenyan digital ecosystem that functions between the macro and the micro level due to knowledge and technology flows in both directions (similar to ICTs as the meso level Sect. 4.2, Sect. 2.2), the actor perspective is likewise crucial with regard to the role of knowledge for sustainable digital development. Hence, a further distinction between knowledgeable and less knowledgeable actors in the Kenyan digital ecosystem emerged from the data analysis. Beginning with the question which actors are assessed as having the required knowledge, according to interviewed experts, the private sector actors are argued to be the most knowledgeable ones in the Kenyan digital ecosystem – with one exception: academia. Academia is mentioned by one expert representative (not by the expert representative of academia) as the actor group that has the knowledge in specific fields concerning digital transformation and with regard to interactions of multi-stakeholders (I5). Nonetheless, expert representatives see the private tech sector actors as most knowledgeable with regard to building systems and providing knowledge around the implementation of technologies (I3). According to I22, this can for instance be attributed to their agility. The fact that knowledge around tech is inherent in the private tech sector may seem obvious, for instance since this group is labeled as the actor group with the "apparent knowledge " (I2). This, however, is not without implications for sustainable digital development, especially with regard to the role of the private sector in the Kenyan digital ecosystem as discussed in Sect. 4.4.5 and Sect. 4.3.4.

On the contrary, the data analysis identified public sector actors, especially the Kenyan government, as the actor group with less knowledge around digital transformation and digital development (than other multi-stakeholders such as private sector actors). The level or role of knowledge around the digital transformation of the Kenyan government is described by expert interviewees as "there's no capacity in government knowledge it doesn't really have to have it [...]" (I22) or "it [the Kenyan government] doesn't necessarily have the knowledge [...]" (I22). Expanding the discussion around less knowledgeable public sector actors, spanning from the Kenyan government to other foreign public sector agencies that are engaged in the Kenyan digital ecosystem, one representative of a multilateral organization likewise admitted that the digital transformation "it is not something a public public sector you know has a lot of expertise in" (I17) and also is "[...] not particularly the (..) the most competent let's say on on the innovation space" (I17). This is important when discussing multilateral organizations' roles in the Kenyan digital ecosystem with regard to digital development cooperation efforts and will be discussed in more detail in Sect. 4.4.4, Sect. 4.4.5 and Sect. 4.3.4.

The distinction between knowledgeable and less knowledgeable actors, illustrates an imbalance in knowledge, similarly to an imbalance in digital capacities (Sect. 4.4.4) and power asymmetries (Sect. 4.3.4).

Transfer of Knowledge

Besides a distinction of who possesses knowledge in the Kenyan digital ecosystem, knowledge flows between multi-stakeholders with different levels of knowledge are further highlighted as crucial for coordinating multi-stakeholder activities. According to I1, there is a lot of knowledge transfer between the international private sector and the startup scene in the Kenyan digital ecosystem that potentially spreads into public sector institutions. More importantly for this dissertation, knowledge exchange is argued to be needed for a sustainable digital development (I6). Sustainability in the context of knowledge transfer is further highlighted as the essence of sustainable development cooperation in a transfer of knowledge that prevails even after a partner has left the project (I20, see Sect. 4.3.2). Here, information flows and knowledge transfer not only from donor organizations to beneficiaries such as Kenya but vice versa are important (I12). Nonetheless, there are also barriers to knowledge transfer in the Kenyan digital ecosystem such as "[...] transmission of knowledge, training, affordability of or the the infrastructure itself so if internet infrastructure is not affordable enough [background noise] is readily spread all over the place" (I1). This specifically includes digital skills and digital capabilities, which are discussed in more detail in Sect. 4.4.4.

The importance of knowledge transfer is further relevant for this dissertation since organizations have been formed with the responsibility to coordinate or govern knowledge transfer in the digital transformation. For instance, this can be explained by the increasing emergence of knowledge hubs (I9; I17) or also knowledge accelerators or knowledge labs (I14). Knowledge transfer between multi-stakeholders in ICT projects, especially local partners, is further highlighted as one key determinant for sustainability and durability of the projects (I18). The role of local and foreign knowledge in the Kenyan digital ecosystem for sustainable digital development will be further discussed in the following.

Foreign Knowledge

Further expanding the discussion around knowledge flows between multi-stakeholders, it is important for this dissertation's research goals to analyze knowledge flows from outside of the ecosystem (foreign) and inside it (internally). For this purpose, this dissertation further distinguishes between the role of local and foreign knowledge in the Kenyan digital ecosystem for sustainable digital development. This likewise functions as another area that needs to be addressed in discussions around implications of the digital transformation and with regard to coordination (of action). The close link between transfer of knowledge and transfer of technologies in the Kenyan digital ecosystem is further underlined through one expert arguing that a lot of knowledge (especially in form of learning) comes into Kenya from outside of the country through digital technologies (I23). At the same time, the role of local knowledge is not to be underestimated, which will be discussed in the following subsection and in reference to brain-drain (Sect. 4.3.4). Nonetheless, both local and foreign knowledge are crucial in the Kenyan digital ecosystem, which is likewise characterized as a "hybrid of both" (I23).

With regard to implications of the digital transformation on Kenya's development, expert representatives firstly highlight the importance of foreign knowledge for digital development (without negating the need for internal knowledge). This discussion here is similar to the development paradigms in development theory with regard to endogenous or exogenous development theory discussions in Sect. 2.1.1. However, the focus here lies on digital transformation and more concretely on the origin of knowledge. I12 and I2 highlight the importance of foreign knowledge, with the latter arguing that

> "[...] in the digital space is so fast moving that it would be it would be very wrong decision to try to to do everything yourself within your community or within your own country I think all of this will only work better with collaboration and and coordination and partnership of some sort."

Moreover, digital technologies and services come with great complexities as previously discussed in Sect. 4.2, so that foreign knowledge here is additionally required (but is not without challenges as discussed in Sect. 4.3.4 and Sect. 4.4.4).

The data analysis of this dissertation identifies a specific form and activity of foreign knowledge in the Kenyan digital ecosystem: the role of foreign founders. In this dissertation, foreign founders are not only understood as an external source of knowledge, but also as an internal one given that they are likewise a part of the actor group of Kenyan startups and hence interact in the Kenyan digital ecosystem. With regard to foreign founders, expert representatives share the experience that it is common for Kenyan (tech) startups to have a mix of Kenyan and foreign founders (I9; I24). One expert interviewee representing a Kenyan startup further assesses that not only the flow of foreign knowledge, but also of experience and investment through networks offered by foreign founders (especially from the Silicon Valley) is influencing the Kenyan digital ecosystem. Hence, foreign founders have an influence on the Kenyan digital ecosystem, for instance with regard to the narrative of Kenya as the 'Silicon Savannah': "[...] driven for by people who are coming to the into the country and I would say also they found people who are ready in in the country to be able to absorb and able to come up together and build some of the solutions." (I24) The value of collaboration and cooperation between foreign and local founders in the Kenyan digital ecosystem is hence highlighted.

Internal Knowledge (and Investment)
After analyzing the role of external or foreign knowledge in the Kenyan digital ecosystem, it is similarly important to discuss the role of local or internal developments. This includes local knowledge as part of internal activities driven by the digital transformation for development, for instance in terms of investments. On the one hand, the need for foreign engagements is assessed as "[...] more of the investment in tech is is coming from outside than it is coming from inside" (I21). One reason behind this is seen in local investments that are perceived as rather directed to real estate projects instead of towards tech companies (I21). Hence the challenge, especially for the Kenyan government, is to find a balance between allowing foreign investment and fostering local digital development, which is one of their roles in the ecosystem as further discussed in Sect. 4.4.5. Balancing the need for both internal and external support, I2 shares a diplomatic response in that "it should be in the interest to to support the growth of local markets from the inside but also from the outside".

Expanding the analysis towards arguments against the need or involvement of external actions and in favor of local development in the Kenyan digital ecosystem, the following arguments for internal digital development are shared by expert interviewees: While I13 (external: bilateral donor organization) argues in favor of digital development driven from the grassroots level instead of by macro interventions, I20 (multilateral donor organization) underlines this position in arguing that the country knows best about their internal processes and needs. More specifically, these expert perceptions on the one hand stem from the value of "[...] local knowledge which is created by local content" (I23). This is assessed as (more) difficult for external actors to meet since they have to adjust their solutions to local contexts that they are not necessarily familiar with (I3). Here it is worth referring to the crucial role of the local context for sustainable digital development as argued in Sect. 4.1 and Sect. 2.1 of this dissertation.

On the other hand, local digital solutions are argued to potentially decrease dependencies on external actors and on external digital technologies and services: "[...] if you want to move away from depending on the developed North and just what can we do for ourselves and how can we tailor make digital transformation for for Kenya by Kenyans" (I11). Further roles local knowledge plays for decreasing dependencies are discussed in Sect. 4.4.4. Local digital knowledge is even ascribed the potential to offer solutions for coordination in form of balancing notions of digital colonialism (Sect. 4.3.7) if knowledge transfer of technology development and project management are kept within the country and local organizations (I18). Synthesizing the role of external and internal knowledge in the Kenyan digital ecosystem, knowledge is further determined by learning and experience, which will be discussed in the following.

Learning and Experience

Besides the distinction between foreign and local knowledge, this dissertation identifies another form of knowledge that is crucial for the Kenyan digital ecosystem: learning. Learning here includes components such as learning by doing, learning from past experience of others or learning from demand (e.g., I19). Hence, experience is an important form of knowledge that correlates with learning. In the theoretical background, learning is one theme in development theories such as modernization theory (Sect. 2.1.1), but also in special regard to governing knowledge in the digital transformation (Sect. 2.3.6).

Firstly, learning from past experience of other multi-stakeholders, here especially actors originating from the Global North, includes learning in order to avoid repeating mistakes and also learning what works. This is highlighted by interviewed experts especially with regard tor the political level (e.g., I2). More specifically, learning allows optimizing to "[...] achieve this kind of sustainable development" (I5). This is especially relevant for sustainable digital development since learning from past experience saves transaction costs through avoided mistakes, learning what has worked in the digital transformation of other countries (I11; I14) and also in avoiding duplications by leveraging synergies reusing existing solutions that are applied to local contexts (I2).

Similar to the previously mentioned discussion around information flows into and within the ecosystem, interviewed expert representatives allude to knowledge and experiences flowing from the *Global North's experiences as rolemodels for learning* to countries such as Kenya. I2 for instance shares that "the Global South is learning from the Global North but luckily we see examples where they're learning from the good and the bad", which is further confirmed by I5. In the area of e-governance, Estonia is for instance mentioned from which learnings can be drawn for Kenya (I18). This is also where experience (and knowledge) of development cooperation, i.e., bilateral donor organizations, comes into consideration since development cooperation likewise 'brings' experience in the digital space to countries such as Kenya (I20). Nonetheless, digital transformation is special here, especially with regard to donor organizations' capabilities and knowledge in the digital transformation as discussed in Sect. 4.3.2 and Sect. 4.4.4. The reason why the experience from the Global North is highlighted here lies in the associated origin of technology (Sect. 4.2) and assessment of the Global North as having more experience in the digital transformation (I18).

However, comparing countries from the Global North and from the Global South in the digital transformation is multifaceted and extends the scope and goal of this research project. Nonetheless, the data analysis raises one crucial difference in the digital transformation journey of both regions with regard to experience: given that infrastructual issues still needed to be addressed, it gave countries such as Kenya time to learn from successes and mistakes of the Global North (I2). This is one component that further underlines the following expert argument that, while countries such as Kenya can learn from experiences of other countries, these still need to be adapted to local contexts since not all experiences can (or should) be replaced (I13). This is especially relevant when considering a shorter span of time: "[...] we cannot really then expect from from the Kenyan side to to have that within the in a shorter amount of time." (I13) Instead, the leapfrogging potential in learning from past experiences of other countries (Sect. 4.2.1, Sect. 2.2.2) can be achieved by "addressing the whole picture" (I2) and through learning from both mistakes and best practices, i.e., "learning from both ends" (I2). This approach is likewise associated with sustainable digital development (I3).

Secondly, this leads over to the dimension of learning and experience in the Kenyan digital ecosystem, in that learning is necessary in a setting of actors lacking experience in digital development (including the external ones from the Global North). This is perceived among expert interviewees as one hindrance to sustainable digital development in Kenya, especially for a long-lasting experience (I18). Another factor hindering sustainable digital development with regard to experience is seen in knowledgeable and experienced actors, especially in the Kenyan public sector: "[...] a way where governments can get into key key strategy where they get somebody who's a digital native person who's an expert in tech really driving the skills from the beginning but that isn't happening here" (I21). Nonetheless, there are exceptions, which highlight the benefit of past experiences of public sector actors that have previously worked in the (international) tech sector (I17).

However, the lack of experience in digital transformation in the Kenyan digital ecosystem is likewise perceived by other multi-stakeholder expert representatives as well, specifically with regard to donor organizations. While digital development has already been identified as a 'hype' topic for donor proliferation (Sect. 4.3.1, Sect. 2.2.3), digital experience is likewise perceived as difficult for bilateral donor organizations themselves:

> "And that's something we we always have to consider in in what we're doing and I think we are also very at the beginning because it on the one hand scientific evidence is still lacking and we're not sure what what actually is the impact of digitalization right?" (I13).

This is specifically relevant for this dissertation's analysis of sustainable digital development, since development cooperation efforts are supposed to make an effective contribution towards sustainable development, and the question is, how can that be ensured if digital experience of actors is questioned? Against this background, a lack of digital experience is further shared by a expert representative of a multilateral donor organization, asserting that: "The the digital focus is so new that we don't even have anybody who is a fully technically qualified in digitalization" (I17) as they "[...] just started working in the sector [...]" (I17). Not only is a lack of experience expressed, but also that the learning process is still ongoing ("learning by doing" I17). Again the importance of experience and (continuous) learning in the digital transformation is essential for every multi-stakeholder in the Kenyan digital ecosystem, not just the Kenyan government, but likewise donor organizations as (sustainable) capacity building measures require knowledge and experience. This is underlined by I1 highlighting the need of having development actors with digital knowledge, skills and experience (see also Sect. 4.4.4).

Moreover, the findings demonstrate that gathering experience through learning and conducting ICT projects seem to happen simultaneously, which poses questions in terms of a sustainable outcome. Certainly, this cannot be generalized with regard to all development cooperation efforts in the digital space or all ICT projects, but it is expressed for some as discussed e.g, in Sect. 4.4.4. Synthesizing these knowledge asymmetries, these likewise create settings of dependence which will be discussed in more detail in Sect. 4.4.4.

4.3.6 Digital Divide in Kenya

Another area where these asymmetries in power, knowledge and experience (among other factors) manifest in, is the digital divide in Kenya. Similarly to Sect. 2.3.4, digital divides hinder sustainable digital development as well as participation in the Kenyan digital ecosystem (see Sect. 4.4.3). More specifically, according to the GSMA (2020) report, top barriers to owning a mobile phone and to using mobile internet in Kenya are affordability and skills (closely linked to Sect. 4.4.4). Similarly, the urban-rural divide "[...] also dominates Kenya's digital economy landscape, where rural communities remain undeserved by existing digital infrastructure, digital skills initiatives and digital entrepreneurship support networks" (The World Bank 2019a:13). Moreover, women remain underrepresented in tech with persistent gender gaps in relation to Kenyan mobile internet users (gender gap of 39 percent, reference size: total adult population) or mobile owners (gender gap of 6 percent, reference size: total population) (GSMA 2019:15). These divides emphasize an example of exclusion in participating in the digital transformation in Kenya and hence an element of unsustainable digital development.

The importance of the digital divide leading to unsustainable digital development and also to exclusion within the Kenyan digital ecosystem are further emphasized by interviewed expert representatives, especially with regard to the role of the Kenyan government in their role as a 'manager' (Sect. 4.4.5). As a first example, the expert interviewees highlight that the digital skills gap in Kenya in combination with the other digital divides (gender, regional etc.) hinder an inclusive participation in the Kenyan digital ecosystem:

> "On the other hand you have a lot of digital skills gaps which I think prevents a lot of actors and especially certain groups at risks or marginalized groups of society to participate in that ecosystem talking about women, talking about people with disabilities [unk] in rural areas who just don't have the the the means and possibilities to acquire the necessary skills to be part of that ecosystem and and thrive in it so maybe these are actor [unk] participation" (I13).

According to I10, the digital skills gap is for instance reflected in having "[...] techies but we also have lots of regions within Kenya or even outside Kenya where we do not have particular skills in particular solutions". Given the discussed importance of digital skills (e.g., in Sect. 4.4.4), this field is one area where external actors such as development cooperation (and private sector) organizations engage the most (I13). The impact on sustainable digital development of these external ICT projects in Kenya focusing on digital skills is further discussed in Sect. 4.3.2. In addition to digital divides in terms of skills, the digital gender gap (e.g., I17) as well as urban-rural divides (I19; I22) in Kenya are highlighted as further examples of hindered participation. Lastly, and similarly to Sect. 4.1.3 and Sect. 4.4.3, infrastructure is highlighted as crucial for participation in the Kenyan digital ecosystem and as 'the basis': "[...] if you don't expand the basis of people who receive or who have access to the basis, you're in fact increasing the digital gap" (I17). For I1, infrastructure and digital skills are the two most important divides where "[...] there's not enough done towards spreading these two elements to marginalized communities". This is similarly reflected in a regional digital divide in comparison to countries of the Global North in terms of equity (I17).

Additionally, and more importantly for this dissertation's governance angle, the expert interviewees highlight the role of the Kenyan government in the digital divide. Here, the government is seen as responsible for closing the digital divide in Kenya (I19). However, the data analysis identified that the Kenyan government is discussed as contributing (instead of addressing) to the digital divide in focussing primarily on Nairobi and elite population groups (I10). This misled focus is argued to be reflected for instance in policies that contribute to the digital divide. I2 here describes notions of digital divide "[...] that's unfortunately also very often due to badly phrased policies". This is similarly confirmed by I1 arguing that ICT policies in Kenya "[...] should be designed more inclusive in the sense of or more targeting in the first place maybe those who so those who are marginalized". The role of the Kenyan government in the digital ecosystem and especially in ensuring sustainable digital development will be discussed in more detail in Sect. 4.4.5.

4.3.7 Digital Colonialism

Manifestations of these asymmetries can also be described by the concept of digital colonialism as discussed in the theoretical background (Sect. 2.1.4, Sect. 2.3.2, Sect. 2.3.6). In these instances, instead of achievements of leapfrogging (Sect. 2.2.2) a mere colonialized setting emerges. With reference to African countries, Wakunuma (2019) and Swart (2011) agree in part with Kwet (2019), arguing that there is an element of 'use, don't own' "due to a lack of security, poor governance, and poor internal continental relations which can stifle innovative use of technologies" (Wakunuma 2019:33).

Among interviewed expert representatives, notions of digital colonialism in the Kenyan digital ecosystem are discussed controversially. On the one hand, experts I5, I1, I6 and I22 negate experiencing digital colonialism in Kenya, with the latter expert seeing differences in the digital transformation here as "[...] it's not quite defined like land and the physical sense." (ibid.). While digital colonialism is mostly referred to regarding donor organizations and the private sector (Sect. 2.3.6), the here quoted expert representatives negating digital colonialism stem from a diverse multi-stakeholder setting of donor organizations, NGOs and academia.

On the other hand, however, digital colonialism in the Kenyan digital ecosystem is experienced by other interviewed experts with regard to the impact of Kenya's colonial past and competition, especially around data. The inheritances from Kenya's colonial past are argued to still have an impact on Kenya's (digital) development, since the country has "[...] taken a lot of time to adjust our laws, to adjust our policies to accommodate some of the changes that are happening and also trying to catch up with what's happening of in the developing world." (I11). This is similarly discussed in Sect. 4.2.1. The impact of external engagements is seen not only in ripple effects of the past, but also in form of current involvement of external or foreign actors influencing development. This is for instance experienced in cases where activities are "[...] not all the time let's say pretty much that what is needed or that what is maybe bought or set up if they could have decided solely on their own." (I5).

Moreover, the data analysis identifies that, of these external engagements in the Kenyan digital ecosystem, competition between multinational private tech companies (see also Sect. 4.3.4) is compared with notions of 'scramble' or competition for business in Kenya (I22). This is especially highlighted with regard to the role and value of data in the digital transformation as discussed in Sect. 4.2.2. More concretely, digital colonialism is further experienced with regard to *data colonialism* (see Sect. 2.3.6) in the Kenyan digital ecosystem as data here is a market where mainly foreign businesses compete (I15). Most importantly for this dissertation's focus on implications for a sustainable digital development, this is a problematic development since data in the context of data colonialism in Kenya is currently perceived as "[...] it's just being taken somewhere else and the value is not remaining with the digital economy of the country" (I13), hence not benefiting the digital development of Kenya. This is further supported by the assessment of I10, sharing the experience that "I can say there's a bit of colonialism in that case cause the they have access to our data and we are kind of helpless to some extent". This 'helplessness' further underlines that tendencies of digital colonialism hinder ownership of Kenyan actors as it "[...] may also destroy or replace or or hinder the the ownership and development on their own for a country" (I5). This underlines the importance of internal digital development as discussed in Sect. 4.3.5. Here, the Kenyan government has a unique role in regulating data protection issues, for instance through legislation such as the Kenyan data protection act (Sect. 4.1.2). Moreover it is worth noting that this issue around data applies to other countries as well.

4.4 Framework for Governance of Inner Workings in the Kenyan Digital Ecosystem (Micro Level)

After discussing the macro (Sect. 4.1) and meso level (Sect. 4.2) of analyzing the Kenyan digital ecosystem, this section will analyze the inner workings of the ecosystem. Moreover, it transfers the theoretically developed framework (Sect. 2.5) to the Kenyan case. For this purpose, expert insights into the characteristics of defining a digital ecosystem as well as a brief introduction into the Kenyan multi-stakeholders will be provided. Ultimately, the framework with its four determinants for governing the Kenyan digital ecosystem (preconditions, challenges, governance roles and governance formats) will be discussed. Generally, networks function as coordination mechanisms, especially with regard to the multi-stakeholders in the Kenyan digital ecosystem as highlighted by interviewed experts such as I5. With regard to the academic literature on the Kenyan digital ecosystem, there seems to be no project delivery framework aimed at optimizing or creating value of ICT projects in the country (Ndegwa, Kiriri and Achoki 2017; Ominde et al. 2020). Tarus, Gichoya and Muumbo (2015) further suggest that the country lacks a systematic framework not only for the delivery of ICT projects, but for the overall management of these projects, which in this dissertation are argued to be expanded from project to country level (ecosystem). The interviewed expert representatives likewise emphasize the need for a framework on how to operate and govern stakeholder engagement in the Kenyan digital ecosystem (I15). Both perspectives further highlight the importance of this dissertation and especially the derived framework. Before analyzing this in more detail, the Kenyan digital ecosystem firstly needs to be defined, serving the dual purpose of highlighting how this dissertation understands and operationalizes the digital ecosystem, but secondly showcasing an area where common understanding of the term 'digital ecosystem' is not clear-cut – both in academic literature as discussed in Sect. 2.2.5 and among interviewed experts.

4.4.1 Nature of the Kenyan Digital Ecosystem as a Network

Firstly, this dissertation understands digital ecosystems as a form of network that is influenced by the digital transformation. Networks have been previously defined in Sect. 2.4.3 while highlighting the chosen theoretical angle of network governance as a coordination mechanism for sustainable digital development. Moreover, this dissertation argues that networks are crucial in the digital transformation (Sect. 2.4.3). Against this background, the data analysis likewise highlights the role of networks and their correlation with digital ecosystems. With regard to the Kenyan digital ecosystem, the interviewed experts further underline the importance of networks in the digital transformation: "[...] the digital sphere is all about networks" (I1). This statement emphasizes that networks and digital (transformation) are associated with each other (not only as discussed in the literature review but also among interviewed expert representatives). Similarly, networks are also used synonymously with digital networks such as social media platforms, as indicated for example by "I go on networks" (I10). This makes the understanding and analysis of networks challenging as different understandings of what a network is and what it entails are identified in the data analysis. Such definitions of networks are likewise identified as challenging in the theoretical background and literature review (e.g., Sect. 2.4.3). Similarly, the network logic of digital technologies, especially of platforms, is highlighted as "[...] the digital tool can facilitate the setup and the administration of of networks" (I5). Hence, digital technologies potentially serve the role of a facilitator, which will be discussed in more detail with regard to the Kenyan digital ecosystem in Sect. 2.5.3.

Moreover, when defining the nature of the Kenyan digital ecosystem as a network, it is important to emphasize the different network formats e.g., ranging from Business-to-Customer (B2C) to Business-to-Business (B2B) to PPPs, especially in development cooperation (I6). The correlation between networks and partnerships is further emphasized by I9 asserting the setup of partnerships through networks in that networks here function as enablers (Sect. 4.4.5). Besides networks as enablers, another governance format mentioned by expert interviewees is characterized in the network form of an association of actors (I13), which is further discussed in Sect. 4.4.6. Moreover, the role of ICTs in governing networks will be further discussed in Sect. 4.4.6. With regard to this dissertation's research focus on digital transformation in Kenya from a development perspective, it is important to highlight that the Kenyan digital ecosystem is comprised of further sub-networks, so that it is important to be connected to these different sub-ecosystems (I1).

Secondly, networks and especially the participation in a network is crucial for digital development. Here, the data analysis further identifies networks as a resource (similarly to Sect. 4.4.4). With regard to the digital transformation in general and especially to the participation in the digital ecosystem, having and being in networks is important (I11). Thereby networks of actors likewise are a resource in that they are, on the one hand, a format in which resource exchange takes place (also between actor groups with diverse resources bases) that is crucial especially for startups that otherwise often lack funding (Sect. 4.4.4). But also, networks of actors are used for pursuing the actors' interests (I24). Network as a resource is further confirmed by I24 (Kenyan startup) for knowledge and resource transfer.

Thirdly, as elaborated in Sect. 2.4.3, a network is comprised of actors. In this dissertation, the actors refer to the identified multi-stakeholder groups in the Kenyan digital ecosystem as members of the network that interact with each other (I5). These multi-stakeholder interactions build "[...] a interconnected group of organizations" (I9) that "[...] build different connections between actors" (I18). What ideally connects these multi-stakeholders, is sharing and working towards a common goal (I9). In this dissertation this common goal is understood to be sustainable digital development in Kenya.

Synthesizing these perspectives on networks as a resource and networks comprising of actors, the fourth characteristic of networks for analyzing the Kenyan digital ecosystem are collaborations between multi-stakeholders. The collaboration between multi-stakeholders in the ecosystem (in the network) is assessed as essential, especially for the Kenyan government (I4). Collaboration here is also understood as "[...] exchanging within a network" (I4). Hence, networks are also understood as "the interplay among the the stakeholders [...]" (I5).

Collaboration in networks is likewise used synonymously with the act of *networking* among multi-stakeholders. A trend towards more networking here is highlighted as evolving with the digital transformation (I1). The impact of networking in the Kenyan digital ecosystem is assessed as crucial, especially for beneficiaries and the grass roots level (I6). Collaboration in the network is further defined by the degree of formality (similarly to formalized and less formalized governance, see Sect. 4.4.6). For defining networks, I13 distinguishes between informal and formal networks. Informal networks in the Kenyan digital ecosystem are for instance characterized by collaboration via word-of-mouth (I10). Similarly to discussion around self-regulation (Sect. 2.5.4, Sect. 4.4.6), I13 highlights that "[...] lot of things happen informally" in innovation ecosystems. The actor angle on networks will be further complemented with the following analysis of multi-stakeholders in the Kenyan digital ecosystem

4.4.2 Multi-Stakeholders in the Kenyan Digital Ecosystem

As illustrated in Fig. 4.1 and in line with Sect. 2.1.4 and Fig. 2.1, in this dissertation the key stakeholder groups in the Kenyan digital ecosystem include the Kenyan government, the (inter-)national private sector including Kenyan startups, bilateral and multilateral donor organizations, NGOs (including hubs) and academia. A similar overview is provided for instance by ITU (2019a). In this dissertation, these multi-stakeholders interact on the micro level of this dissertation's framework and will here be introduced briefly before analyzing their interests and interactions. In Fig. 4.1, a selection of stakeholders represented by individual organizations of each actor group are illustrated as an example for the Kenyan digital ecosystem (without claim of completeness). Some of these organizations belong to more than one actor group, for instance the C4D Lab (hub) is a innovation hub at the University of Nairobi (academia).

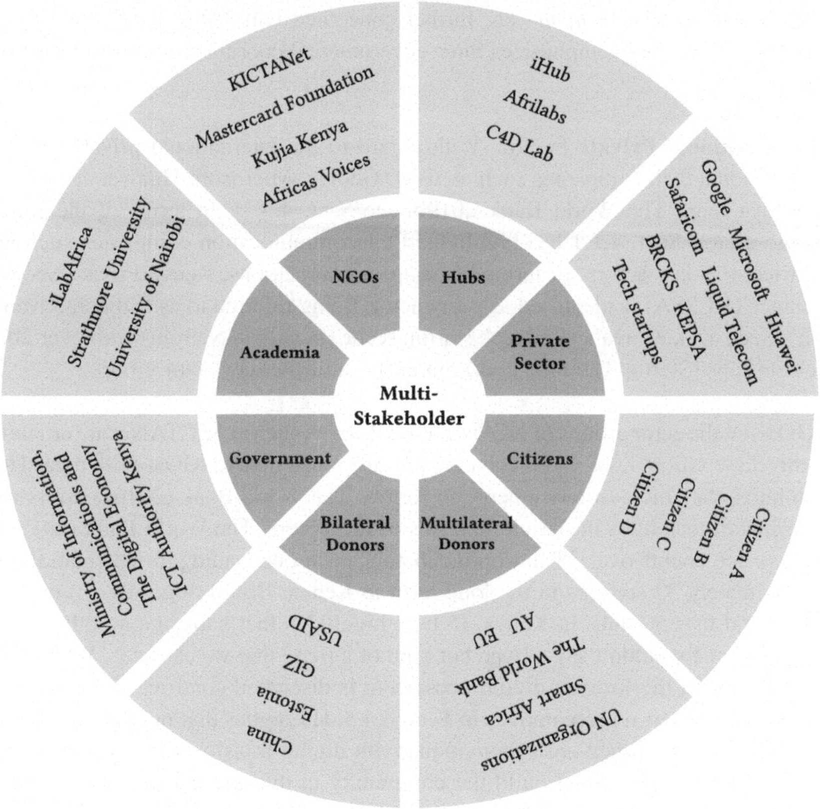

Fig. 4.1 Selection of key multi-stakeholders in the Kenyan digital ecosystem (own illustration)

Kenyan Government Beginning with the Kenyan government, this stakeholder group includes governmental institutions, with the *Ministry of Information, Communications and The Digital Economy* as the focal ministry. The role of the Kenyan government is not only crucial in the Kenyan digital ecosystem, but especially for the political system and development of the country. The government is for instance described as the "absolute ruler" I21. Due to the cross-sectoral nature of the digital transformation, other ministries such as the *Ministry of Industrialization, Trade* *and Enterprise Development*[7] are further concerned with 'governing' the digital transformation which emphasizes inter-governmental coordination as discussed in Sect. 4.1.6.

(Inter-)national Private Sector With regard to the international private sector, multinational tech companies such as IBM, Google, Microsoft, Huawei are active in Kenya (e.g., The World Bank 2019b, see Sect. 4.4.3). In terms of the local private sector, Sect. 4.1.1 has highlighted telecommunication companies such as Safaricom as key actors. In terms of business associations, Kenya Private Sector Alliance (KEPSA) is identified as a key actor. Being referred to as 'Silicon Savannah' (Sect. 4.1), Kenya's vivid tech startup scene (mostly in Nairobi) with over 200 tech startups (Startup Genome n. d.) represents a further key actor group.

NGOs Of the actor groups of NGOs, national actors such as KICTANet are crucial. In this dissertation, hubs and incubators are subsumed under NGOs. With regard to tech hubs, they increasingly emerge in Kenya. The World Bank (2016a) offers an overview of tech hubs in Africa. In terms of incubators, The World Bank (2019b) for instance counts over 38 startup incubators, each one leading to the creation of approximately 32 startups in the long term in Kenya. Briterbridges (n. d.) on the other hand lists 50 hubs in Kenya. I5 here highlights that hubs among others are "[...] kind of I wouldn't say a hype but kind of a trend that we can see". While the role of hubs in the Kenyan digital ecosystem is discussed controversially among interviewed expert representatives in Sect. 4.4.5, I11 argues that not only are hubs important but "[...] they are synonymous with digital transformation because we [unk] because hubs helped build the community in the first place". Hence, hubs here are discussed as an actor group.

Academia Of the actor group academia, Strathmore University has been identified as a key actor. Other universities focusing on digital transformation for instance include the University of Nairobi.

Bilateral and Multilateral Donor Organizations As discussed in Sect. 2.1.4, especially in terms of their ODA to Kenya, bilateral donor organizations from Germany, the US and China are emphasized. With regard to the digital transformation, Scandinavian donors are also identified as important. On a multilateral level, Smart Africa Alliance, UN Organizations and the EU are identified as key actors. Kenya was one of the seven countries who first joined the Smart Africa Alliance that is part

[7] For further information see https://www.industrialization.go.ke/.

of the African Union's Digital Transformation Strategy for Africa, which focuses on the creation of "more homegrown digital solutions" (African Union 2019).

4.4.3 Determinant 1: Preconditions for Governance in the Kenyan Digital Ecosystem

After this introduction into the framework for governance of inner workings in the Kenyan digital ecosystem (micro level) and beginning with the first determinant of this dissertation's framework, there are preconditions for governance of the digital ecosystem in Kenya. In line with this dissertation's theoretical framework, the following preconditions derived from Sect. 2.5.1 have been transferred to the Kenyan digital ecosystem and are empirically analyzed through the qualitative data analysis using expert interviews. Accordingly, the following preconditions will be analyzed with regard to the Kenyan digital ecosystem:

- Network size (number of actors)
- Levels of activity
- Visibility
- Transparency
- Voice
- Type of actors
- (Actor's interest, see Sect. 4.3.3)
- Willingness (also motivation)
- Goal consensus
- Perceptions: Awareness (Common understanding see Sect. 4.4.4)
- Trust
- (Mutual dependency/interdependence, see Sect. 4.4.4)
- Network's rationale
- Duration of actors' relationships
- Frequency of interactions
- Intensity (closeness and distance)
- Degree of centralization (also embeddedness)
- Expectations
- Network effectiveness and efficiency

As a variation to the structure of Sect. 2.5.1 where common language and understanding are discussed as precondition for network governance and coordination, the data analysis of the Kenyan digital ecosystem proved as a challenge, since it emerged as an area where efforts often fail. For this reason, common language and understanding will be discussed in Sect. 4.4.4. Similarly, actors' interests are identified as a challenge and hence discussed in Sect. 4.3.3. Hence these are entirely put in brackets in the list above.

Before analyzing each precondition with regard to the research question as to how this ecosystem can be coordinated, it is important to note that these preconditions share overlaps with required bases and contexts similarly to Sect. 4.1 and Fig. 1.1. What is needed for coordination to contribute to sustainable digital development, is seen in having appropriate legislations and policies for digitalization, change management processes, trained and knowledgeable employees around technologies and IT processes as well as choosing and implementing 'the right' technology I18.

Moreover, the preconditions provide insights into not only preconditions for governing the digital ecosystem, but indirectly for a thriving and sustainable ecosystem. Against this backdrop, the correlation between a 'thriving' and 'sustainable' ecosystem is further important since a thriving ecosystem does not automatically lead to or equal a sustainable ecosystem (I17). Hence this dissertation demonstrates that all of these preconditions not only need to be incorporated by multi-stakeholders in their activities, but also to be governed and coordinated.

Network Size / Number of Actors

Beginning with the first precondition and in close relation to identifying and elaborating the different multi-stakeholders present in the digital ecosystem of Kenya (Sect. 2.1.4, Sect. 4.4.2), it is important to analyze the impact and relevance of both the number and the type of actors in the context of governance and coordination. Here the implications of the number of stakeholders in the ecosystem, i.e., the size of the ecosystem, are similarly discussed in the theoretical framework (Sect. 2.5.1). This precondition further emphasizes this dissertation's argument for a need of coordination.

Generally, the perception is shared among the interviewed experts that a high number of actors is involved and present in the Kenyan digital ecosystem, for instance as "lot of various players involved" (I6), or "a lot of organizations setting up their bases here or they're planning to set up [...]" (I10). At this point of the empirical analysis, this fact is analyzed regardless of physical presence, an issue which will be discussed further in Sect. 4.4.3, as the focus of this section lies only on the actual amount of actors. Interestingly, the high amount of actors in the Kenyan digital ecosystem is discussed as both beneficial and challenging for digital development and coordination.

On the one hand, beginning with a further analysis of the positive impact of a high number of multi-stakeholders, the increased amount of actors is associated with simply more interest as indicated for instance by "[...] generally more money going around as well" (I2). Moreover, not only is the benefit of "the more the merrier" (I9) shared, but likewise explained by the need for many actors to address and cope with the vast amount of digital transformation topics where "[...] it's not enough for just one specific group of actors to be involved" (I9). I11) further argues that smaller networks of actors that share capabilities would limit barriers to the digital transformation.

On the other hand however, the influence of the number of actors in the Kenyan digital ecosystem is mostly perceived as a challenge and hence as a negative impact on digital development by expert representatives. Most importantly, the high amount of actors lead to the challenge of fragmentation (I6) and even confusion that may hinder development (I12), which is in line with Sect. 4.1.5. This underlines the importance of coordination of actors and their activities for achieving sustainable digital development. Moreover, there is criticism among interviewed experts not only in regard to "[...] too much involvement of too many players" (I1), but also regarding actors "that may may not be necessary" (ibid.). This includes the role of donors and donor fragmentation with I4 arguing that there already is a vast amount of donor agencies present that makes it difficult for (new) actors to establish their legitimacy. The vast amount of actors also complicates the identification of key ones and might explain the emergence of subnetworks or even silos (Sect. 4.4.3).

Despite benefits or challenges arising from the number of actors, it is also important to discuss the impact of a small amount of actors for the success of ICT projects as illustrated in Sect. 4.3.2. While a limited number of actors can increase the efficiency of projects, it however cannot include all actors, which leads to exclusion instead of inclusion (I12, Sect. 4.4.3, Sect. 4.4.3). This is a balance that needs to be addressed – in projects but also in the development of the digital ecosystem (I12).

These findings suggest that a small amount of stakeholders is easier to coordinate, but leads to exclusiveness even to the degree of silos, less inclusivity and less holistic development, but also have the potential to achieve higher efficiency and impact, thereby a more sustainable digital development. Although this dissertation provides some insight into how this balance can be governed (Sect. 4.4.5, Sect. 4.4.6, Sect. 4.5), further research is needed.

Levels of Activity of Multi-Stakeholders

As elaborated in Sect. 2.5.1, interactions between multi-stakeholders require some degrees of activity that build dynamics. Activity as a metric is also reflected in the degree of participation and collaboration in policy processes (I21) as actors are characterized by degree of actively influencing for instance content, course and outcomes of social interaction processes (Scharpf 1997; Klijn and Koppenjan 2016). Activity in this dissertation hence is understood as a continuum from high levels of activity to low levels of activity, similar to being passive or active. This is operationalized in Table 3.3 and will be completed in Table 4.1 based on the findings of this section on activity levels in the Kenyan digital ecosystem. A 'high' level of activity here is defined as active and visible participation in the Kenyan digital ecosystem (e.g., through participation in policy processes or ICT projects), while a 'low' activity level is defined as the opposite i.e., passive or no (visible) participation (see also Sect. 3.4.2). How visibility is correlated with activity will be discussed in more detail in Sect. 4.4.3.

Transferring this theoretical perspective, data from the conducted interview content analysis firstly highlights the importance of understanding dynamics for a sustainable development, especially with regard to the Kenyan digital ecosystem (e.g., I12). This simultaneously underlines the need for coordination and governance of these multi-stakeholder dynamics, which is especially relevant for the Kenyan digital ecosystem with much dynamism (I20). Hence dynamics likewise address activity levels. For analyzing how digital ecosystems can be coordinated, it needs to be discussed what the actors' activity levels are. Are the active ones also the ones governing? This section will deliver insights into who the active actors are, and as to whether they are also the ones governing will be discussed in more detail in Sect. 4.4.5.

Firstly, activity level is used among expert representatives to define the Kenyan digital ecosystem:

> "[...] what makes a vibrant ecosystem or more how I would define that when you have a lot of activity in that ecosystems so so let's say the the rate of interaction is quite high and also the it's hard to measure but to ja I would say the rate of of activity and output of an ecosystem " (I5).

In line with the perception of the Kenyan digital ecosystem as 'vibrant' (ibid.), it is important to note that the level of activity is generally high. Not only is there "more and more activities in in in this ICT sector" (I5), a lot of drive (I6) and "local activity" (I7). More importantly, this dissertation argues that it is important to analyze the different levels of activity of the multi-stakeholders in the Kenyan digital ecosystem before understanding their roles in governance and coordination, as underlined by the statement that "for each actor group there is some activity" (I5). Which actor group is more active and which is more reactive or passive will be analyzed in the following with regard to the Kenyan digital ecosystem.

High Level of Activity of Actors Secondly, before analyzing the gathered interview data in terms of activity levels, it is important to address possible biases that might be underlying in the responses as most of the interviewees named themselves as the most active actors. Through the methodological approach of this dissertation, this possible bias is balanced in that every interviewee has been asked the same question so that every actor group has been commented on by others. More specifically, a *high level of activity* in the Kenyan digital ecosystem is emphasized by the role of driving the digital transformation, which will be discussed in more detail in Sect. 4.4.5. Generally, the findings demonstrate that the most active groups are the private sector (both international and local private sector) as well as hubs. I4 for instance suggests hubs as the most active actor group. The dynamic of hubs (and accelerators) is further confirmed by I17. The international private tech companies are further highlighted as the most active group, with international tech companies from the US at the forefront (I19), followed by actors from China, but to a lesser degree (I3). Hence, these findings suggest that private sector actors are perceived as most active, although further research is needed here given this dissertation's limited data.

Less Active Actors Thirdly, on the continuum of levels of activity, NGOs and the Kenyan government are perceived as *less active*, which will be further discussed in Sect. 4.4.5. With regard to NGOs, interviews I1, I3, I21 and I7, assess this actor group as less active. Additionally, the Kenyan government and respective public sector ministries are further assessed as less active (I5), rather reactive (I4) and passive (I24) to the extent of "they're completely not doing anything" (I10). These perceptions are opposed to the ones that see the Kenyan government in the role as driver of digital transformation in Kenya (at least to a certain degree, see Sect. 4.4.5). Similar activity evaluations of the Kenyan governmental institutions such as the Ministry of ICT is shared by I21 with I9 and I4 asserting the potential for more activity of the government. Moreover, a need for governmental institutions in the Kenyan digital ecosystem to be active rather than reactive for ensuring ownership is emphasized (I20).

On the contrary, I1 sees at least some activity on the part of the government. The perceived 'passiveness' of the Kenyan government can be explained by the further findings of power imbalances, digital capacities, and the impact with regard to legitimacy and efficiency – all of which pose challenges to sustainable digital development. Moreover, the Kenyan government as being passive is not exclusively associated with criticism, as it is also highlighted as beneficial. More specifically, the more passive role of the government in the digital transformation is assessed as not leading to stifling innovation (as discussed in Sect. 4.4.6). I1 for instance highlights the governmental passiveness as a successful approach for fostering innovation and digital transformation in creating frameworks without too much active involvement. I16 likewise highlights that only a certain amount of activities, such as in form of regulations, is needed in order to maintain the innovative spirit.

With regard to the activity level of the actor groups of donors and academia, theirs did not emerge from the data analysis in the context of activity. This can be explained by the methodological approach and also with regard to the correlation between presence, visibility and perceived activity. These findings determine the importance of actors' activity levels for coordination in the Kenyan digital ecosystem as it firstly needs to be established who the active actors are. For this, the presence and visibility of actors is likewise important, which will be discussed in more detail in Sect. 4.4.3 and in the following section.

Visibility of Multi-Stakeholders

Following up on the precondition of multi-stakeholders' activities, visibility is a crucial part of interactions in the digital ecosystem (Sect. 2.5.1. Here, visibility is another factor that explains if and how coordination of the ecosystem for sustainable digital development is possible. Firstly, the interview data suggests that the digital transformation enhances visibility of action (e.g. also in sense of transparency). The global interest in the Kenyan digital ecosystem is likewise visible and simultaneously demonstrates the necessity for actors to be visible, especially for participation in the digital ecosystem (I2). Thereby visibility is closely linked to awareness, since only if something is visible, awareness can be reached (Sect. 4.4.3). This correlation between awareness and visibility is for instance illustrated by I6: "[...] but I'm also not aware of all planned or pipeline projects of all donors or development partners". Hence, if actions or actors are not visible, they are not recognized, which hinders interaction (and provides no connecting factor for coordination).

One experience from the Kenyan digital ecosystem is shared by I9 asserting that it is not always visible what the dynamics in the Kenyan digital ecosystem are:

> "I also think in terms of visibility sometimes it's not always clear what exactly is going on and who's doing what in the ecosystem so then you are you kind of have to do a lot of research to find out what someone is working on, how long they have been working on it and the most difficult one to get news on is something that isn't implemented yet you know maybe it's in the discussion stages or they're planning because often we tend to hear about something when it's been launched or in the doing progress reports or when they're closing so then it's a little bit difficult to have visibility on that."

As the findings suggest, one way to identify or measure visibility is to look for participation in policy making processes which is a similar method as proposed in Sect. 4.4.3. This approach is named by experts as enabling them to see who is present or engaged in specific policy-making (I13). The consequence of this approach however is almost a vicious circle that feeds into power imbalances (Sect. 4.3.4), since those who are not recognized, are also not involved in the processes in the Kenyan digital ecosystem leading to exclusion (Sect. 4.4.3). The impact here alludes to unsustainable digital development and highlights a need for coordination to achieve inclusion.

Secondly, expanding on the previous discussion around activity levels in the Kenyan digital ecosystem, a correlations can be drawn between activity and visibility, i.e., actors who are perceived as active are also the ones who are most visible in the Kenyan digital ecosystem. This is for instance highlighted among interviewed experts with regard to China being perceived as very active and also "very present and very visible in in Kenya" (I6, see also Sect. 4.4.3). With regard to the private sector and specifically startups, the ones who are visible are the ones who get supported (I3). Thereby visibility also determines the potential and degree of participation and can lead to exclusivity instead of inclusivity (Sect. 4.4.3). What is likewise demonstrated here is the argument around the role of development cooperation in the digital ecosystem of Kenya, as this discussion around visibility highlights that donor darlings prevent others (e.g., startups) from becoming visible (I13). This is an instance where visibility correlates with increasing the exclusivity of the digital ecosystem through multi-stakeholders' actions and interests. Hence, this leads to the assumption that visibility is in the interest of actors (Sect. 4.3.3) as it is crucial for recognition and explains actors' drive for proliferation in digital development (Sect. 4.3.1) so that there is almost a race for visibility.

Thirdly, a further finding of the qualitative interview analysis is that activity does not necessarily correlate with or equal impact. This is supported by expert I21 referring to multinational private tech companies being perceived as more visible (here also indicated by voice, see also Sect. 4.4.3) than impactful: "[...] the reality is that that of the big tech companies probably they're making more noise about what they're doing than that than that then they're really doing." This dualism between visibility and impact is similarly experienced by expert interviewees with regard to bilateral donor organizations as the most visible ones are not necessarily the most impactful ones (I6). I19 further acknowledges the activity of donors and their visibility given their requirement to transparently fund projects (e.g., ODA), which makes their visibility possibly different to other external actors. Here further analysis is required as to whether donors' interests or their funding is (more) transparent, which will be discussed in more detail in Sect. 4.4.3.

Finally, regarding the remaining actor groups of NGOs and government, these have not been specifically mentioned as being visible by expert interviews but will be discussed further in the precondition of voice (Sect. 4.4.3).

Presence of (foreign) Multi-Stakeholders Visibility is further manifested in this dissertation's framework in the form of presence. Presence here is understood as physical presence as only one part of determining actors' visibility. This dissertation argues that another explanatory factor for perceived levels of activity of actors is the role of presence. However, the importance of physical presence is contested and perceived differently among interviewed experts. Generally, I10 shares the perception of the Kenyan digital ecosystem as "a lot of organizations setting up their bases here or they're planning to set up [...]." This suggests an activity towards achieving or increasing physical presence for instance through having offices in Kenya. More specifically, I19 confirms this experience with regard to multinational tech companies, which are claimed to all have (regional) headquarters or offices in Kenya: "[...] all the big tech companies Facebook Microsoft Google I believe they have a headquarters there apart from South Africa the next place they usually go is Kenya." At least with regard to Google, this is confirmed by I9.

Hence, physical presence is also a factor determining or expressing visibility that demonstrates interest and power. However, while I16 confirms the presence of international companies in Kenya, the expert claims that none of them have physical presences in the form of offices in Kenya. Multinational tech companies not being physically present in Kenya is further shared by I23 and I23, while highlighting the digital (intangible) presence of their services such as Google, YouTube or Facebook. With regard to the actor group of donor organizations, their physical presence for instance through embassies is highlighted by I18 (bilateral donor) sharing that their country does not have an embassy in Kenya while other countries' donors do. Despite not having a physical presence, this particular donor organization is nonetheless active and visible in Kenya (see e.g., I6 and I18 referring to the country). This might lead to the conclusion that physical presence in the digital ecosystem is not as important and is rather manifested by services offered through the digital transformation. With regard to multilateral organizations, especially UN organizations are highlighted as being present with offices in the Kenyan digital ecosystem as well as various (international) NGOs (I21).

Furthermore, presence cannot only be analyzed through distinguishing between presence or no presence, but also in terms of the size of presence i.e., determined by number of staff (bigger or smaller presences). Some multinationals for instance only have smaller physical presence in the Kenyan digital ecosystem (I10). Hence the question is whether it is necessary for organizations' participation in the Kenyan digital ecosystem to have a physical presence in Kenya. The answer depends on the actors' interests and whether the entrance into the digital ecosystem can also be achieved through digital presence (Sect. 4.3.3). For instance, outsourcing enables a digital inclusion of stakeholders in the ecosystem (I21). Another example for actors' interests being the determining factor for (physical) presence is the argument that physical presence increases visibility and is a tool for engaging in competition, especially with regard to multinationals. Here, correlations to their motivation can be drawn, for instance relating back the discussed actor proliferation (Sect. 4.3.1). Thereby presence can also be an indicator for the degree of closeness or distance (here physically) between multi-stakeholders in the Kenyan digital ecosystem, which will be discussed further in Sect. 4.4.3.

Transparency

As mentioned in the beginning of the previous precondition visibility, another related precondition for sustainable digital development through governance and coordination is transparency. The role of transparency in network governance and in ICT4D for coordination has been derived in the theoretical background (Sect. 2.5.1). Firstly, the need for and importance of transparency for a sustainable digital development especially in development cooperation is stressed by I4 arguing that not enough efforts are made towards transparent structures and capacity building. Notions of unsustainable digital development are here determined by a lack of transparency that especially in ICT context, for instance driven by hidden agendas (I3, Sect. 4.3.3 and Sect. 4.3.2), or a lack of willingness to be transparent (I2). A lack of transparency can hence also be traced back as barrier to sustainable digital development with regard to the discourse around ICTs that is inherent in 'hype' (Sect. 4.3.1) as illustrated by: "I think you know even as we hype the usefulness of digital technologies we we sort of have not also been very open in in in all it's aspects" (I15). This specifically applies to this dissertation's analysis of the Kenyan digital ecosystem and the Kenyan ICT sector, which according to I7 needs to be more transparent for a sustainable digital development, especially with regard to preconditions for participation.

Secondly, transparency is further essential for sustainable digital development with regard to multi-stakeholder activities with this dissertation arguing that transparency can be achieved through coordination. More crucially for this research project, transparency and coordination are closely aligned with regard to multi-stakeholder activities in the Kenyan digital ecosystem in order "to figure out what are the actual synergies and that needs some kind of of coordination and and transparency about who's doing what and why" (I5). This is further highlighted by I20 emphasizing the need for a holistic frame in which interactions are guided sustainably. This dissertation will deliver further insights into how transparency can be achieved for sustainable digital development through a coordinative governance means in more detail in Sect. 4.5. Transparency in this research project has a further component in sustainable digital development and coordination since transparency here is closely linked to trust with transparency as "[...] a key factor to overcoming the the trust issue that I've mentioned before so if nobody trusts the other" (I2). Transparency in the intersection to trust hence is another factor that is crucial when analyzing multi-stakeholder interactions in the Kenyan digital ecosystem. This is likewise experienced by expert representatives with regard to the interplay of transparency, trust (Sect. 2.5.1, Sect. 4.4.3) and funding (Sect. 4.4.4), which is crucial for innovation and digital development. The interviewed expert representatives in this research project highlight the need for transparency in both decision making around funding (I7) and for attracting funding (I20). More importantly with regard to this dissertation's focus on the different multi-stakeholder groups in the Kenyan digital ecosystem, NGOs are assessed as the most transparent actor by I3 (bilateral donor organization).

Thirdly, this dissertation's findings suggest the following solutions to achieve coordination in a transparent way: standards (e.g., I7 and further discussed in Sect. 4.4.6), digital technologies' potential to enhance transparency as a step towards alignment of actors (e.g., I4, further discussed in Sect. 4.4.3), transparency against corruption and for efficiency (e.g., I18, see Sect. 4.4.3), and digital platforms as the ultimate combination of these three. Digital platforms are one main proposal of this dissertation that will be discussed separately in Sect. 4.5. In terms of governance form, transparency is another indicator for the need for digital platforms as a governance format as it combines these transparency arguments (I1). Here also the governance form of self-governance and digital platform are combined (see further Sect. 4.4.6). Moreover, a digital platform coordinating activities in the Kenyan digital ecosystem could likewise potentially enhance transparency of multi-stakeholder activities (I2). This dissertation thereby takes up the argument of digital technologies enhancing transparency (e.g., transparency in code highlighted by I7, I18, see also Sect. 2.2).

Voice

Another determinant that loosely falls into the scope of activity, presence and visibility of multi-stakeholders in the Kenyan digital ecosystem from a coordination point of view, is identified in 'voice'. As derived from the theoretical section (Sect. 2.5.1), using voice in interactions, especially in multi-stakeholder policy consultations, is a further governance determinant. More specifically, voice is understood also as a form of power (Sect. 4.3.4) and as a means to further distinguish between levels of activity and inclusion as some voices of actors are heard better than others. Hence voice functions as an indicator for influence in this dissertation as illustrated in Table 3.3 and Table 4.1.

Firstly, voice as a determinant for influence is highlighted in the context of policy making processes. Here voice is understood by distinguishing between who is heard in policy consultation, and who is not. With regard to multi-stakeholders in the Kenyan digital ecosystem, the importance of voice in these processes is for instance highlighted by the following expert experience:

> "I think if you want to look for the major in of civil society for instance you look at the policy-making process of new regulations for instance a startup act, data protection act and you look who has voiced who has put effort into it and has been heard. I think that's a good source maybe to to see who is who you see and and maybe that is also the case for for certain private sector initiatives." (I13)

Therefore voice as an indicator further allows for distinguishing actor groups that are 'loud', here meaning (more) powerful in influencing policy (I19), from ones that are less loud, meaning less powerful. Consequently, the following role of voice can be derived: one, the impact of having and/or using voice, and two, voice as an indicator for equal participation or inclusion. The latter is further confirmed by I10 arguing that "some voices are heard better because of their positions, because of who they are [...]". Another example for this is the 2016 ICT policy process, where the expert representative of a Kenyan NGO shares the experiences of formal inclusion in the process, but a lack of recognition in the final document (i.e., no impact) (I15).

Secondly, as illustrated in the expert statement, there is a gap between (formal) involvement or the possibility to use voice and the actual recognition of voices in implementation. On the one hand, involving and recognizing differing voices of various multi-stakeholders is the essence of sustainable digital development (Sect. 2.1.3). This is further emphasized by I15 highlighting the importance of equal representation of voices, such as of civil society, academia and 'techies' in (digital) development issues. However, notions of a 'proforma' inclusion of multi-stakeholder representatives is experienced by I21: "[...] just to maybe will feel good that their views have been heard but no one ever listens or takes into account". Experiences of voices being included in formal multi-stakeholder processes but not having found their ways into implementation (e.g., of policies), is additionally shared by I15 highlighting (general) disadvantages of multi-stakeholder processes (for further discussion see Sect. 4.4.6). Another reason for failed recognition is likewise experienced in actors being "[...] onboarded way too late in certain processes and it's more like 'this is our decision' and they don't really get to design it rather than just say 'yeah with we like it or not' " (I13). These experiences emphasize unsustainable developments that need to be coordinated (better).

Thirdly, the data analysis highlights that it is preconditional for sustainable digital development and coordination for actors to speak with one joint voice, which is closely linked to common language and understanding (Sect. 2.5.1, Sect. 4.4.4). More specifically, the data analysis emphasizes the importance of having a joint voice for sustainable digital development as it is one area where coordination is already observed. For instance expert I20 (multilateral organization) shares the experience of their organization trying to have joint advocacy with others. Another expert of a different multilateral organization likewise highlights efforts of EU member states (*Team Europe Initiative*) to "combine their strengths and come up with a joint voice" (I6) in supporting the Kenyan government. While this may lead to the impression that multilateral organizations are the most likely to achieve a joint voice given their supranational nature, it needs to be noted that bilateral donor organizations are (mostly) part of these multilateral organization. Hence, some credit for achieving joint voices may likewise apply to them. One donor organization for instance reports efforts of their organization to "[...] speak in one voice because one of the challenges in development cooperation is that you have so many partners that you overwhelm the government." (I17) This emphasizes the goal of combing multi-stakeholders voices in one joint voice as an coordinative effort to not overwhelm the government, meaning not to overstretch their capacities which ultimately would otherwise manifest in unsustainable digital development (as similarly argued in Sect. 4.1.5 and Sect. 4.4.4). These findings likewise underline this dissertation's argued importance of coordination as also highlighted in Sect. 4.1.6.

Fourthly, these experiences emphasize that there is a difference in having, using and recognizing voices of multi-stakeholders in the Kenyan digital ecosystem. At first, actor groups that are perceived as not being heard are discussed before switching to actor groups perceived as being heard. Beginning with the first group, the data analysis of the conducted expert interviews identifies two actor groups assessed as not being heard: NGOs and Kenyan startups, hence demonstrating inequalities. In addition to findings on NGOs in the previous passage, experts such as I13 highlight the lack of acknowledgment of NGOs in the Kenyan digital ecosystem and especially in policy making processes. I15 on the other hand emphasizes the underlying danger here since NGO are the actor group whose systematic function is to counterbalance (similarly discussed in Sect. 2.1.4). With regard to the second group of Kenyan startups, they are argued to be weaker in policy processes (I21), for instance due to not raising their voice (I21). This might be due to a lack of a joint voice represented by an organizational startup association or community where voices are pooled (I2). Similarly, I19 highlights that there is no advocacy, especially for rural tech startups, i.e., no one speaks for them, which emphasizes a need to form startup alliances across the country. The role of such associations in governing the Kenyan digital ecosystem will be discussed in more detail in Sect. 4.4.6 as it is argued to be a solution for this unequal recognition of voices.

Switching to the second actor groups in the Kenyan digital ecosystem being perceived as being heard and recognized i.e., whose voices find recognition and influence, are mostly identified as (multinational) tech companies. This finding here is closely linked to visibility (Sect. 4.4.3) and power (Sect. 4.3.4). In connection to the previous finding of NGOs as a lesser heard actor group, I15 further states that the government listens (may want to listen) more to the private sector than NGOs and "[...] their views are given more I think that's a univers disadvantage you find that their views are given more weight than probably other stakeholders." In the same vein, international tech companies are argued to come with strong voice (I20) that underlines their power (Sect. 4.3.4).

In sum, one (governance) solution to overcome these asymmetries is seen in the value of NGOs that need to be considered more by the Kenyan government as it would provide them with the opportunity to decrease dependencies on certain actors such as private sector actors: "I think how the government of Kenya could be strengthened is working with the civil society. Strengthening the civil societies in Kenya to work more closely with the government so they bring their voice they bring their voice there." (I19) Thereby the discussed unsustainable outcomes of imbalances in voice can be counteracted for achieving sustainable digital development. Moreover, the 'lesser heard' actor groups need to be encouraged and empowered to use their own voices which can likewise be achieved through governance and contribute to sustainable digital development, especially considering fairness, inclusion and equality (I7). Synthesizing these preconditions of voice, activity and visibility, this leads to a classification of multi-stakeholder groups in the Kenyan digital ecosystem according to their level of activity and influence as illustrated in the following matrix as proposed in Sect. 3.4.2. The findings of these subsections can be summarized as illustrated in Table 4.1.

Table 4.1 Activity and influence matrix of multi-stakeholders in the Kenyan digital ecosystem (own illustration)

Actor Group	Activity	Influence
Academia	? Unknown	? Unknown
Donor Organizations	▲ Mostly Active	~ Debated
Hubs	▲ Active	▼ Moderately Low
Kenyan Government	▼ Mostly Latent	~ Debated
Kenyan Startups	▲ Active	▼ Moderately Low
NGOs	▼ Mostly Latent	▼ Moderately Low
Private Tech Sector	▲ Active	▲ High

Type of Actors

In reference to the previous precondition of the network size, not only is an increasing number of actors in the Kenyan digital ecosystem observed by interviewed experts, but also an emergence of new types of actors as derived from the theoretical section (Sect. 2.5.1). Here, the 'new' stakeholders pose a specific challenge for sustainable digital development in Kenya as supported by: "[...] on one hand you have the usual challenges and then on top the are very specific challenges with regard to the new stakeholders and and the thematic scope of digital transformation" (I5). New actors are especially seen in the form of "a new kind of ministry or the digitalization or innovation hubs" (I5), such as the Kenyan Ministry of ICT (I6) or innovation hubs (I5). Their emergence is a result of the ongoing developments of the digital transformation (I5). Besides hubs and ministries focusing on digital transformation in Kenya, multinational tech companies are not necessarily new in the ecosystem, but their importance grows due to their engagement in the digital transformation (see also Sect. 2.5.1). This is for instance mirrored in their growing number and also in their increasing presence in the ecosystem (I10), which is expected to further grow in the future (I3). Here the preconditions of number and type of actors are connected. With further regard to actor groups in the digital transformation, ICTs are discussed as actors themselves, for instance in Sect. 2.2.1, the role of ICTs in the Kenyan digital ecosystem will be discussed in more detail in Sect. 4.4.5.

Willingness

A further precondition identified in the theoretical background and interview analysis is willingness. While willingness is also discussed as a part of interests (Sect. 2.3.3, Sect. 4.3.3), it here is understood more in the sense of motivation as identified in the theoretical background (Sect. 2.5.1). Willingness of multi-stakeholders is further closely linked with the precondition transparency (Sect. 2.5.1, Sect. 4.4.3), for instance willingness to be and act transparent. More specifically, transparency about failures by multi-stakeholders in the ecosystem is perceived as limited by expert I2. This dissertation argues that the potential for coordination depends on the willingness of multi-stakeholders, which is likewise confirmed by I3 and I2 as "it's not necessarily the form of of structure that would lead to the right answer, it's more the the the the willingness of the participants to achieve the right result" (I2). More specifically, willingness is a crucial determinant for collaboration, for implementation or/and using digital technologies.

Firstly, willingness is a determinant relevant for analyzing collaboration between multi-stakeholders. The Kenyan digital ecosystem and ICT sector here has a special role in achieving sustainable digital development since the willingness for collaboration in the digital sector is assessed as greater than in 'traditional sectors': "[...] this willingness there's appetite for working together is not like other sectors where maybe the political economy has already been you know has established who are the major power actors and they want to keep things as the are for a few decades and they don't want to change." (I17)

Secondly, willingness in the Kenyan digital ecosystem is crucial with regard to digital development in terms of implementing and using digital technologies. Here expert representatives' perspectives on the willingness of multi-stakeholders in the Kenyan digital ecosystem to implement digital technologies in processes or projects differ, which is discussed in further detail in Sect. 4.4.4. Willingness as a precondition for ecosystem governance is of further interest at this point with regard to willingness of multi-stakeholders towards the suitability of coordinative digital platforms as a governance form (Sect. 4.5). More specifically, I2 argues that the success of the platform is dependent on the actors' willingness, which is further confirmed by I3.

Goal Consensus

As elaborated in Sect. 2.5.1, goal consensus as a precondition for network/ecosystem governance is expressed in facets of common language and understanding and closely linked to network rationale (see further Sect. 4.4.3, Sect. 4.4.4). The empirical interview analysis suggests that goal consensus serves the purpose of an indicator challenging cooperation and coordination rather than a sole precondition. Instead, it revolves around the question of multi-stakeholders having the same (aligned) goals that is prerequisite for cooperation and coordination. As the interview data analysis suggests, shared objectives are challenged in the Kenyan digital ecosystem by diverging interests of multi-stakeholders (see e.g., Sect. 4.3.3) and by different powers of multi-stakeholders in communicating (as discussed in Sect. 4.4.3). More specifically, the interviewed expert representatives highlight the importance of a goal consensus for achieving sustainable digital development in the Kenyan digital ecosystem as it facilitates multi-stakeholder cooperation (e.g., I18). Similarly, goal consensus is seen as crucial for the performance of partnership formats, while highlighting that goal consensus is a theoretical precondition that is more difficult to achieve practically (I5).

Perceptions: Level of Awareness

As a further precondition for governance and coordination, this dissertation identifies perceptions and more specifically the level of awareness within the ecosystem. This empirical analysis of perceptions as a precondition differs from the one derived in the theoretical background (Sect. 2.5.1) in that common language and understanding, as an original part of perceptions, are here discussed as a challenge (Sect. 4.4.3, Sect. 4.4.4). Awareness in this dissertation is analyzed with regard to awareness of implications of the digital transformation as well as awareness of multi-stakeholders of each other in order to identify coordination potentials. Awareness is insofar a precondition for governance with regard to multi-stakeholders as issues can only be addressed if there is a level of awareness before actions are required. Awareness is likewise the starting point for (digital) innovation, because only when awareness of problems exists, can (digital) solutions be found as argued by a Kenyan startup representative (I24). I15 (NGO) on the contrary highlights the role of awareness for decision making in the Kenyan digital ecosystem with regard to sustainable digital development arguing that recognition and awareness of multi-stakeholders views and especially awareness of issues are crucial.

In addition to the importance of awareness for sustainable digital development, awareness in this dissertation is further identified as a starting point for all multi-stakeholder interactions in the Kenyan digital ecosystem, since multi-stakeholders need to be aware of each other in order to interact with each other. However, awareness of each other is challenging, for instance due to the number (and type) of actors as discussed in the beginning of this subsection. I2 further shares the experience of actors in the Kenyan digital ecosystem not being aware of each other. This emphasizes the correlation between a lack of awareness of each other and the fragmentation or lack of interaction (as discussed below in the subsection on duplications). With regard to the multi-stakeholder groups, a lack of awareness is for instance shared in the interaction between a bilateral donor organization and multinational tech companies. Here the expert highlights that in their function as a representative of a bilateral donor organization they are not aware of multinational tech companies such as Facebook or Google and their activities in the Kenyan digital ecosystem (I4).

Additionally, the data analysis identifies a lack of awareness among actor groups themselves, for instance I13 (bilateral donor organization) shares the experience of bilateral donor organizations not being aware of each other in the Kenyan digital ecosystem:

> "I think when when I started we had three different donor coordination groups in ICTs that I I was aware of maybe there are more maybe there are still some out there that we are unaware of right, this is always a little bit confusing when suddenly there's a new actor on the stage and 'oh oh okay you also work in ICT, good'."

This is another instance of both aid fragmentation and the role of awareness hindering coordination for sustainable digital development.

Moreover, the digital transformation emphasizes the importance of awareness with regard to the usage of digital technologies. Expert representatives here highlight the example of awareness around using certain applications such as WhatsApp: On the one hand this is an example where a user lock-in (Sect. 4.4.4) is created since "[...] almost everybody knows what WhatsApp is but if you look at other platforms people are not aware." (I10) Hence, awareness (and usage) of alternative communication applications is one area that could prevent dependencies such as lock-ins (Sect. 4.4.4). Awareness around apps likewise has the component of data protection, which is one area where the level of awareness of multi-stakeholders varies greatly (see e.g., Sect. 4.1.2). More specifically, I3 for instance perceives a difference in awareness around data protection between the Kenyan government and EU governments. Similarly I6 argues that "the awareness and sensitization in terms of data protection and privacy rights is is way lower than in the in the North or in in Europe". This lack of awareness can possibly endanger sustainable digital development if concerns around data protection are not fully addressed. To address this, various awareness raising initiatives are highlighted by expert representatives such as of the AU (I3) and bilateral donor organizations (I2; I1), but also of the private sector, especially towards the Kenyan government (I10). Awareness raising measures here are also highlighted as a solution for overcoming a lack of awareness. Expanding on the impact of awareness on sustainable digital development, this further strengthens the argument that awareness is needed to pursue and reach an (estimated) outcome: "[...] why would people engage in and put resources into something when they are not aware of what will be the outcome?" (I12).

Conclusively, awareness throughout digital development efforts ranges from being a precondition for innovation to interactions and to the awareness around ICT. Hence, these dimensions of awareness need to be considered in digital (development) processes for achieving and fostering a sustainable digital development.

Trust

As an additional precondition, the role of trust in the digital ecosystem of Kenya is crucial for governance towards a sustainable digital development. While trust has already been established as a preconditional determinant in the theoretical background (Sect. 2.5.1), the empirical findings of the interview analysis assess the role of trust as follows. Most importantly for this dissertation's research question, I5 supports the importance of trust insofar as that coordination is dependent on trustworthy actors. Trust is further highlighted for the functioning of the Kenyan digital ecosystem, especially in terms of funding for startups (I23) or for sustainable (in the sense of long-term) investments from the private sector (I12). Hence, a lack of trust is a factor that impedes the sustainable digital development as it weakens the ecosystem's performance. In the context of the Kenyan digital ecosystem, trust can be further determined by political culture (I18) and more specifically trust towards the government (I2), but also trust in the multi-stakeholder relationships. Here trust in the relation between the government and the private sector is of specific importance (I2) also considering the discussed knowledge and power asymmetries (Sect. 4.3.5, Sect. 4.3.4) between these two that are impacting sustainable digital development. Digital technologies in settings of trust are additionally important. Using technologies is argued to achieve trust and to address trust issues (I19). Another solution to the trust issue is transparency (I2) as previously discussed in the precondition of transparency.

Expectations

Closely linked to the precondition of trust are expectations of multi-stakeholders (and their expectation management). More specifically, coordinating expectations is here emphasized as a precondition for sustainable digital development. As derived in Sect. 2.5.1, multi-stakeholders in the ecosystem generally have certain expectations similar to having differing interests. This is supported by experiences of I6 in the Kenyan digital ecosystem. Most importantly for this dissertation, one component are expectations concerning digital technologies, which are argued to be accompanied by notions of 'overexpectations' (Sect. 4.3.1). Similarly to the discussion in Sect. 4.3.1, misdirected expectations around digital technologies are likewise highlighted in the data analysis, for instance arguing that expectations of any digitalization program can be achieved without using AI (I18). Nonetheless, some expert interviewees admit to having flawed expectations, such as I5 highlighting the challenging experience of stakeholder management. More specifically, a gap between expectations and reality are discussed with regard to the success of ICT projects where "[...] there are cases in which in theory in theory you have that but during the project you realize that somehow the fit is not as good as expected" (I5). This likewise applies to (over-)expectations around the speed of results in digital development, as argued by I13: "So I think we can't expect this from happening from one day to the other, it will it will take some time in terms of data protection [...]".

This is why, in addition to the network governance perspective on expectations (Sect. 2.5.1), expectations have been further derived in the empirical data analysis as an important precondition for coordination in the Kenyan digital ecosystem. Specifically the need for expectation management, for instance in form of initial communication of expectations for achieving coordination is highlighted with regard to the success of MSPs (I3). This falls within the scope of expectation management that here serves the purpose of coordination. The need for expectation management for sustainable digital development in the Kenyan digital ecosystem is further emphasized by I12 and I19. Expectation management is also experienced as a challenge in the Kenyan digital ecosystem by I12:

> "[...] the challenge is to to really find the right to to have the right setup for the engagement and the right objectives and the right communication to everyone to be clear we are all on the same page to overcome also different institutional and cultural influences and have where people may understand things differently and then really manage expectations right, channel the dynamic right?"

Hence, if expectations are not addressed and managed, they can potentially contribute to unsustainable digital development, especially in terms of exclusion of actors (I13).

Frequency of Interactions

Frequency of interactions is another precondition for understanding the multi-stakeholder interactions in the Kenyan digital ecosystem (see Sect. 2.5.1). With regard to the Kenyan digital ecosystem, frequency is difficult to establish since the gathered data only provides a limited insight. Of the expert interviewees that did provide insights into their frequency of interaction with other multi-stakeholders, they stated differences: On the one hand, I21 describes the frequency of their organization's interactions with other actors rather vaguely as "some we are some were not". On the other hand, multi-stakeholder dialogues on a regular basis are identified by I5 referring to general multi-stakeholder dialogues. I12 for instance states regular internal meetings (here between EU member states and D4D Hub). I20 describes regular meetings between UN organizations and the Kenyan government and other multi-stakeholders. A frequency of annual meetings, for instance of multilateral organizations and the Kenyan government is shared by I19.

Moreover, the frequency of interactions can also be explained by the size of multi-stakeholder gatherings as meetings with more participants occur less frequently than the ones in smaller groups (I3). Other experts report interactions on a rather adhoc basis (I12) or on a short-term basis (I11). Hence frequency of interactions between multi-stakeholders as an determinant focussing on relational metrics has an impact on sustainable digital development. Frequency here is closely linked to the further relational determinant of duration of interaction, which will be discussed in the next subsection.

Duration of (Inter-)actions

Closely linked to frequencies of interactions in the Kenyan digital ecosystem are durations of interactions (Sect. 2.5.1) that serve as indicators for sustainable digital development. While short term projects can certainly be impactful, the focus here lies on sustainable digital development specifically aiming at long-term impact and longevity (of activities and also ICT projects). While regularity as a metric has already been subsumed under the above mentioned frequency, duration here is further determined by metrics of short-term and long-term bases.

Firstly, the duration of multi-stakeholder interaction in the Kenyan digital ecosystem is characterized by political rhythms. Usually the government in Kenya is elected every five years (Sect. 4.1.2), so that the actor group of the Kenyan government follows this cycle and determines the term of the elected public sector actors. Based on the interview data analysis, this political cycle has implications for the duration of multi-stakeholders with the Kenyan government. With regard to sustainable digital development, I19 expresses experiencing challenges of this political cycle, since the political context potentially changes after five years creating a volatile setting for sustainability (as opposed to stability as discussed for instance in Sect. 2.1): "[...] by the time I implement it, the next government comes in, the priorities shift and then the next the government comes in says 'well I don't think the priorities is in e-government we need to focus on water water resources management'." More specifically with regard to other multi-stakeholders, the duration of policy making processes and multi-stakeholder processes, is argued to exceed the resource scope of actors such as NGOs facing challenges in funding their work or involvement in years long processes (I13). Funding over a longer period of time that allows participation in the digital ecosystem and hence interaction is correlated to the duration and sustainability of digital development and may be a reason for the weaker position of NGOs (as discussed in Sect. 4.3.4 and will be addressed in more detail in Sect. 4.4.4).

Secondly, the duration of multi-stakeholder interactions in the Kenyan digital ecosystem depends on both the history of relations between stakeholders (especially between the Kenyan government and donors), and also the duration of collaborations in form of (ICT) projects or working groups (Sect. 4.4.6). The latter are argued to be mostly set up "over weeks and months" (I12). Similarly to the implications of political rhythms, this by its very nature often short-term based involvement of the Kenyan government in ICT projects, "[...] sometimes prevent projects from being implemented or carried out successfully" (I3). More specifically, in Sect. 4.3.2 the durability of ICT projects has already been addressed while noting that projects are mostly implemented on a short term basis. With regard to the actor group of donor organizations, the duration of interactions in the Kenyan digital ecosystem is likewise determined by their presence or engagement in the country (see precondition visibility above). While some donors have longstanding development relations with Kenya (such as the UK or US), the field of digital development is somewhat different since the donor involvement in digital development in Kenya is rather recent (see Sect. 4.3.5, Sect. 4.3.1). While one multilateral institution for instance has a longstanding relation to and presence in Kenya, their digital development involvement only started in 2019 (I17). Hence the duration of the involvement in digital development in Kenya is another angle determining sustainability of digital development.

Ease of Entering the Ecosystem: Exclusivity vs. Inclusivity

Considering this dissertation's focus on sustainable digital development, inclusivity is not only a part of the development goal (Sect. 2.1.3), but likewise serves as its indicator. Inclusivity here is operationalized as a further relational precondition derived in Sect. 2.5.1 that helps analyzing the degree of inclusiveness or exclusiveness of the ecosystem. More specifically, it aims at determining whether the Kenyan digital ecosystem is perceived as open or closed among expert interviewees. Hence this dissertation analyses the ease of entry in the Kenyan digital ecosystem using the continuum of exclusion and inclusion as elaborated in Sect. 2.5.1. The goal is to further identify factors hindering participation of actors in the ecosystem (beyond mere factors of the digital divide) by focusing on the interaction level and requirements of participation. In sum, the ease of entering the Kenyan digital ecosystem here is measured as 'exclusive' (closed ecosystem) vs. 'inclusive' (open ecosystem), which will be elaborated in more detail in the following.

Generally, determinants for entry into the ecosystem are identified in the data analysis with regard to digital skills, access to hardware, software or infrastructure (Sect. 4.2, Sect. 4.1.3) as well as to resources (Sect. 4.4.4). These are summarized by I9 as follows:

> "I think one thing that a case that we consider in digital transformation is exactly that, how do we get you know the populations that may not be or you know have access to a computer or they may not have gone to school or maybe they don't have a smartphone what solutions can you provide for them to get them to a point where they are more connected to the digital transformation".

Resources here are specifically highlighted as a precondition for entering the ecosystem (I10), especially financial resources or funding (I13), which will be further discussed in Sect. 4.4.4. When analyzing the ease of entering the Kenyan digital ecosystem, this reminds of the ease of doing business (measured by the *Ease of Doing Business Index*) as discussed in Sect. 4.1.1. Whether it is easy for startups to enter the Kenyan digital ecosystem is experienced differently among interviewed experts (including startup representatives). On the one hand, one expert representative of a Kenyan startup perceives it to be easy to start a business in Kenya and thereby to enter the ecosystem (I24). On the other hand, policies and legislations (also the lack thereof) pose barriers for digital development and thereby also for the development of the digital ecosystem. This is emphasized in I11 (hub) highlighting obstacles for investments: "[...] barriers to entry for investments which have been stifled due to again policies and the laws governing us through policymakers" (see further Sect. 4.1.2).

Further factors that aim at determining the Kenyan digital ecosystem as inclusive/open and exclusive/closed will be discussed in the following.

Degree of Openness: Inclusive Ecosystem Beginning with the analysis of factors determining the Kenyan digital ecosystem as inclusive, digital ecosystems in general and including the Kenyan one, are perceived as being inclusive by expert representatives. Moreover, the tech scene is described as very open (I7) and "very meritocratic" (I7). Additionally the "digital world" (I7) as well as "digital networks" are perceived as highly inclusive (I1).

Firstly, inclusiveness of the Kenyan digital ecosystem in the sense of ease of entry is determined by an openness to include other actors. With regard to the Kenyan digital ecosystem, this is described as an "[...] openness of the tech ecosystem in terms of the events they used to go on, they were all very open, people could attend" (I7), which is similarly confirmed by I10. Openness here is similarly confirmed by I22 sharing the experience that "[...] it's really easy to come in, depends on what area of the market one [unk] but in the overall I think the entry point is still open". Expanding the perspective of entering the ecosystem to external actors, the expert representative of a Kenyan startup likewise assesses an ease of entry for foreign actors in describing the Kenyan digital ecosystem as "one of the easiest" (I24).

Another example for the inclusivity and openness of the Kenyan digital ecosystem is seen in the ease of starting a business in Kenya for locals and foreigners (I24).

Secondly, another factor for explaining the openness and inclusivity of the Kenyan digital ecosystem is identified in the data analysis with regard to the role of open structures. Open structures here include governance structures such as hubs and also events, which are used to define the Kenyan digital ecosystem (I1; see also Sect. 4.4.6). Additionally, openness and inclusivity is associated with the increased interest and investment in the Kenyan digital ecosystem (I13). Similarly, openness is likewise associated with an openness specifically of Kenyans towards digital technologies and services. The open mindset towards digital technologies in Kenya is seen as an determinant for assessing the Kenyan digital ecosystem as inclusive as argued by experts I20, I24 and I6.

Degree of Exclusion: Exclusive Ecosystem In contrast to these aforementioned perceptions by expert representatives towards the Kenyan digital ecosystem as inclusive and open, factors of exclusion and exclusivity likewise emerged from the data analysis. Exclusive here is understood in terms of a closed ecosystem and barriers to entry. In the data analysis, degrees of exclusion emerged to the extent of silo creation, which is discussed in more detail in the respective precondition further below. Generally, the Kenyan digital ecosystem is perceived as rather exclusive by I6 and I2. Factors for this perceived exclusion will be elaborated as follows:

Firstly, I7 argues that structural barriers are factors contributing to hurdles of entry or notions of exclusion. More specifically, such 'structural barriers' are determined by notions of *elitism* which by definition demonstrates notions of exclusivity. I1 for instance argues that the Kenyan digital ecosystem is not inclusive for all multi-stakeholders, but for "[...] those who are already advantaged in the centers of commerce and tech". This is likewise supported by I7 arguing that actors who do not belong to the elite are excluded from participation in the digital ecosystem, especially with regard to digital entrepreneurship.

Secondly, and closely linked to the previous passage, resources emerged from this dissertation's data analysis as factors linked to exclusion, which will be discussed in more detail in Sect. 4.4.4. I15 assesses resources as hurdles for entering the ecosystem and factors of exclusion. Resources here are emphasized with regard to (digital) skills and financial resources. Beginning with digital skills, the correlation between financial resources and education is highlighted as one barrier for entry since "[...] you can't even get your foot in the door without a degree and it's really hard to get a degree without money" (I7). Moreover, (the lack of) digital skills is a further factor that determines notions of exclusion and barriers to entry which prevents actors, especially marginalized groups from entry (I13). This is reminiscent of the digital divide (Sect. 2.3.4). Financial resources are further determinants for exclusion as they are needed to participate in the digital transformation and hence the ecosystem as it "still boils down to who has the resources" (I10). According to I7 this especially applies to actors who cannot afford financial risk taking, such as young Kenyans that do not belong to the before mentioned elite, which creates a "[...] sort of self-excluding situation." Financial resources are likewise required for using hardware or digital infrastructure (I20), thereby further determining exclusion. In sum, the barriers of entry into the Kenyan digital ecosystem and participation in the digital transformation is closely linked to the digital divide (Sect. 2.3.4) as an example for exclusion of marginalized groups (I9).

Thirdly, access to informal networks is a further determinant for exclusive notions in the Kenyan digital ecosystem. Informal networks are perceived as a further form of resource, i.e., the resource of contacts that enables entry into the digital ecosystem (I10). The importance of contacts as enablers into the ecosystem are likewise highlighted by I15. Similarly, I13 perceives informal networks as rather unofficial channels, also in terms of access to decision makers. The discussion whether informal networks correlates with corruption expands the focus on exclusion in the Kenyan digital ecosystem and is instead discussed in more detail in Sect. 4.1.2. More importantly for this dissertation's analysis of an exclusive Kenyan digital ecosystem, one expert representative of a Kenyan startup for instance shares the experiences that it is difficult to enter the ecosystem if actors do not have a network of contacts in the ecosystem that can introduce them (I23). Hence, a lack of access to contacts in the ecosystem is assessed as one factor impeding participation and inclusion (I7).

These factors lead to a setting where multi-stakeholders are excluded from the Kenyan digital ecosystem (in addition e.g., to the digital divide). Here, differences between included and excluded actor groups emerged from the data analysis. For instance, the experience of exclusion is shared by the expert representative of a Kenyan NGO in rural Kenya, sharing the experiences of feeling excluded from the ecosystem as "we are so far off the epicenter of everything" (I1). In this regard, exclusion here correlates with the determinant of degree of centrality, which will be discussed further below. In addition to the experience shared by I1, NGOs are also assessed as "[...] quite a ex excluded most of the time" (I3) and "[...] civil society is not included enough" (I13). These instances are especially highlighted in the context of policy making processes (e.g., I15), emphasizing the relevance of power asymmetries as factors contributing to unsustainable digital development as discussed in Sect. 4.4.3 and Sect. 4.3.4. In the policy making processes in the Kenyan digital ecosystems, one expert representative likewise mentions Kenyan startups as excluded from support and recognition (by the Kenyan government):

"[...] the government almost pushed it out and and gave so many advantages to the foreign private sector to come in so that that was one thing that that was quite exclusive" (I2). This however needs further research since this is the only interview data here available for startups in the discussion around exclusion at this point.

In addition to the actor group of NGOs, further actor groups experiencing exclusivity instead of inclusivity, are seen in hubs. This applies specifically to hubs that only have a specific target group, which is emphasized as contributing to exclusion (I6). More specifically, this mismatch in terms of target groups is experienced with regard to "those traditional or economic MSMEs as well as tech startups what about young enterprises coming from the manufacturing sector or coming from the agricultural sector with innovative ideas with with innovative potential they don't find support services at innovation hubs" (I6).

Additionally, donor organizations are criticized as contributing to exclusion in the Kenyan digital ecosystem due to a Western (representation) bias in ICT donor coordination groups (I13) and their expectations (I13). At this point of analysis this is not discussed in more detail, since it is closely linked to findings in Sect. 4.1.5 and Sect. 4.3.2.

Degree of Closeness between Multi-Stakeholders

For governing the Kenyan digital ecosystem and its multi-stakeholder activities sustainably, the developed framework further argues that an assessment of the degree of closeness (and distance) between the actors is crucial. Closeness here means strong ties and relations between actors, whereas distance refers to weak ties between actors in line with Sect. 2.5.1. As discussed in the theoretical framework (Sect. 2.5.1), closeness can also be explained through other factors such as similarity between actors. Another factor that emerged from the data analysis that correlates with the operationalized degree of closeness can be seen in reference to 'voice' as discussed in Sect. 4.4.3, where close ties are observed between actors that are recognized. Closeness has also been discussed in Sect. 4.3.4 focusing on the role of private sector actors and especially their relation to the Kenyan government in policy making processes. The closest relation between multi-stakeholders is seen in a symbiotic relationship that is likewise characterized by interdependence (Sect. 2.5.1), which can likewise be transferred to the Kenyan digital ecosystem: "it's mostly a kind of symbiotic relationship we need each other" (I23). Hence, the degree of closeness can be further distinguished between very strong ties such as alignment, streamlining, harmonization and joint programs, which will be discussed in the following. The degree of distance is to be understood as opposite to the degree of closeness and here specifically entails contracting, duplicative working and silos, which will be discussed afterwards.

Alignment, Streamlining and Harmonization Firstly, the data analysis identified the experts expressed need for *streamlining* and *harmonization* as forms of coordination efforts in the Kenyan digital ecosystem. Alignment, streamlining and harmonization represent a close relationship between actors. The importance of these three is especially highlighted with regard to donor development activities, for instance: "[...] there is a tremendous need of coordination, streamlining and harmonization of donor activities and development activities hundert percent" (I6). This partly roots from the discussed aid fragmentation as discussed in Sect. 4.1.5. I6 further highlights this especially with regard to sustainable digital development that is characterized by a further need for sustainable solutions in terms coordination among multi-stakeholders. The need is further seen in the possibility of alignment in terms of orchestrating each actor's capabilities and competencies in the digital transformation (I5), meaning using each other's synergies. Further taking an ecosystem perspective on this potential for alignment, I5 highlights the potential for alignment in the ecosystem. Using these potentials of alignment for a sustainable digital development in the Kenyan ecosystem is further expressed in the actors' goals of alignment: One goal is seen in an alignment between regions in the digital transformation (I2) as part of a global sustainable development. Another is seen in future alignment of multi-stakeholder actions for achieving sustainable digital development (I13). For this, coordination plays an essential role since it would not only allow to streamline actors' activities, but also to simultaneously avoid duplication, especially with regard to donor activities (and fragmentation) (I6).

One specific area where alignment of multi-stakeholders' activities is already experienced, is highlighted by expert interviewees with regard to alignment on the level of agendas. On a regional level, one example of such alignments of agendas is the EU–AU digital single market (I2). On a country level, one approach is the alignment between digital agendas of the EU and the Kenyan government (I17). More specifically, the alignment of digital agendas between EU actors and the Kenyan government is emphasized with regard to the 'Kenyan Digital Economy Blueprint' (I6, see Sect. 4.1.2). Here, the experts refer to an *overlap* between the strategies of these two actors (overlap here not meant in terms of duplication but an alignment of agendas (I17).

Synthesizing these arguments, alignment of actors in their activities and goals is a crucial determinant for sustainable digital development embodying coordinative efforts since it likewise serves as an indicator for the success (or failure) of ICT projects (Sect. 4.3.2). Regarding alignment on a project level, one expert shares the experience that projects with an alignment of actors' agendas are more likely to be implemented (I3). What this demonstrates however is that alignment of agendas of these diverse multi-stakeholders is not without challenges (e.g., due to differing interests Sect. 4.3.3).

Secondly, alignment of agendas in the interactions poses challenges to coordination and governance (I14). Factors that are impeding an alignment are for instance actors' interests and objectives (I11, Sect. 4.3.3) and different understandings (Sect. 4.4.4). The latter is specifically highlighted with regard to different understandings of what harmonization entails, who is harmonizing and how (I20). This is further expressed in sensitivities and carefulness around using the term 'harmonization' (I20, see Sect. 4.4.4).

Thirdly, the question of how coordination through alignment can be steered is raised in the interview analysis. The previous passage already addressed upon the role of standards for harmonization and alignment (e.g., I20) and will be further discussed in Sect. 4.4.6). More importantly, the role of transparency for enabling alignment (e.g., I4) is further highlighted (see precondition transparency). The data analysis of this dissertation further emphasizes the role of governance structures to streamline and align. On the one hand, expert representative I5 emphasizes the need for a governance structure to align multi-stakeholder activities in the Kenyan digital ecosystem. On the other hand, experts likewise highlight the role of governance formats for alignment which is ascribed to hubs (I11). Additionally, a digital transformation center as part of a donor ICT project is ascribed the role of streamlining and coordinating (European) donor activities (I6). Further roles of streamlining and harmonizing is discussed in Sect. 4.4.5.

In conclusion, the degree of closeness between actors in coordination is expressed within *alignment*, *harmonization*, *streamlining* and *overlap* of multi-stakeholder activities in the Kenyan digital ecosystem. The ultimate organizational closeness is an institutionalized or formalized partnership, which will be discussed in separate sections due to their importance for this research project (Sect. 4.4.6, Sect. 4.4.6).

Joint Programs Another indicator for degree of closeness between actors are joint programs. These joint programs are formed in order counteract fragmentation, whereby coordinative efforts are made by multi-stakeholders to collaborate jointly. For instance, bilateral donor organizations have joint acceleration programs (Sect. 4.3.2) with private companies such as in the Gesellschaft für Internationale Zusammenarbeit (GIZ)'s program 'Make-IT in Africa' (I4). Moreover, the interviewed experts highlight joint programs between bilateral donors, especially between European donors as well as between bilateral donors and the EU (I17). Another joint program is mentioned between private tech companies and multilateral organizations (I20) and the EU (I12). Nonetheless, these are to be understood as only a selection of joint program examples without claiming completeness.

Summary: Degree of Closeness Synthesizing these findings on the degrees of closeness between multi-stakeholders in the Kenyan digital ecosystem, the following close ties between multi-stakeholders can be summarized:

Beginning with the stakeholder group of *startups*, according to expert interviewees there are close ties between founders, investors and elite (I7) as well as with academia "[...] because the lot of ideas are generated by also organization close to the university" (I5, bilateral donor organization). An assessment of the ITU however contradicts this by stating that in Kenya the entrepreneur community is not linked with the academia and that academic institutions are not successfully creating spinoffs (ITU 2019a:53). Startups are further perceived as close to the (potential) needs of the users (I5). As demonstrated in this chapter's section on competition (Sect. 4.3.4), there are close ties between (multinational) private tech companies and startups with regard to funding and strategic purposes. One indicator for close relations of startups is for instance observed in their participation in acceleration programs of various multi-stakeholders (Sect. 4.3.2). With regard to ties between donor organizations and startups, the data of this dissertation is limited due to the fact that none of the interview startups were in contact with donors as they did not participate in donor funded acceleration programs.

With regard to the next actor group, bilateral and multilateral *donor organizations*, a closeness between European bilateral donors and the EU institutions are mentioned by expert representatives (I6). Close ties are further identified between multilateral donor organizations such as the UN and the EU (I14). Bilateral donor organizations are likewise in close interaction with the Kenyan government, for instance with the Kenyan Ministry of ICT as one main interlocutor (I17). Although bilateral donors further cooperate with private sector actors, e.g., in joint programs, their relation in terms of closeness is for instance noted as "[...] there's a raised interest to include them" I3, which is discussed in more detail in Sect. 4.4.4. Due to the rapid evolvement of the digital transformation in Kenya, I17 expects that the collaborations with the private sector will further increase in the future. The exchange between bilateral, multilateral donor organizations and NGOs in Kenya is perceived as close, but not as much as with private sector actors (due to financial resources of NGO, I3). With regard to the actor group of NGOs, they are perceived as closer to "[...] the needs on the ground" (I5). However, the closeness between donors and NGOs (as well as between the Kenyan government and NGOs) is rather distant, which is analyzed in more detail in the sections on voice and power asymmetries (Sect. 4.4.3).

Expanding the perspective of close ties between multi-stakeholders and the private sector, the private tech sector has rather close ties with Kenyan startups (I24). Similar to donor organizations, private sector actors are perceived as having close ties with the government, which is for instance confirmed by I15. From these statements the conclusion can be drawn that the Kenyan government has close ties with the actor groups of donor organizations and private sector organizations. This will be further discussed in Sect. 4.3.4 since these close relationships are an essential component of analyzing (un-)sustainable digital development in this research project.

Degree of Distance between Multi-Stakeholders

In contrast to the before mentioned degree of closeness, the data analysis of this dissertation further demonstrates empirical insights into degrees of distance in the interactions between multi-stakeholders in the Kenyan digital ecosystem. More specifically, this is determined by relations of *contracting*, *parallel working*, *doing 'own thing'* and *silos*, structured by an increasing degree of distance (i.e., contracting as less distant to silos as most distant). These determinants are derived from Sect. 2.5.1 and are complemented by an analysis of the gathered expert interview data in the Kenyan digital ecosystem. The identified degrees of distance can be further complemented by other previously elaborated preconditions such as the level of awareness (i.e., ultimate distance is assessed if actors are not even aware of each other, see Sect. 4.4.3). This dissertation argues that these degrees of distance pose challenges to collaboration and sustainable development as they impede coordination of activities. Hence this subsection not only characterizes the multi-stakeholder interactions in the Kenyan digital ecosystem, but also emphasizes the need for coordinating these degrees of distance for a sustainable digital development.

Contracting Firstly, the degree of distance is determined by a relation between multi-stakeholders in the form of contracting. Contracting here is assessed as less distant in comparison to silos or a total lack of interaction (as elaborated in the following passages). While the degree of distance derived in the theoretical background (Sect. 2.5.1) does not specifically address relations of contracting between multi-stakeholders, contracting is analyzed in the context of PPPs. Contracting emerged as an individual theme in the data analysis, describing the multi-stakeholder relations in the Kenyan digital ecosystem (beyond formats such as PPP).

More specifically, contracting especially refers to the private (tech) sector where contracts are a formalized mode of interaction that can be extend to partnerships based on contracts (I10). Collaboration through contracts designed as client or business relations is also highlighted by expert interviewees as a mode for interaction between Kenyan tech startups and private tech companies. Here contracts are used for strategic purposes such as entry to private company networks (I24). Adding to this perspective of a Kenyan startup, an expert representative of a multinational tech company further shares the experience of contractual interaction between the company and its customers (including the Kenyan government), which they regard as a longer-term partnership to offer continuing support to the organization (I10). Two expert representatives of bilateral donor organizations deliver further insights into their perceptions of the interaction between the Kenyan government and the private tech sector in terms of contractual relations: I5 and I3 assess that the relationship between the Kenyan government and the private tech sector is more of a contractual nature than an actual partnership. Hence this is a counterexample to the perceived closeness between the Kenyan government and private sector discussed in the precondition 'voice' (Sect. 4.4.3). Donor organizations themselves further interact with the private tech sector (e.g., as customers and partners). Their interaction with the private sector is intended as a partnership due to the 'leave no-one behind' goal as opposed to contracts with their foundations or philanthropy departments (I12). While these relations follow a market logic, it also creates dependencies and potentially lock-ins for customers (as discussed in Sect. 4.4.4). This is supported by another potential danger for sustainable digital development that is experienced by a representative of a multilateral donor organization reporting instances of contracts between governments and private sector companies that failed (I20).

Synthesizing these arguments, the contractual settings of multi-stakeholder interactions classify as one degree of a rather distant relation, but can be assessed as a (short-term) partnership form, which will be discussed further in Sect. 4.4.6.

Own Thing and Parallel Working Secondly, and more importantly for the perspective of sustainable digital development, some expert representatives shared the experience of a lack of collaboration between multi-stakeholders in the Kenyan digital ecosystem, where they instead are "doing their own thing" contributing to parallel and duplicative activities (Sect. 2.5.2). This fragmentation of action and lack of coordination (similar to Sect. 2.1.5, Sect. 2.1.6, Sect. 4.1.5) is for instance emphasized by expert I21 in the space of digital skills where various Kenyan ministries, multinational tech companies and donor organizations are all engaged, "but it is just not really a cohesive". This illustrates that various multi-stakeholders interact in a specific space, but with less alignment or cohesion in action. This dissertation argues that is one instance where coordination can contribute to a sustainable digital development. In sum, parallel working of multi-stakeholders in the ecosystem correlates with a *lack of alignment* (opposed to Sect. 4.4.3).

Notions of 'doing their own thing' in the Kenyan digital ecosystem are further highlighted by the following expert statement: "I can't say that as a network we are or as an ecosystem we we are in the same place. I think everybody is doing their own stuff you know [...]" (I10). These can be interpreted as instances of actors pursuing their own interests (also in the context of hyping digital topics, see Sect. 4.3.3, Sect. 4.3.1) that likewise leads to multi-stakeholders not necessarily interacting or communicating with each other (I12). Here, another explanatory factor identified in the data analysis can be seen in the discussed notions of *competition* between multi-stakeholders in the digital sector in Kenya that here may serve as motivational factors (Sect. 4.3.4). Consequently, the strongest indication for distance mentioned by expert representatives is emphasized in a *lack of collaboration*, that can be explained due to competitive motivations. Specific examples mentioned here are for instance Microsoft and Google, whose collaboration in the Kenyan digital ecosystem might not be close due to individual interests and competition (I9). This is similarly confirmed with regard to the relation between other multinational tech companies (I10). Against this background, actors' interests likewise explain notions of exclusion in the Kenyan digital ecosystem (Sect. 4.4.3).

Moreover, a lack of communication leads to parallel working and *duplication* of activities that is experienced in the Kenyan digital ecosystem by experts such as I12. This further emphasizes an area where coordination of (development cooperation) activities can contribute to sustainable digital development. The discussed aid fragmentation instead of actor alignment further leads to doubling of donor activities and possible unsustainable ICT project outcomes (I4, see further Sect. 4.1.5, Sect. 4.3.2). More concretely, this argument is expanded by one expert representative highlighting the need to further align or streamline development cooperation activities with priorities of the Kenyan government in order to avoid duplication of donor activities (I6). This dissertation argues that one way to streamline these multi-stakeholder activities through coordination efforts are governance forms, which will be discussed in more detail in Sect. 4.4.6 and Sect. 4.5.

Besides the discussed duplication of donor activities, duplication and parallel working also in the public sector in Kenya highlight a *lack of intergovernmental coordination* and is identified as an impeding factor for sustainable digital development in the data analysis. Digital transformation contributes to the issue of intergovernmental coordination insofar as it is a cross-cutting development that affects and is handled by various ministries (I13), which can cause overlaps or parallel working. While I19 highlights intergovernmental coordination as a key factor for sustainable development in the digital transformation of Kenya, I21 delivers further insights into the (unsustainable) implications of this lack of coordination:

> "[...] you got policy coming from ICT ministry but then you're having implemented a implemented things that are that are not that that are not in line with the policy and that comes from a you know from the ministry of finance who says we're going put twenty percent broadband excise duty on on all on on all broadband services which doesn't

really add up to the policy you you're trying to increase your digital development because you you're putting the cost out of people's reach."

This further emphasizes the counter-productive impact of parallel working by Kenyan ministries on sustainable digital development highlighting that parallel working of multi-stakeholders in the Kenyan digital ecosystem is a consequence of a lack of coordination and can hence be addressed through coordination. Moreover, a lack of intergovernmental coordination in the Kenyan digital ecosystem likewise has implications for other multi-stakeholders interacting with the Kenyan government due to the cross-cutting responsibilities over various ministries that lead to a fragmented interaction (I17).

Synthesizing these arguments, parallel working of multi-stakeholders leads to duplication of activities, which poses a challenge for sustainable development and highlights the importance of coordination.

Silos Thirdly, the ultimate degree of distance between multi-stakeholders is identified in the form of silos (Sect. 2.5.1). In the Kenyan digital ecosystem, silos of multi-stakeholders are experienced by expert representatives such as I11, I2 and I20. More concretely, I4 delivers further insights into silos in the ecosystem in reference to competitive interests: "[...] what many ecosystems are coming from including Kenya is very much a silo thinking quite a competitive thinking rather than you know acknowledging that it's the ecosystem that builds an entrepreneur". Silos in the Kenyan digital ecosystem are further explained with regard to subgroups within the ecosystem that are also labeled as "bubbles" (I5) and a result of uncoordinated interaction exclusively within certain groups (I3). This emphasizes this dissertation's argument around coordination of multi-stakeholder activities for a sustainable digital development.

Expanding on the silos *across* multi-stakeholder groups on the one hand, rather distant relations are observed between donor organizations (especially bilateral ones) and multinational tech companies, which is for instance highlighted by I10 and discussed in Sect. 4.3.4. Another distance in collaboration is expressed between (the interviewed) startups and NGOs: "NGOs NGOs very few like I'm not so much engaged with them" (I24). Thereby silos emphasize a lack of interaction between certain actor groups in the ecosystem. On the other hand however, silos are further experienced even *within* each actor group due to a lack of interaction, which is perceived both in the actor groups of the public and the private sector: "I've seen them in the public and in the private sector that none of them would actually collaborate and and put together they were all pushing themselves into the silos." (I2). Given the discussed challenge of intergovernmental coordination (as discussed in previous subsection), siloed approaches of Kenyan governmental institutions are likewise observed by interviewed experts and are assessed as very costly for the country (I2; I20). Using Fig. 2.7 as a further illustration, each of the multi-stakeholder groups in the Kenyan digital ecosystem that is identified by expert interviewees as operating in silos is active in their own 'bubble' without interacting with other silos (e.g., donor organizations in silo 1, multinational tech companies in silo 2, etc.).

In conclusion, the degree of distance is a further precondition that needs to be assessed in multi-stakeholder network governance for sustainable digital development.

Degree of Centralization

A final relational factor characterizing the network of the Kenyan digital ecosystem is the degree of centralization (Sect. 2.5.1), which is closely linked to the previously discussed degree of exclusion (Sect. 4.4.3). The data analysis resulted in the activities in the Kenyan digital ecosystem of Kenya being quite centralized. Centralized here refers to geographical centralization in Nairobi and centralized in the sense of exclusivity due to having siloed activities (Sect. 4.4.3). Those multi-stakeholders that are not present in Nairobi share experiencing notions of exclusion. More specifically, the centralization of the Kenyan digital ecosystem is shared by expert interviewees I13, I10 and also I1 with the latter arguing "[...] that it is very centristic or very much centered all in the tech community in Nairobi". A NGO based in rural Kenya (I1) perceives being "[...] quite far from those epicenters of power and of also of business accumulation of Nairobi so we don't really get much of that", thereby expressing notions of exclusion. This is similarly confirmed for digital development activities on the startup level since "[...] there is a lot of startup engagements that happen in Nairobi" (I10). I23 (Kenyan startup) similarly shares their experience of having to move from their home county to Nairobi due to their inability to locally attract investment. Therefore, one explaining factor for the (physical) centralization of the Kenyan digital ecosystem on Nairobi is the high concentration of interest and action "[...] from also donor organizations or other partners also from from the private sector, IT companies, tech companies" (I3) and their physical presence mostly in the capital Nairobi (see further precondition visibility above). Moreover, this centralization of the digital ecosystem on Nairobi can be further explained with regard to the highlighted contextual factor of Nairobi as the center of most economic and political development (I22), simply as "that's where everything is" (I22, see further Sect. 4.1).

In sum, this centralization of activities in the Kenyan digital ecosystem on Nairobi correlates with fragmentation (see Sect. 4.1.5), that is likewise highlighted by I13: "Because I was saying it's fragmented, there's so many actors we were talking about it so many actors also focused on Nairobi, so as well I mean we we're struggling to to try to [unk] find connections to to other actors beyond the capital." This is a further example demonstrating difficulties in finding interactions and also alludes to difficulties to enter and participate in the Kenyan digital ecosystem (Sect. 4.4.3). Hence, this centralization on Nairobi poses a challenge to inclusive and sustainable digital development as it not only emphasizes the digital divide (urban-rural, Sect. 2.3.4), but also the need for coordination for sustainable digital development. A governance solution addressing this challenge is decentralization, which is long happening in Kenya (I22), but has likewise impact on the digital ecosystem. In the conducted interviews, a decentralized governance approach is mentioned interchangeably with decentralized coordination which can be facilitated by ICTs (e.g., I7; I13). This finding, in line with further findings from Sect. 2.5.4, has led to this dissertation's proposal of a technical coordinative solution in form of a coordinative digital platform (Sect. 4.5). These results are similar to and support the self-governance approach to governing the Kenyan digital ecosystem for sustainable development, which will be discussed in Sect. 4.4.6.

Network Rationale: Sustainable Digital Development
Sustainable digital development is the argued network rationale of the Kenyan digital ecosystem in this dissertation. In the theoretical framework, the network rationale is hence understood both as a general goal (Sect. 2.1.3) and also as a precondition for network governance (Sect. 2.5.1). The data analysis has identified three main dimensions of sustainable digital development in the Kenyan digital ecosystem: *sustainability*, *fairness and equality* and *inclusion*. Moreover, the network rationale is different from the analysis of multi-stakeholders' interests as discussed in Sect. 2.3.3 and Sect. 4.3.3 as the network rationale here is understood as an ultimate goal that is above individual actors' interests. As phrased by I9, the network rationale in the Kenyan digital ecosystem is more directed "[...] to simultaneously do something for community while achieving their [actors'] own internal goals". This is supported by I14 arguing that sustainability is ensured through need-based digital technologies (i.e., human centered digitalization), especially of European actors (I17), as opposed to solutions "geared towards the masses" (ibid.). This stands in contrast to the discussion around hype and actors' interests (Sect. 4.3.1; Sect. 4.3.3). Hence, sustainability here not only entails a technological component (Sect. 4.2), but also an actor component that is crucial for this dissertation. Here "the goal is not to digitize everything" (I18), but the other way round (i.e., asking: this is the specific problem, which technology can help best?). Hence the network rationale aims at need-based solutions for a sustainable digital development that cover all three dimensions of sustainability (equality and fairness as well as inclusion). This

example further emphasizes the correlation between these three dimensions culminating in the goal of a *holistic* sustainable digital development. More specifically, I6 highlights this argument by drawing on the ecosystem's rationale of "[...] more holistic solutions, more sustainable solutions and allows probably long term impact and really address the identified obstacles and challenges."

Ultimately, in order to achieve a sustainable development of the digital ecosystem, it is preconditional that multi-stakeholders have at least some degree of interest to foster sustainable digital development (Sect. 4.3.3, Sect. 4.4.3). With regard to the Kenyan digital ecosystem, I5 asserts that " [...] more inclusive and more sustainable way of using digitalization for development" is not only a goal but also a trend. For further clarification of sustainable digital development as the network's goal, the network rationale here is similar to the SDGs (Sect. 2.1.3). The SDGs on the one hand address sustainable development of countries, including digital development, for instance with regard to addressing aid fragmentation (I2). This global goal or global consensus building on the other hand influences multi-stakeholders' activities in the Kenyan digital ecosystem, especially with regard to development cooperation. One expert representative of a multilateral organization shares that their work is guided by the SDGs, with digital transformation as an enabler for implementing those (I20). This is similarly confirmed with regard to other donor organization's activities that are (should) be aligned along the SDGs (I2).

Hence, *sustainability* is a crucial dimension of sustainable development as a network rationale. Sustainability here is not analyzed with regard to environmental impact, although this association is observed as one area where the common understanding of multi-stakeholders (and interviewed experts) differs (Sect. 4.4.4). Sustainability is a further example where a lack of common language and understanding among multi-stakeholders poses a challenge to sustainable digital development. In line with the discussion in the theoretical background (Sect. 2.1.3), sustainability is here understood with regard to the *durability* of actions (long lasting perspective on development and transformation), which is similarly influenced by the duration of interactions between multi-stakeholders (Sect. 4.4.3, Sect. 4.3.2). This is further confirmed by the empirical data as for instance argued by I12, for whom sustainable digital development entails multi-stakeholders' collaborations and building "something on the long term", similarly to I5 stating "it's also about long lasting impact".

A second dimension of sustainable digital development as the ecosystem rationale includes *fairness and equality*. Equality is not only part of sustainable development (e.g., SDGs, especially SDG 10, Sect. 2.1.3), but especially relevant with regard to digital development. For this dissertation's analysis of the Kenyan digital ecosystem, equality is essential as demonstrated by the following expert statement: "You can have a thriving ecosystem here but I think there's a lot about equity and social inclusion that needs to happen in a society is a very unequal society [...]" (I17). Hence the development of a thriving ecosystem can only be sustained if a focus on equality and social inclusion is provided.[8] On the one hand, this includes the argument that all multi-stakeholders in the Kenyan digital ecosystem should be treated *equally*, which is challenged as argued by this dissertation for instance by power and knowledge asymmetries (Sect. 4.3.4, Sect. 4.3.5, Sect. 4.4.3). Moreover, equality as part of sustainable digital development is closely linked to fairness, which is a further network rationale in the context of sustainable digital development, especially with regard to "[...] fair values where everybody understands entirely what is at stake, what is what are the costs, what are the benefits and everybody gets their equal share of both ends." (I2) Hence, fairness also aims at including 'weaker' actors, weaker for instance in terms of financial resources (I15, Sect. 4.4.4) and voice (Sect. 4.4.3).

On the other hand, equality and fairness further imply the *inclusion* of all multi-stakeholders in the Kenyan digital ecosystem as a third dimension of sustainable digital development. I6 shares the observation that the goal of inclusion is palpable in the Kenyan digital ecosystem as the "intention is there to have an inclusive one [...]" (a gap between intention and reality however is discussed in Sect. 4.4.3). Inclusion here has two components: one, the role of inclusion in multi-stakeholders' activities and two, inclusion of all multi-stakeholders in the ecosystem (i.e., an inclusive ecosystem, especially in political decision making processes). The first dimension is especially reflected with regard to *inclusion in multi-stakeholder activities*, and also inclusion between Kenya and other countries (regional digital divide; see e.g., I2 and in Sect. 2.3.4). For I5, sustainable digital development is synonymous with inclusion. Inclusion in the sustainability context for instance entails a focus on the Kenyan (rural) youth (I19; I19), women (e.g., I4; I13), people with disabilities (I13), rural areas (I13) and language (I4). If these are not included, this is an instance where the digital transformation is an obstacle to development (e.g., I4) and to entry into the Kenyan digital ecosystem (Sect. 4.4.3).

The second dimension of inclusion in the context of the network rationale is the importance of *inclusion of multi-stakeholders* in the ecosystem, especially in decision making processes. Expert interviewees here highlight the rationale to involve all multi-stakeholders in (and for building) the Kenyan digital ecosystem (I12), where goals "can only be achieved with a concerted effort" (I13), Moreover, inclusion can be achieved by operating holistically for instance in considering challenges and opportunities (I12). The 'concerted effort' of multi-stakeholders here refers to the stakeholder groups of civil society organizations, academia, private tech sector

[8] Especially relevant for this research project is the fair design of digital policies recognizing inclusion, equity and fairness, especially of marginalized groups (e.g. I1, and see Sect. 4.1.2).

organizations including startups (I15), but also donors and the Kenyan government as well as further associations. The inclusion of private sector actors in the multi-stakeholder approach of the Kenyan digital ecosystem is crucial, although this is discussed critically among expert representatives in terms of power asymmetries and compatibility of interests (Sect. 4.3.4, Sect. 4.3.3). Nonetheless, inclusion of private sector actors for sustainable digital development is highlighted by I3 while I17 acknowledges the need "[...] to keep commercial actors to restack certain principles of inclusion" without neglecting private sector interests in long-term investments (see further Sect. 4.3.3). Another multi-stakeholder group that experts identify as crucial to be included in the decision-making in the Kenyan digital ecosystem are (end-) users. While users need to be included in both dimensions (in multi-stakeholders actions and in the ecosystem decision making processes), the interview data here suggests the role of users is crucial from an actor perspective. I20 and I3 highlight the need to include users in the ecosystem decision-making processes.

How this network rationale is achieved (from the perspective of an outcome) will be analyzed in this dissertation following a governance perspective arguing that it can be achieved through coordination, which is discussed in more detail in Sect. 4.4.5, Sect. 4.4.6 and ultimately in the proposed outcome Sect. 4.5.

Ecosystem Efficiency and Effectiveness

Efficiency and effectiveness of multi-stakeholder activities in the Kenyan digital ecosystem are not only essential for achieving sustainable digital development (Sect. 4.4.3, Sect. 2.1.3) and influential for the performance of ICT projects (Sect. 4.3.2), but can also be achieved through both digital technologies and coordination. Thereby this section synthesizes some of the previous findings on preconditions and context, especially efficiency from a development cooperation perspective (see e.g., Sect. 4.1.6, Sect. 2.1.7, Sect. 2.1.6). Thereby findings of this subsection are similarly derived as in the theoretical background (Sect. 2.5.1), from a technical perspective, a development perspective and also a governance perspective. Ecosystem efficiency and effectiveness are often used interchangeably.

Beginning with the technical perspective regarding the potential of digital technologies and ICTs as drivers for development (Sect. 4.2), the interviewed experts highlight the potential of ICTs for increasing efficiency and effectiveness in the Kenyan digital ecosystem (and in general). The general potential of ICTs to enable more efficiency through digital technologies is highlighted by expert interviewees (e.g., I16; I2). More specifically, increased efficiency and effectiveness through digital technologies and services is emphasized with regard to processes, for instance by I5 and I18. The latter expert interviewee gives a further example of the role of ICTs in making tax collection and prevention of fraud processes more efficient, e.g., through e-tax systems (I18). Digital technologies are even discussed by experts in the context of using digital technologies for increased or more efficient aid deliveries and allocations (I2), which underlines this dissertation's proposal of a coordinative digital platform as one solution (Sect. 4.5). In sum, these arguments underline the discussion around the role of ICTs as an enabler, in this context an enabler for increased efficiency and effectiveness of processes (e.g., I18), which will be further discussed in more detail in Sect. 4.4.5. The proclaimed potential of ICTs for increasing efficiency and effectiveness is likewise mirrored in their need, meaning to address observed inefficiencies in the Kenyan digital ecosystem, especially inefficiencies in development cooperation and in ICT4D projects (see e.g., Sect. 4.1.6, Sect. 2.1.7, Sect. 2.1.6, Sect. 4.3).

Firstly, aid effectiveness has already been highlighted as a challenge to sustainable digital development that needs to be further analyzed from an ecosystem effectiveness and efficiency perspective (see e.g., I10 and Sect. 2.1.7). Challenges that emerge from inefficiencies and ineffectiveness of digital development activities are highlighted for instance with regard to duplications of activities (Sect. 4.4.3) and an unsustainable design or implementation of ICT projects. As discussed in Sect. 4.3.2, 'quick fixes' in projects are assessed as inefficient from a sustainable digital development perspective since their value is generated on a short term basis (I10). Moreover, taking a macro perspective on the Kenyan digital ecosystem, inefficiencies of multi-stakeholder interactions in working groups (Sect. 4.4.6) are further highlighted by I7 citing experiences of "wasted time" and "things just don't get move forward [...] in these multi stakeholder groups". While the importance of inclusion of all multi-stakeholders in the Kenyan digital ecosystem from a sustainability perspective has been elaborated, the downside of the multi-stakeholder process is highlighted as often inefficient.

Secondly, this directly leads over to digital technologies and services coordinating in the context of increasing inefficiencies and ineffectivenesses. Potentials of ICTs for efficiency and effectiveness purposes in the ecosystem are further highlighted with I1 arguing that the digital transformation might not lead to more coordination but better information flows. I15 further confirms that coordination would enhance ecosystem effectiveness and efficiency similarity to I5. How coordination in the Kenyan digital ecosystem for increased effectiveness and efficiency can be achieved will be discussed in more detail in the following determinants (Sect. 4.4.5, Sect. 4.4.5 and Sect. 4.5).

Summarizing these findings, preconditions can likewise function as specific challenges in the Kenyan digital ecosystem from the synthesis of development cooperation, digital transformation and governance challenges for achieving sustainable digital development. This will be discussed in more detail in the following section (Sect. 4.4.4).

4.4.4 Determinant 2: Challenges for Governance and Coordination of the Kenyan Digital Ecosystem

As a second determinant, following the structure of the proposed framework (Sect. 2.5.2), this subsection will analyze challenges for coordination through governance of the Kenyan digital ecosystem incorporating the macro, meso and micro level. Similarly to Sect. 4.3, challenges are derived from the synthesis of development theory and cooperation, digital transformation and governance. Building on the previous determinant 1 (Sect. 4.4.3), if preconditions are not considered or fulfilled in the Kenyan digital ecosystem, then they pose challenges to coordination for sustainable digital development. In contrast to the theoretical background, which derives common language and understanding as preconditions (Sect. 2.5.1), the empirical data analysis identified this as a major challenge (see also beginning of Sect. 4.4.3) and hence subsumed under determinant 2. Similarly to Sect. 2.5.2, specific digital development challenges from the empirical data insights are (analogue) bureaucracy and traditional structures, digital complexity as well as digital literacy and skills, (financial) resources, common language and understanding, dependencies (e.g., on protocols, on private sector) as well as interdependencies and lock-ins. These will be discussed in the following.

(Analogue) Bureaucracy and Traditional Structures

Beginning with the first challenge and in correspondence to the theoretical section (Sect. 2.5.2), this dissertation emphasizes that bureaucracy in (mostly public) institutions as well as a focus on analogue instruments both impede flexibility and agility required to respond to digital development challenges. In sum, these three components pose a challenge to coordination in the Kenyan digital ecosystem for achieving sustainable digital development. While Sect. 2.5.2 already highlighted theoretical implications, this section will elaborate on these three factors using insights from the interviewed expert representatives.

Analogue Firstly, implications of analogue instruments, processes and mindsets of multi-stakeholders are identified as challenges in the data analysis. In contrast to the following passage on bureaucracy, a focus on analogue processes in both public and private sector groups is highlighted. More specifically, a preference "to do a lot of things manually" (I10) is experienced with regard to both companies and public sector in Kenya. Here a dualism is experienced between 'traditional' businesses and 'newer' or younger businesses, especially tech businesses. On the one hand, the tech scene in Kenya is described as "[...] more open than a lot of the traditional existing businesses" (I7). 'Open' here does not refer to inclusion (Sect. 4.4.3), but to open mindsets around (digital) innovations (as discussed in precondition ease of entry, Sect. 4.4.3). More specifically, reasons behind this are seen in the fact that in certain sectors, analogue physical innovations can be far more valuable to the user (business) than digital innovations (e.g., I21). Similar experiences with regard to 'traditional' sectors or businesses being influenced by the digital transformation in Kenya are shared with regard to investments of 'traditional' Kenyan investors that rather invest in physical objects such as real estate than into tech startups (I21). Consequences of the discussed opportunities of using digital technologies and services also for Kenyan businesses (Sect. 4.2, Sect. 4.1.1), are also observed with I11 arguing that "it's so much harder to do business traditionally as it was done".

Reasons or rationales behind this preference of using analogue techniques instead of digital ones are further identified in the data analysis: ('traditional') mindsets of decision makers, cultural factors and incentives of corruption. Analogue activities in the Kenyan digital ecosystem are especially highlighted with regard to public sector actors working "paper based" (I10; I9). More importantly, a lack of willingness and even resistance to adopt digital technologies or services is experienced by an expert representative of a multinational tech company when working with public sector actors in Kenya (I10, I11). Not only is a correlation to the challenge of multi-stakeholders' interests noted (Sect. 4.3.3), but also the precondition of willingness (Sect. 4.4.3). Reasons behind this resistance on behalf of public sector actors can for instance be explained by a lack of adequate digital equipment (and its usage) in public sector organizations (I10), Hence a lack of capabilities both in terms of digital infrastructure (hardware, software etc.) and of digital skills and knowhow is seen as one reason behind the public sector working analogous. This is similarly confirmed by I9 sharing the experience of public sector actors that "[...] don't really think about how to integrate technology to make work easier so so really at a government level we are quite behind and I think it now affects other areas"—here again maybe due to limited digital capacities and capabilities. These arguments are only based on a brief sample that are only used as an example with missing governmental interview data and will be further discussed in Sect. 4.4.4.

Hence a preference of decision makers to do things analogous is associated both with a lack of capability, interest and also with the discussed perception of the digital transformation as a hype (Sect. 4.3.1). The lack of digital skills and capability, especially of public sector actors, will be discussed further in Sect. 4.4.4.

Combining these arguments with the associated underlying interests as one factor explaining the preference of analogue instead of digital techniques, a further reason here is seen in the possibility of corruption, which are emphasized by I10 and I21 since analogous or manual work is argued to offer ways for conducting activities less transparently than digital ways (Sect. 2.1.2, Sect. 4.1.2)

Expanding on these assessed notions of 'traditional' mindsets and a perceived preference of analogue activities, experts I10 and I6 further highlight a correlation with the role of (political) culture in Kenya. This is not only associated with public sector actors, but with multi-stakeholder across the board including users. Taking end users as an example representing the Kenyan society, I10 indicates Kenya is a "collective culture" where "we still want to do some things traditionally instead of digitizing" (I10). One example here is mentioned regarding online grocery shopping where "we want that interaction instead of buying online [...] it's that cultural aspect of talking to somebody." (I10) This argument stands in contrast to the discussion in Sect. 4.4.3, where an open mindset and willingness towards (digital) innovations was ascribed to the Kenyan society. Since the data analysis of this dissertation provides only limited insight into the cultural angle on the digital transformation in Kenya and since this exceeds the scope of its research goals, this is an area for further research.

Bureaucracy Closely linked to the findings on analogue processes of some multi-stakeholders in the Kenyan ecosystem, this dissertation likewise argues that bureaucracies or bureaucratic activities pose challenges to coordination and sustainable digital development. On the one hand, bureaucratic processes are associated with the public sector in Kenya, especially the government that is described as "a big administrative bureaucratic monster" (I2). More specifically bureaucracy here is attributed with "very slow and and small steps" (I2). Similarly, governmental institutions are described as "mammoths by themselves they're very bureaucratic" (I19) and also associated with slow process of decision and policy making (e.g., I21; see also Sect. 4.4.3). Slow bureaucratic processes however are not only highlighted as a barrier to sustainable digital development and impeding coordination, but also a barrier for other public sector actor groups such as donor organizations. One multilateral donor organization, the UN, is mentioned as an example where bureaucratic activities pose a challenge to sustainable development (e.g., I2). Thereby inner-organizational bureaucratic processes contribute to ecosystem inefficiencies and ineffectiveness in the sense of unsustainable digital development as previously discussed in Sect. 4.4.3. This is also manifested in a lack of flexibility and agility of bureaucracies, which will be discussed in the following.

Flexibility Synthesizing analogous approaches and bureaucracy, these lead to a third factor that hinder sustainable digital development and also pose challenges to coordination of activities: a lack of flexibility and also lack of agility of multi-stakeholders. Inflexibility and lack of agility of multi-stakeholders in the Kenyan digital ecosystem are for instance experienced by expert interviewees with regard to the Kenyan government. More specifically, it is noted that the digital transformation poses a challenge for regulation and legislation by the government (see also Sect. 4.1.2) due to a level of speed that is not aligned with the speedy development of digital technologies and services:"[...] digitalization or innovation just moves way quicker than law does. [S]o that always creates this kind of impression that you lagging behind and trying to regulate something which is already out of date." (I13) Besides legislation processes by the actor group of the Kenyan government, other (public sector) actors and their bureaucratic nature such as the UN are described as being "[...] risk averse for one it's also slow, it's also bloated and bureaucratic" (I7). According to this expert, this is an argument against the UN in the role of a coordinator in the Kenyan digital ecosystem (Sect. 4.4.5). Nonetheless, these insights emphasize that fast (and efficient) responses to the digital transformation are needed simultaneously (e.g., I20). The data analysis shows that more agile approaches are needed for governing digital development activities in the Kenyan digital ecosystem.

In sum, these three factors of analogue processes, bureaucracy and flexibility simultaneously correlate with the complexity of the digital transformation and a lack of capacity and skills, which will be elaborated in the following subsection.

Digital Complexity

As emphasized in the previous subsection and in the theoretical background (Sect. 2.5.2, Sect. 2.2), the digital transformation and emerging digital technologies are highly complex. This is further emphasized in the expert interviews. In line with this dissertation, complexity is not only perceived on a technological level, but also on an actor level. I5 for instance associates challenges in the digital transformation with the emergence of new actors that add to the complexity (similarly argued in Sect. 4.4.3 and Sect. 2.5.1). To further emphasize the role of digital complexity in (sustainable) digital development, it is important to note that not only digital transformation is perceived as complex, but also ICT projects themselves are complex. Consequently, this makes engagements in digital development increasingly complex as for instance shared by one expert representative of a bilateral donor organization (I5). Besides the shared complexity challenges for donor organizations, this is likewise expressed with regard to beneficiaries i.e., partner governments such as Kenya (I5) and for the private tech sector. With regard to the latter, complexity here is for instance seen in adapting digital technologies to local contexts

(I24, see also Sect. 4.1.4). In sum, digital complexity requires digital literacy and skills as well as a digitally skilled workforce (I5), but also governance in form of regulation and legislation. The digital complexity is thus identified as a barrier for all multi-stakeholders in the ecosystem. It highlights the need not only for collaboration and cooperation among them for using each others' strengths and knowledge (Sect. 4.3.5), but also for coordination (e.g., given the number and type of actors, their capabilities, etc.).

Shifting to the technological side of digital complexities, the data analysis identified one specific area of challenge, especially for governments: *cybersecurity*. Firstly, the interview analysis provided further empirical insights into cybersecurity as highly complex that requires a great amount of knowledge and skills (I2). Secondly, cybersecurity affects not only the national level of countries such as Kenya, but also the global level as it challenges almost every country (I2; I22) and every sector (I22). However, the vulnerability due to cybersecurity issues is greater for countries such as Kenya where the legal infrastructure, especially legislation and regulation around digital transformation, is not fully implemented as discussed in Sect. 4.1.2. This is further emphasized by I22 and I2, with the latter arguing that cybersecurity goes hand in hand with digital transformation and "[...] it can result in the totally wrong decision to go with digital means if you have fundamental flaws in your cybersecurity setup it can destroy pretty much everything and it can also trigger long lasting damage." Hence this is an area where governance is needed to also secure political stability that in turn serves stability of the ecosystem (see e.g., Sect. 2.1.2).

Thirdly, this further highlights the need for digital policies (see e.g., Sect. 4.1.2). Here, I10 emphasizes the need for a separate organizational body in the Kenyan government for areas around data protection and cybersecurity. Whether this is necessary is debatable and discussed in more detail in Sect. 4.5. One specific governance approach for coordinating cybersecurity efforts that is already used are thematic committees (I15), which are similarly discussed in Sect. 4.4.6. In sum, governing digital complexity is hence an important area for achieving and securing sustainable digital development, especially for the Kenyan government. For governing such complex digital areas such as cybersecurity, respective knowledge and skills are required, which this dissertation demonstrates as being further challenges that will be analyzed with regard to the Kenyan digital ecosystem in the following.

Digital Literacy First of all, digital literacy is a requirement for acquiring and executing digital skills. This is underlined by I19 and I5, with the latter emphasizing "the people need to be digitally literate". Hence, digital literacy is a precondition for digital development, similar to "hard" factors such as hardware, software or electricity, as it enables access to digital technologies and thereby digital development (I14). According to I19, digital literacy is assessed as even more important in Kenya in contrast to the Global North (see also Sect. 4.1.4): "I wanna say I think even in Africa it's needed more than even in the West because we are dealing with the largely digitally illiterate pop population." Hence, the level of literacy is an important factor for participation in digital development and in the digital ecosystem that is identified as an area that needs governance. More specifically, this for instance applies to the education systems and the digital equipment of schools as emphasized in Sect. 4.1.4 and Sect. 4.1.3. I19 further highlights the importance of digital literacy in Kenya, but also in Africa: "two hundred sixty million children go to schools in Africa and only s ten percent of these have access to a computer. So what it means it means that once this children finish school they don't have any literacy skills [...]". This is one example for the importance of building digital literacy capacities at early stages that have implications for future digital development. Moreover, these findings suggests a correlation between general levels of literacy and digital literacy, which cannot be fully verified nor falsified in this dissertation, which leaves room for further research.

Digital Skills As discussed above, digital literacy is one determinant and precondition for digital skills. Before analyzing how and where digital skills pose a challenge not only to coordination in the Kenyan digital ecosystem, but also to sustainable digital development, digital skills need to be further defined. As elaborated in the theoretical background of this dissertation (e.g., Sect. 2.5.2, Sect. 2.3.4), different degrees of digital skills can be distinguished (basic and advanced skills). The data analysis of this dissertation highlights the correlation of digital literacy ('basic') and advanced digital skills: "[...] certain number of people having skills to consume ICT services whether that be basic digital literacy to use apps and UBER and this and that whether it be sort of more advanced medium level skills to use word excel and and and databases and things like this." (I21) Hence digital skills are also closely linked to knowledge in the digital transformation as discussed in Sect. 4.3.5.

Firstly, the importance of digital skills in the Kenyan digital ecosystem is emphasized among expert interviewees in that enhancing digital skills is likewise framed as a goal. The need for more digitally skilled people, especially employees across sectors, including the 'traditional' economy (I20), is highlighted for instance by I9 and I5. Similarly to different degrees of digital skills, I10 here highlights gradations in which levels of digital skills are needed: "Yes we do not want everybody to sort of become an ICT specialist [...]". This again highlights the focus and the goal of digital skills ICT projects of multi-stakeholders (Sect. 4.3.2). Similarly, the interlinkage between digital infrastructure and digital skills is emphasized by expert interviewees. I12 for instance argues that enhancement of both infrastructure and skills should happen simultaneously for a sustainable digital development, which will be discussed further in the following.

Secondly, while statistical data provided one indication of the state of digital skills levels in Kenya (Sect. 4.1.4), expert representatives here have different perceptions. According to the *Global Gender Gap Report 2021*, Science, Technology, Engineering and Mathematics (STEM) attainment in Kenya among females is at 11.19 percent and among males at 20.84 percent. Moreover, ICT attainment is only distributed among 3.50 percent female and 6.71 percent male, thereby indicating quite low levels of digital skills (World Economic Forum 2021:240). With regard to the interview analysis, I10 on the one hand assesses Kenya as "[...] one of those countries with a lot of a lot of skill set especially in the tech space." On the other hand, weak digital skill levels are attested, for instance by I6 sharing the perception that "[...] at the moment still quite low digitization level of local companies and another obstacle is is limited digital entrepreneurial skills". I21 similarly highlights digital skills distribution in Kenya as only a smaller group of people have skills to program and create software in comparison to countries such as the US. Digital skill distribution on a national level in Kenya is moreover associated with the digital divide (Sect. 2.3.4). More specifically, digital skills gaps between rural and urban areas are highlighted as a further challenge (I9). Moreover, this emphasizes digital skills as a further precondition for the participation in the digital ecosystem (Sect. 4.4.3), in that a lack thereof prevents entry, especially with regard to marginalized groups and the digital divides (I13, see Sect. 4.3.6). Hence experts highlight the specific challenge of a lack of digitally skilled labor force for a sustainable digital development. The attested low digital skills level likewise has implications for the performance of the ecosystem as well as its governance with regard to multi-stakeholder actor groups and their roles (Sect. 4.4.5).

A third challenge in the context of digital skills is the mentioned interrelation with (digital) infrastructure. Not only are these two highlighted as most important barriers to sustainable digital development that need to be (further) considered by multi-stakeholders (e.g., I19), but their interlinkage is also emphasized (e.g., I14). This is similarly confirmed by I1 further arguing that "[...] there's not enough done towards spreading these two elements to marginalized communities". More specifically, the interrelation between digital skills and infrastructure is discussed among expert interviewees with regard to Kenyan schools. According to I20, schools in Kenya are not equipped for distributing, enhancing and upskilling in terms of digital skills (e.g., due to lack of connectivity, electricity, hardware but also digitally skilled teachers, see also Sect. 4.1.3).

This stands in contrast to the assessment of the educational system as "quite good in Kenya" with "very very well renowned universities". However, the experience of going abroad for training (I20) reminds of the discussion around brain-drain (Sect. 4.3.4). While the innovative and entrepreneurial spirit of Kenya has been highlighted in terms of leapfrogging potential (Sect. 4.1.1, Sect. 4.2.1), one expert interviewee further raises the issue of formal educational (ICT) degrees in general, but also in regards to digital skills as one area that poses a challenge to sustainable digital development. Especially the requirements of foreign multinational companies are one example where certified digital skills are required (I7, see Sect. 4.4.3). This is similarly reflected in the STEM and ICT attainment rates as discussed in the previous passage and in Sect. 4.1.4.

In sum, expert interviewees emphasize the importance of digital skills in the Kenyan digital ecosystem. For a sustainable digital development, further upskilling across sectors and institutions is needed to also compete on an international level, see e.g., I1. While this is not a new finding, the ways in which this can be achieved are proposed in this dissertation. Here, governance plays a crucial role, especially with regard to governance of multi-stakeholders, as the digital skills gap can for instance be overcome through closer collaboration among multi-stakeholders in the ecosystem (I6). Thereby issues of dependencies on the knowledgeable and digitally skilled actors as well as power asymmetries (as further discussed in Sect. 4.3.5 Sect. 4.3.4) can be minimized. For this purpose, this dissertation offers not only the potential of achieving this through governance roles of actors in the ecosystem (Sect. 4.4.5), but also through governance formats (Sect. 4.4.6) and especially in the context of organizations or digital platform as a solution (Sect. 4.5). More importantly, the synthesis of the passages on digital literacy and skills further culminate in digital competencies and capacities of multi-stakeholders, which will be discussed in more detail in the following subsection.

Digital Capacity (of Public Sector Actors)

Expanding on the previously discussed importance of digital literacy and digital skills in the context of digital complexities (Sect. 4.4.4), this section analyzes the capacities of public sector actors in the Kenyan digital ecosystem: the Kenyan government and donor organizations. The capacities and capabilities of the remaining multi-stakeholder groups have been previously discussed in Sect. 4.3.3, Sect. 4.3.4 and Sect. 4.3.5. In this section, digital capacities are understood especially regarding political capacities in line with the theoretical background analysis in Sect. 2.1.2. Here, the functioning of governmental institutions was determined by state authority, state capacity and state legitimacy (Grävingholt et al. 2015:1286). Hence, digital capacities here are understood as both capabilities of political actors, but also in terms of their competencies. In the conducted expert interviews, the term 'competencies' was used interchangeably with 'know-how', so that political competencies in this dissertation are subsumed under legitimacy (similar to Sect. 2.5.2). In this, digital capacities, skills and literacy are both preconditions as well as obstacles (if not achieved) with regard to coordination for a sustainable digital development in Kenya. This dissertation argues that digital competencies not only refer to digital literacy and skills, but also to general understandings around digital technologies. However, these understandings have been identified as a major challenge where multi-stakeholders fail (Sect. 2.5.1). This will be discussed with regard to the Kenyan digital ecosystem in more detail in Sect. 4.4.4. Similarly, the issue of digital capacities in terms of understanding is further linked to the discussion around hype (Sect. 2.2.3, Sect. 4.3.1), especially in the context of AI capabilities, which has been identified as correlating to the failures of digital skills ICT projects (Sect. 4.3.2).

Synthesizing these findings further highlights the importance of both the Kenyan government as the political decision maker as well as of donor organizations with the mandate to enhance the capacities of the former, which will be discussed as follows.

Digital Capacities of the Kenyan Government Beginning with the digital capacities of the Kenyan government, regulation in the digital transformation is highlighted by I13 as a "question that applies to all governments worldwide". Generally, but also with reference to the Kenyan case, the public sector is perceived as less experienced with regard to digital transformation issues (I17). This is for instance explained by limited capacities of governmental staff as argued by I4, which is also discussed as one reason behind the government's perceived passivity in Sect. 4.4.3. Limited digital capacities of the governmental decision makers is likewise emphasized by I21. A shortage of digital skills and capacities among governmental staff has been elaborated with regard to competition around employees with the private sector and issue of brain-drain (Sect. 4.3.4), but is also associated with the bureaucratic nature of governmental institutions (Sect. 4.4.4).

Further explanations behind limited digital capacities of Kenyan governmental institutions are seen in limited interest or willingness to use digital technologies and services, which is different from having the capacity to do so (e.g., I10 and further in Sect. 4.4.3). Moreover, constrained financial capacities and resources are identified as a further reason, since they are closely related to the development context of the lower-middle income country (Sect. 4.1). As one specific example, financial capacities and associated costs for training employees in enhancing their digital skills are highlighted by I19, especially with regard to 'advanced' technologies such as AI. The role of financial resources as a challenge for coordination (here coordination by the Kenyan government) and also for sustainable digital development as an outcome will be discussed in more detail in Sect. 4.4.4.

Synthesizing these findings, limited digital capacities of Kenyan governmental institutions and their employees likewise hinder the government in enforcing laws that aim at regulating the digital transformation. This is for instance emphasized by I19 sharing the experience of authorities lacking knowledge around issues such as cybersecurity. This is similarly supported by I6 (bilateral donor): "[...] the main issue I see is the lack of capacity so we have a very good data protection act but as I said in the beginning limited capacities for the implementation [...]". In sum, this leads to questions around the Kenyan government's digital capacities with regard to their governance roles, which will be discussed in more detail in Sect. 4.4.5. Moreover, this illustrates donors' engagements in enhancing these capabilities and capacities through capacity building measures. However, this is discussed critically among expert interviewees, as emphasized in the following.

Digital Capacities of Donor Organizations Moving the analysis of digital capacities from the Kenyan government to donor organizations, the role of donor organizations in supporting the Kenyan government will be discussed in the context of (donors') digital capacities and capacity building measures. Firstly, the digital development activities of donor organizations are aimed at providing capacity building on higher political levels (I3); I6). Further donor engagements and specific ICT areas are discussed in Sect. 4.3.2.

Secondly, however, digital capacities of the donor organizations themselves are identified as challenging in the data analysis. Ironically, expert interviewees of both bilateral and multilateral donor organizations state that in digital development activities (in Kenya) they 'start from scratch' (I18), digital development activities are in their "early stages" (I18) and "[...] this is completely new" (I17). More importantly the expert interviewee further highlights a lack of experience, digital capacity and qualified personnel among their donor organization (I17).

Similar experiences are shared by a representative of both a multilateral and bilateral donor organization sharing the experience of a lack of capable experts on partner sides in ICT projects (I12). Other interviewees did not specifically address their own digital capacities so it cannot be fully clarified or falsified that every actor is 'learning by doing' in the digital transformation given the cross-sectoral nature (see Sect. 2.1.3, Sect. 2.2.1, Sect. 2.3.1, Sect. 4.1.1).

Besides a lack of (finding and building) workforce for digital capacities, underfunding of (multilateral) organizations is mentioned as one reason behind this (I17). Although these are only a few expert insights of donor representatives, these findings lead to a question of their digital capacities and thereby their digital development efforts: If donors themselves struggle with digital capacities, what effect does this have on the performance and especially on the sustainability and suitability of their digital capacity building measures? Do cases of their limited experience and capacities strengthen or weaken the Kenyan government's digital capacities? Phrased differently: Are these development efforts leading to 'capacity building' or rather 'capacity draining'? This will be elaborated in the following passage.

(Digital) Capacities of the Kenyan Government Overstretched by Donor Organizations Another instance where capacities, especially capacities in the digital transformation, are overstretched by donors is the number of donors and aid fragmentation leading to increased transaction costs for the Kenyan government (Sect. 4.1.5, Sect. 4.4.3). Empirically, this is for instance highlighted by I13. As a further example of this development cooperation setting, the Ministry of ICT has to manage all donor relations with their different requirements and regulations which binds the ministry's capacities (I13). This barrier to sustainable digital development and also further implications for the Kenyan government's role in the digital ecosystem will be discussed in more detail in Sect. 4.4.4 and Sect. 4.4.5.

Similarly, and inherent in the development cooperation relationship between Kenya and donor organizations, Kenya is a recipient and "[..] ultimately they are at the receiving end" (I2) with "hardly any capacity to do their actual work this is my longterm experience" (I12). Hence they are dependent on donors which demonstrates a further imbalance (Sect. 4.3.4, Sect. 4.4.4).

Synthesizing these arguments cumulates in the experience shared by expert interviewees of donor organizations overstretching capacities of the Kenyan digital ecosystem, including governmental institutions such as the Ministry of ICT. These findings further underline this dissertation's argument that coordination, here donor coordination, is needed for sustainable digital development. In some (donor) cases, tendencies are already found for pooling resources, knowledge and experience as done on an European donor level (I18). These findings likewise underline the importance of coordination as argued by this dissertation and also highlighted in Sect. 4.1.6.

Financing, Funding, Financial Resources and Costs

As a further barrier to sustainable digital development and coordination, this dissertation emphasizes the role of resources, especially financial resources that are associated with costs as proposed in the corresponding theoretical background (Sect. 2.5.2). Firstly the 'general' importance of resources for sustainable digital development and for coordination in the Kenyan digital ecosystem will be discussed before focusing specifically on financial resources and funding. The analysis is complemented by a discussion around the role of costs and affordability in the Kenyan digital transformation context.

(Non-financial) Resources Beginning with an analysis of resources in the Kenyan digital ecosystem, resources have been derived in the theoretical background as essential for development of countries in general, but specifically in the context of the digital transformation (see e.g., Sect. 2.3.6).

Firstly, resources in the digital transformation are manifold. On the one hand, the digital transformation is assessed as a resource itself (I12), especially in terms of leapfrogging (Sect. 2.2.2, Sect. 4.2.1). Similarly, digital transformation as a potential for leapfrogging and as a resource is seen with regard to the tech sector being "a low asset industry" (I7). This is consistent with the perception of I24 highlighting that the digital industry is not as resource intensive as for instance the construction industry. Whether the digital transformation, on the other hand, contributes to a more equal distribution of resources in Kenya is however discussed controversially among expert interviewees. While the digital transformation contributes to resources being distributed more broadly as "resources are rather getting broader than narrow" (I13), I10 emphasizes that resources are (still) not being equally distributed. Sustainable development here is dependent on a time "till people who have who do not have access to resources" (I10), which corresponds to discussed notions of imbalances (Sect. 4.3.4, Sect. 4.3.5) and digital divide (Sect. 2.3.4).

Secondly, this dissertation has highlighted that resources are a precondition for entering the Kenyan digital ecosystem (Sect. 4.4.3). Resources are highlighted not only as crucial for entering the ecosystem (e.g., I10), participation in the ecosystem (e.g., I13) and potentially posing barriers to entry if resources are lacking as "[...] it's it's really it's about resources" (I15). Resources in the cited expert interviews here refer to both hardware (see further Sect. 4.2.2) and infrastructure (e.g. Sect. 4.4.3, Sect. 4.1.3) as well as knowledge and digital skills (Sect. 4.3.5, Sect. 4.4.4). Moreover, entry into the Kenyan digital ecosystem is highlighted as enabled through having networks with I15 assessing networks (in terms of contact with present actors) as a resource. In addition to requiring resources for entering the ecosystem, a relating question refers to requirements for continuing participation in the ecosystem. The data analysis states that gaps also prevail in having enough resources (both financial, but also in terms of manpower) for continued participation, especially in multi-stakeholder interactions and processes that are discussed as resource draining (I13, see Sect. 4.4.3) This alludes to the role of financial resources and distribution among actor groups in the Kenyan digital ecosystem, which will be discussed in more detail in the following paragraph.

Thirdly, and similarly, those who have resources can enter the ecosystem, which simultaneously is an indicator for dominance since the ones who have resources are more dominant or powerful than the others, as emphasized in Sect. 4.3.4. Initially this implies a resource gap, which has been highlighted in the beginning of this paragraph as well as by I14, arguing that there is a resource gap between foreign and local tech companies as well as between private sector and public sector actors as discussed in Sect. 4.3.4. Generally, the data analysis highlights that actors who control resources are the dominant ones (I14), are the ones who exert power ("armstrong" I11) or "who calls the shots" (I11) as well as the ones who "sometimes might not align truly" (ibid., see further Sect. 4.4.3). Besides dominance through power and knowledge (Sect. 4.3.4, Sect. 4.3.5) in the digital transformation context, financial resources are highlighted as essential determinants for dominance (I15). Given their importance, financial resources will be discussed separately in the next paragraph.

Synthesizing these three arguments, this hence underlines the value of coordination of resources for sustainable digital development, as highlighted for instance in Sect. 4.1.6 and Sect. 4.4.4 that could contribute to sustainable digital development in addressing resource imbalances.

Financial Resources, Financing and Funding Focussing more specifically on financial resources, these were also referenced in the data analysis in the context of 'financing' or 'funding', further including funding possibilities and availabilities. More specifically, this paragraph analyzes which of the actor groups are assessed by interviewed experts as having or lacking financial resources in the Kenyan digital ecosystem.

Firstly, expert interviewees highlight that there is a lot of investment in the Kenyan digital ecosystem (I7) and also a lot of interest to invest there (I20), especially in terms of venture capital that is for instance acquired through foreign founders (I7). Nonetheless "access to markets, access to finance" (I13), especially for Kenyan tech startups, is assessed as a factor hindering digital development (similarly discussed in Sect. 4.4.3).

Secondly, the role of foreign founders in the context of financial resources in the Kenyan digital ecosystem is emphasized by various expert interviewees. While foreign founders have already been highlighted as bringing external knowledge into the Kenyan digital ecosystem and having more experience (e.g., I23, Sect. 4.3.5), they are also associated with financial strength (I21) that enables Kenyan startups' access to funding (e.g., through co-foundership or access to networks, I24). Moreover, it is claimed by I10 that "most of the successful startups in the country are actually somehow foreign funded". Hence foreign investments, especially of foreign founders, can be classified as having a strong position regarding financial resources in the Kenyan digital ecosystem.

Thirdly, and on the contrary, Kenyan (tech) startups are highlighted among experts as one actor group with limited financial resources. Access to financial resources in the Kenyan digital ecosystem is on the one hand assessed as not easy, for instance by a Kenyan startup (I23), especially for tech startups not based in Nairobi, i.e., rural startups (I23), as similarly discussed in Sect. 4.4.3. This struggle for access to financial resources and funding is further confirmed by I10 and I14. For progress in digital development and innovation, funding is crucial (e.g., I13, see also Sect. 2.3.5).

With regard to sources of funding (besides foreign founders), funding sources are highlighted with regard to private sector actors and donor organizations. One interviewed representative of a Kenyan startup reports their individual experience of receiving private funding instead of donor organizations' funding (I24). This is closely linked to the discussion around (donor) funded ICT projects focussing on digital entrepreneurship, for instance with regard to flawed accelerator programs (e.g., I24, as alluded to in Sect. 4.3.2). Moreover, experts share the possibility for startups to miss out on funding if they do not comply with rules, protocols and requirements of both public and private actors (e.g., I4), which will be discussed in more detail in Sect. 4.4.4. Additionally, and closely linked to Sect. 4.3.1, one representative of a Kenyan startup shared the experience that certain topics pursued by startups are easier to get funding for than others (not specified whether funded by private or public sector): "[...] it depends on the domain you gain you gain social social enterprise and then [unk] trying to solve a social problem with digital technology then it's much easier to get get get funding" (I23).

This, fourthly, directly leads over to the financial capacities of the remaining actor groups in the Kenyan digital ecosystem. With regard to donor organizations, the data analysis only provided limited insights. Although I19 notes financial flows from donor organizations, also in reference to financial aid provision to Kenya (I11), their financial resource strength is nonetheless limited given their nature of a public sector agency (e.g., see discussion in Sect. 4.4.4). Similarly, data on the financial resource provision of NGOs are limited, but have been alluded to for instance in Sect. 4.4.3 with the conclusion of limited financial capabilities. The Kenyan government is highlighted as enabling financial investments in the digital transformation context for instance through foreign funding by means of PPPs or MSPs (Sect. 4.4.6) or other partnerships with development partners, especially in ICT infrastructure projects. In this context, foreign funding is constituted as the largest percentage in terms of technology (King'ori Ndegwa et al. 2016:3). Considering the contextual factors surrounding the Kenyan government (Sect. 4.1) and their capacities (Sect. 4.4.4), leads to the hypothetical assessment of the Kenyan government as having rather limited financial resources. This dissertation argues that a lack of financial resources impedes multi-stakeholders in their governance roles (Sect. 4.4.5) and is a further indicator for power asymmetries and dependencies (Sect. 4.3.4, Sect. 4.4.4).

Costs and Affordability As a third element of financial resources, costs and affordability of digital technologies and services are further identified by this dissertation as crucial for achieving sustainable digital development and explaining implications for governance in the Kenyan digital ecosystem. In the interview data analysis, affordability of digital technologies and services including costs for license fees limiting financial resources of actors such as the Kenyan government is highlighted as a core theme. While implications of affordability in the context of sustainable digital development have been previously discussed with regard to access to digital infrastructure, hardware and software (Sect. 4.1.3, Sect. 4.2), here the implications of license fee costs that are payed for using the software of multinational tech companies is further discussed among expert interviewees.

Generally, I22 shares the perception that these multinational software licenses offered by Oracle or Microsoft among others are too expensive. More concretely, the cost of license fees is further shared by I19 with regard to small Kenyan businesses: "[...] you cannot afford a hundred dollar if you are a small business to pay for license fees." More importantly for this dissertation's research focus on governance, the costs of license fees for digital technologies and services are further emphasized with regard to financial resources of the Kenyan government. This is further confirmed by I19 with the expert arguing that this is likewise a result of policies that are "skewed towards private sector" (as further discussed in Sect. 4.3.4). This statement demonstrates that having these software license can create lock-ins for users such as the Kenyan government, which will be discussed in more detail in Sect. 4.4.4.

One solution in the context of financial resources and dependencies are open source software offerings (as discussed in Sect. 2.2.4 and Sect. 4.2.2). I22 here shares the experience of "[...] there's always a struggle to see whether you really need to have this licenses or [unk] go open source". More specifically, I19 (multilateral donor) here shares the experience of the Kenyan government being (unsuccessfully) consulted to adopt open source software as "I think the government cannot afford this huge bill to pay license, why don't we adopt open licenses [...]." Given the limited data on the impact of license fee costs for digital services and also on open source as a means to reduce financial costs, no clear result is observed here as to whether or not open source is a solution.

Common Language and Understanding

As a next challenge in the context of coordination for sustainable digital development, common language and understanding of multi-stakeholders in the Kenyan digital ecosystem are identified by this dissertation. In contrast to the theoretical background where common language and understanding are derived as preconditions for governance (Sect. 2.5.1), the data analysis identified it as a challenge rather than a precondition as the data suggests a lack thereof impedes sustainable governance. While the theoretical background combines perceptions (common language and understanding) with awareness, the derived challenge of common language and understanding is analyzed individually as a challenge. Awareness is discussed as a precondition similarly to the theoretical background (Sect. 2.5.1, and also in Sect. 4.4.3). Common language and understanding is also closely related to the roles of harmonization and alignment as discussed in Sect. 4.4.3.

The data analysis of the conducted expert interviews in the Kenyan digital ecosystem highlights common language and understanding among multi-stakeholders in the ecosystem as a challenge that questions the sustainability of ICT projects and measures. Moreover, it emphasizes a lack of knowledge of digital technologies also in the sense of hype (Sect. 4.3.1), and highlights self-interests of stakeholders as further reasons behind a lack of common understanding and language of multi-stakeholders. Common language and understanding can likewise allude to vocal activity of stakeholders (Sect. 4.4.3). In contrast to Sect. 4.4.3, common language and understanding is here understood as alignment of knowledge and similar understandings of multi-stakeholders. Hence common language and understanding are in that sense a precondition for speaking with one voice that goes beyond using voice as a preconditional indicator for identifying (power) relations between multi-stakeholders. In the conducted expert interviews, differences in understanding of certain terms are identified, for instance expressed in using terms interchangeably or synonymously. In sum, common language and understanding (or the lack thereof) is highlighted among expert interviewees as major challenge that impedes coordination of multi-stakeholder activities in the Kenyan digital ecosystem due to differences in their organizational languages and understanding of sustainable digital development topics. From this, the question arises how sustainable digital development can be conducted or even governed if it is not understood properly or vastly differently? In the same vein, the lack of shared understandings among multi-stakeholders simultaneously poses challenges for this research project and its data quality, whose implications will be discussed in Sect. 6.3.

Firstly, expert interviewees discuss the role of common language and understanding for sustainable digital development in the Kenyan digital ecosystem. More specifically, I18 argues that cooperation[9] among multi-stakeholders is easier if they share a common language and understanding, which also facilitates sharing a common vision. Communication in the sense of having a common language associated with common goals is seen as a further challenge by I12 (see also Sect. 4.4.3). Most importantly for this dissertation, the data analysis emphasizes the importance of a common language and understanding for coordination. Common language and understanding for coordination here refers to "[...] how do you eventually get everybody on the same table and find an agreement that is truly based on fair values where everybody understands entirely what is at stake, what is what are the costs, what are the benefits and everybody gets their equal share of both ends" (I2). The data analysis highlights this as not only a challenge, but also as a 'vicious circle'. Understanding further refers to recognizing the roles that partners play (I16), which is discussed in more detail in Sect. 4.4.5. Moreover, I7 highlights that as a first step before coordination can be achieved, it is necessary for multi-stakeholders to "come together and come up with with first they understand the problem properly which is very difficult to do for the people who are in these organizations, come to a shared understanding of the problem". In specific reference to the digital transformation in the coordination context, I11 highlights the necessity to "just understanding if we all get to understand what coming together and coordinating might do for the for digital transformation then we we could move this much faster." Finding a common language that results in structured outcomes (I2) is not only perceived as a challenge, but also can be achieved through defining terms and agreeing on their meaning (ibid.).

Secondly, for achieving a sustainable digital development a common understanding among multi-stakeholders is further highlighted with regard to understanding the context in which the Kenyan digital ecosystem operates (similarly to Sect. 4.1). The importance of 'understanding' is especially highlighted with regard to the (roles of) the Kenyan government. For the actors who conduct and foster sustainable digital development, "[...] the government has to be very understanding of what is happening on the ground so it cannot just sit ah the top of the hill and just come up with the regulation that couldn't could regulating digital space so startup space without understanding what is happening." (I24) This would likewise sharpen sustainable legislation for digital transformation contexts and the role of the government, which will be further discussed in Sect. 4.4.5. This is similar to the findings in Sect. 4.3.4 highlighting the importance of recognizing and understanding power asymmetries in the Kenyan digital ecosystem.

[9] Note that cooperation, collaboration and coordination are often used interchangeably among expert interviewees as discussed in Sect. 4.1.6.

Thirdly, after highlighting the importance of a common language and understanding for coordination for a sustainable digital development, a lack of understanding is emphasized in the data analysis. More specifically, the data analysis highlights a lack of understanding and hence a lack of a shared understanding of the term sustainability. This can for instance be derived from queries of experts in the interview setting such as: "Sustainability now not in an environmental way?" (I1). This question indicates that sustainable (digital) development, although it is globally proclaimed as a goal, is not clear and interpreted in various dimensions by interviewed experts.[10] With regard to the correlation between sustainability and ICTs in the context of common language and understanding, a common understanding of sustainability is potentially achieved through the SDGs in that SDG can facilitate to create a common understanding and the way ICT components are handled and managed (I5). Besides facilitating the achievement of common language and understanding through the SDGs, partnership formats are additionally highlighted for achieving a better (common) understanding (I9). Such impacts on coordination in the partnership setting will be further discussed in more detail in Sect. 4.4.6.

Fourthly, and most importantly for this dissertation, a lack of common language and understanding is highlighted by expert interviewees with regard to the digital transformation. This is closely linked to findings in Sect. 4.2 highlighting a lack of common understanding of the term 'digital'. This is underlined by I18 as "the danger of the term and then the process and discussion been very vague because digital digital is very broad term and the broad statement". The difficulty of a shared understanding of 'everything digital' is further confirmed by I6. More importantly, implications of digital technologies "if it's not understood properly" (I22), can lead to the discussed notions of hype (Sect. 4.3.1) rather than to sustainable digital development. Experiences of failing to understand what ICTs mean are further expressed by I17, similar to the discussion around ICTs as a 'means to an end' (Sect. 2.2.1, Sect. 4.4.5).

Adding to complexity, the interviewed experts highlight the importance of understanding the context of the Kenyan digital ecosystem as well as putting digital transformation in the specific (country) contexts (I18; I11). For achieving these understandings, the role of development cooperation and the private sector is highlighted. Here, it is important to draw links to awareness raising efforts of development cooperation actors (Sect. 4.4.3) and to the previous findings relating to the state of digital capacities of donor organizations themselves (Sect. 4.4.4). With regard to the role

[10] This can also be traced back to the interview setting and requirement of clarification on behalf of the researcher in an open line of questioning that was pursued for gathering authentic associations of the interviewees.

of the private sector in the context of common language and understanding, this is seen in helping clients such as the government "understand why they need to go digital" (I10), which however is perceived as a struggle. Here also the understanding of collaboration with certain actors, especially the digitally strong private sector actors, includes understanding its benefits and limits (I17) for also decreasing possible dependencies or lock-ins. Interestingly in this context, donors are identified as overtaking the wordings of the private sector such as "I don't think all donors see how that like how they could use such an approach for their own storyline and their own KPI's amongst each other donors" (I4).[11] This is similar to the argument of non-profits and international donors "adopting some of the language, and perhaps even the beliefs, of the for-profit sector." (Marchant 2017:323) Given the limited data of this dissertation, further research is needed here.

In addition to the highlighted lack of common understanding and language of sustainability and digital transformation, the following concepts and terms are further examples:

Common language and understanding: Both digital transformation and development cooperation The data analysis resulted in a lack of common language and understanding in this dissertation's setting of the Kenyan digital ecosystem and sustainable digital development with interview questions aimed at both digital aspects as well as development aspects. Although experts in respective fields were chosen for this qualitative research, it was difficult for some interviewees to answer questions of both sides. This is for instance highlighted by "your questions are hard" (I4, bilateral donor) or also "that's a difficult question" (I10, multinational tech company). Further examples are provided in the previously discussed lack of knowledge around digital transformation (beginning of this section). With regard to development cooperation, a similar effect is observed where terms relating to the field of development cooperation and aid where unclear or unknown to interviewed expert representatives. This is for instance illustrated in a lack of understanding the development discourse terminology such as 'Global North' and 'Global South' (I10, multinational tech company). Similarly, 'aid fragmentation' (Sect. 2.1.5, Sect. 4.1.5) is unclear as expressed by I11 (hub) and I14 (multilateral donor). As a further example, a lack of knowledge and understanding of the term 'multi-stakeholder' was identified (I1, NGO).

[11] Key Performance Indicators (KPI)s mostly measure the business objects such as activities or value propositions among others and more specifically the organization's performance and progress (Camarinha-Matos et al. 2019). This business logic is also transferred to other contexts, in this case development cooperation.

Common language and understanding: Human centered digital transformation
The concept of human-centered digital transformation was identified as only noted in interviews with mostly European multilateral donors. More specifically, only two statements of multilateral donor organizations referring to human-centered digital transformation were noted. For instance, by "we have another model to offer to our partners in the Global South where digitalization will be built right from the beginning on a human centric basis" (I12). Similarly the expert representative of another multilateral donor organization highlights "the dig human centered digitalization has become one of the key priorities." (I17)

Common language and understanding: Network and ecosystem As highlighted in Sect. 4.4.1, network and ecosystem were used interchangeably by expert interviewees without recognizing the differences (see also Chap. 4). This is a further implication for this research project as it is not always clear which of those is meant and what is understood by it (see further Sect. 6.3)

Common language and understanding: Governance Similarly to a lack of common language and understanding of network or ecosystem, governance is likewise understood differently among expert interviewees. More specifically, governance was sometimes used in the sense of 'regulation', sometimes as 'coordination' (see further Sect. 4.4.6, Sect. 4.4.6). This is for instance highlighted by expert statement "I don't maybe we have different understandings of of of governance" (I14).

Common language and understanding: Harmonization Different understandings of harmonization are identified by I20.

Sensitivities
Closely linked to common language and understanding and also to awareness (Sect. 4.4.3, Sect. 2.5.1) the data analysis identified sensitivities of multi-stakeholders as challenges to coordination and sustainable digital development. Sensitivities here refer to sensitivities of interviewees in answering, for instance due to diplomatic reasons (see Chap. 6). More concretely, sensitives are emphasized by interviewed expert representatives around certain digital topics and also towards certain multi-stakeholders in the Kenyan digital ecosystem. Both are also expressed in a limitation to comment for sensitivity reasons that lead to methodological research limitations (see also Sect. 6.2). Identified sensitivities function as a challenge for governance and are highlighted in the context of partnership formats. I20 for instance highlights the important role of sensitivities in multi-stakeholder interactions and in partner-

ships for the purpose of sustainability and inclusivity, especially in development cooperation partnerships.

Firstly, the data analysis identifies digital transformation as a "sensitive topic" (I12). Moreover, sensitivities (and awareness) of multi-stakeholders are highlighted in the context of data protection and privacy rights, which are labeled as differing among regions (I6). Additionally, data is emphasized as a sensitive topic (I6, see also Sect. 4.2.2).

Secondly, sensitivities are expressed towards interactions with or engagements of certain multi-stakeholders in the Kenyan digital ecosystem. I10 shares experiencing sensitivities of Kenyans towards the government's close interaction and (financial) dependence on some actors. More concretely, the engagements of certain multinational tech companies are argued to be handled with caution as further highlighted by I21, emphasizing caution around a gap between hype or CSR interests and actual impact (as discussed in Sect. 4.3.3).

Dependencies

As a final challenge that is closely linked to the before mentioned power asymmetries (Sect. 4.3.4), this dissertation further argues that dependencies in the Kenyan digital ecosystem hinder governance for sustainable digital development. Dependencies are presented as a challenge in the theoretical background, for instance in Sect. 2.3.6 and Sect. 2.5.2. More specifically, five dependency categories have been derived from the data analysis: 'new' dependencies, dependencies on protocols and on the private sector as well as interdependencies and lock-ins. These will be discussed in this subsection.

Firstly, the data analysis of this dissertation highlights a possible emergence of 'new dependencies' of and on actors due to the digital transformation. New dependencies here are understood as dependencies that have not existed prior to the emergence of the digital transformation. The interviewed experts experience new dependencies arising from the digital transformation (e.g., I5), especially with regard to software (I5; I1) and (lack of) local digital knowledge and skills (I1). Moreover, dependencies on infrastructure are to be mentioned further in the synthesis with the findings of Sect. 4.1.3. For instance, a (continuing) dependency "[...] on expanding the the significantly fiber optics to then provide other services" (I17) is experienced. Similarly, electricity and connectivity are needed for enabling the digital transformation thus creating dependencies (I2). In turn, infrastructure is also dependent on certain providers, mostly from the private sector (I5). In funding of such digital infrastructures, the Kenyan government is argued to be dependent on either infrastructure provision or financial funding of multi-stakeholders such as private sector and/or development cooperation actors (I3) given the economic and infrastructural contexts (Sect. 4.1.1, Sect. 4.1.3). Dependencies on these two actor groups will be discussed in more detail in the following. In contrast to these arguments, I19 and I20 negate the hypothesis that digital transformation leads to new dependencies in the Kenyan digital ecosystem.

While the question around 'new' dependencies through the digital transformation cannot be answered completely at this point, it will be further discussed and linked to the other dependency forms mentioned above. Additionally, as discussed in Sect. 4.1.5 ad Sect. 4.4.4, aid or development cooperation relations are characterized by dependencies. Similarly, as discussed in Sect. 4.4.3, 'new' actors in digital development such as tech companies might likewise lead to 'new' dependencies.

Dependencies on (Donor) Protocols Secondly, multi-stakeholders' dependencies on protocols or requirements, especially of donor organizations, further challenge sustainable and inclusive digital development in the Kenyan digital ecosystem. More specifically, expert interviewees highlight that there are a multitude of aid organizations (Sect. 4.1.5, Sect. 4.4.3) that have different requirements for collaboration, which are assessed as being difficult for the Kenyan digital ecosystem – on the level of individuals, on a community level and for the public sector (I2). Requirements for instance include internal policies of stakeholders, here especially of donor organizations, that likewise impact collaboration in multi-stakeholder groups (I7). Hence, dependencies on protocols are closely linked to the previously discussed actors' interests (Sect. 4.3.3). In this instance, donor organizations and private sector companies are both similarly discussed in the context of dependencies on protocols with both of their requirements being criticized. This refers to not matching innovative target groups (I4) or protocols leading to exclusion when actors fail to meet or comply with the requirements (Sect. 4.4.3). Dependencies on the private sector will be further discussed in more detail in the next subsection.

Moreover, the impact of dependencies through protocols on the Kenyan government and on Kenyan startups in the context of sustainable digital development are further highlighted by expert interviewees. On the one hand, the impact of dependencies on protocols for the Kenyan government is assessed as being counterproductive and hence serves as a crucial determinant for (un-)successful digital development. For instance, each donor organization has "[...] slightly different rules, terms and conditions, political agendas and the transaction costs can be huge" (I17, similarly stated in I2). Hence these different rules of a multitude of donor organizations are assessed as one factor that poses a specific challenge for the Kenyan government as it impedes their work (I12) and limit their capacities due to fulfilling these requirements. This effect is multiplied considering the number of donor organizations present in the Kenyan digital ecosystem interacting with the Kenyan government

(Sect. 4.1.5, Sect. 4.4.3). As one specific example, the Kenyan Ministry of ICT is mentioned as having "[...] to manage all the different donor systems, everybody has other requirements and rules and regulations." (I13). Conclusively, this is another instance of misdirected (digital) capacities of the Kenyan government in regard to being 'overburdened' by multi-stakeholders as discussed in Sect. 4.4.4.

On the other hand, the impact of dependence on protocols for Kenyan startups, especially the ones that seek donor support e.g., in acceleration programs, is further emphasized (as discussed in Sect. 4.3.2). Here, capacities of this stakeholder group are similarly crucial to the one of the Kenyan government. Expert interviewees assert that startups that are new or inexperienced with donor regulations and processes are disadvantaged in receiving donor support (I4). Therefore the amount of Kenyan startups' experience is not only crucial for receiving donor support, but also an instance where inequalities are created. In the metric of this research project, this would classify as exclusive rather than inclusive digital development effects. Moreover, it has shown that protocols and requirements of donors can potentially hinder participation of (unknown) startups in their donor ICT programs.

Dependencies on the Private Sector (Protocols, Technologies and Knowhow)
Thirdly, dependencies of multi-stakeholders in the Kenyan digital ecosystem are further experienced by expert interviewees with regard to the private tech sector. This especially applies to their protocols, provision of digital technologies and service as well as knowhow that likewise mirror dominance and power asymmetries (see Sect. 4.3.4). Similarly to the impact of protocol dependency of donor organizations in interaction with Kenyan startups, I4 argues that startups and larger corporates have a close(r) relation, because startups know private companies' requirements given their similar nature. Hence, this similarity can be understood as a further indicator for the degree of closeness between the two private sector actor groups in the Kenyan digital ecosystem (Sect. 4.4.3). However, dependencies of multi-stakeholders on the private sector are highlighted with regard to dependency on foreign private tech organizations (I20), on "Westerners or [...] Easterners" (I1), their (foreign) technologies with specific protocols and systems (I17), and their provision of digital infrastructure (I5). Especially with regard to the latter, dependencies to the extent of monopolistic settings of users being dependent on only one provider, are experienced in the Kenyan digital ecosystem (see further Sect. 4.3.4). For instance, Safaricom is mentioned as having a monopoly in the telecommunication sector (I6, see also Sect. 4.1.1), similarly to providers of apps (I17; I2). Since monopolies hinder competition and entry of competitors, this is a further example for exclusive developments in the Kenyan digital ecosystem (Sect. 4.4.3) as they can likewise lead to provider lock-ins (Sect. 4.4.4). A last dimension of dependencies on the private

sector is identified with regard to their knowhow and digital skills (Sect. 4.4.4; Sect. 4.3.5).

Interdependence Fourthly, dependencies between and on multi-stakeholders in the Kenyan digital ecosystem are further manifested in notions of interdependence. Interdependencies are similarly discussed in the theoretical background in Sect. 2.5.1. Interdependence is here understood as a dependency between at least two multi-stakeholders emphasizing that multi-stakeholders need each other for their activities in the Kenyan digital ecosystem (similar to the ecological ecosystem logic). Digital technologies in this context are an exogenous factor that contribute to an interdependence, while the interdependence between multi-stakeholders here is understood as endogenous (internal, in the ecosystem).

The interviewed expert representatives share experiences of interdependence between multi-stakeholders in the Kenyan digital ecosystem for instance in contractual relations for *licensing* between customers and providers of digital technologies and services (I16) as demonstrated in Sect. 4.4.3. More specifically, this interdependence between private sector companies (provider) and public sector actors (customers) is experienced as "[...] the private sector knows that a lot of the possibilities for them depends on public sector investments in subsidies be it on fiber infrastructure for example." (I17) This also demonstrates an interdependence of knowledge and capabilities in the digital transformation (I22). This interdependence between these two actor groups is representative for the entire digital ecosystem in Kenya. For instance, I23 emphasizes the interdependence of multi-stakeholders as needing each other, even to the degree of a symbiotic relation.

Lock-in Lastly, this dissertation emphasizes a final degree of dependencies in the form of lock-ins in the Kenyan digital ecosystem. It argues that lock-in effects between multi-stakeholders and in the digital transformation are a challenge for sustainable digital development and for governing these challenges. In this dissertation, lock-ins are identified as factors that contribute to unsustainable digital development, both due to technological aspects of the digital transformation and due to multi-stakeholder interactions (as also discussed in the theoretical background Sect. 2.3.6). Therefore, lock-ins and interdependencies are closely linked. More specifically, if the interdependence is too strong, switching to another provider becomes uneconomical due to high transaction costs, which leads to the so-called 'lock-ins'. In addition to the before mentioned categories of dependencies and power asymmetries, further factors such as exclusion (as discussed in Sect. 4.4.3), culminate in lock-ins. In sum, the data analysis of this dissertation highlights two specific dimensions of lock-ins in the Kenyan digital ecosystem: a technological and an actor perspective.

On the one hand, lock-ins are experienced by interviewed expert representatives as occurring due to digital technologies and services. Against the background of perceived lock-ins in the digital transformation (I11), technological lock-ins in the Kenyan digital ecosystem are experienced with regard to hardware (e.g., I23, see Sect. 4.2.2), infrastructure and software (I5, see Sect. 4.2.2). Lock-in effects are especially perceived with regard to certain software (providers) such as Microsoft Office, with users being used to a specific software and therefore want to continue its usage. This affects every customer, ranging from individuals to the Kenyan government (I13). These are instances are further emphasized with regard to other software applications such as WhatsApp, where lock-in effects occur due to a lack of awareness of alternatives (I10) and due to network effects (I10). The 'price' is voluntarily providing personal data and is a consequence of such network lock-ins (data further discussed in Sect. 4.2.2).

Moreover, this similarly leads to lock-ins through contracts (Sect. 4.4.3), where for instance contracts are signed for software provision and support where the signing parties are tied to it, because switching providers or systems would mean huge transaction costs that are not sustainable (e.g., I10). These contractual lock-ins are similarly experienced in the Kenyan digital ecosystem with regard to other partnerships, for instance in ICT projects or programs (Sect. 4.3.2). Lock-ins here are experienced during the project: Even if these instances of overlaps or duplications with other projects are recognized by the project partners, the ongoing project cannot be altered during the project execution, so that overlapping activities – even though unsustainable – continue: "But we also realize that a lot of programs are ongoing and that we can certainly have certain lockins already, so we see 'oh actually we have a lot of overlap here but the programs are going, the partnerships are there'." (I13) The effects of such project lock-ins are discussed in more detail in Sect. 4.4.3 as these are instances where this dissertation highlights that (prior) coordination can prevent these duplications for a sustainable digital development (Sect. 4.1.6), for instance by creating awareness (Sect. 4.4.3).

On the other hand, this leads over to the perspective of lock-ins in multi-stakeholder interactions, so not on a technological or procedural level, but on an actor level. More specifically, lock-ins are experienced in the interaction between multi-stakeholders hindering collaboration in the Kenyan digital ecosystem (I13). Synthesizing the previous findings of knowledge transfer between multi-stakeholders (Sect. 4.3.5), digital capacities (Sect. 4.4.4) and dependencies on the private sector, interdependence and also lock-ins demonstrate differing levels and asymmetries in power of multi-stakeholders.

After highlighting these digital development and coordination challenges in the Kenyan digital ecosystem, this dissertation proposes the setup of governance roles and governance formats as an approach to addressing and potentially overcoming these challenges. These will be discussed in the following two subsections in line with the theoretically developed framework (Sect. 2.5).

4.4.5 Determinant 3: Governance Roles in the Kenyan Digital Ecosystem

As third determinant of this dissertation's framework, governance roles are proposed as one governance angle on the Kenyan digital ecosystem in line with the network governance roles for sustainable digital development derived from the theoretical background Sect. 2.5.3. Here, governance roles of *coordinator* (e.g., Raeymaeckers et al. 2020), *facilitator* (e.g., Provan and Kenis 2008) here synonymous with *broker* (e.g., Börzel and Risse 2010; Klijn and Koppenjan 2016; Kenis and Provan 2009, *network manager* (e.g., Klijn and Koppenjan 2016) and *driver (of digital transformation)* are derived from the literature review. All of these are active roles that multi-stakeholders take in interacting and intervening in the ecosystem, so that these roles are taken in governing the ecosystem for sustainable digital development. As elaborated, some roles are more similar than others with definitions provided in Sect. 2.5.3. In this section, governance roles are analyzed using empirical findings from the conducted expert interviews and the respective qualitative data analysis. Here, the roles are identified by expert interviewees through their self-perception of their own roles and through the external perception of other multi-stakeholders. More specifically, I23 underlines the argument that all actor groups in the Kenyan digital ecosystem are important as they actively intervene and shape the Kenyan digital ecosystem where everyone has their role. This further emphasizes the importance of each multi-stakeholder for a sustainable digital development of the ecosystem. In contrast to the governance roles derived in the theoretical background, the empirical data analysis identified additional governance roles in the Kenyan digital ecosystem: advisor, balancer, connector, enabler and supporter.

The identified roles have been ascribed to the stakeholders as illustrated in Table 4.2. These governance roles illustrated in Table 4.2 are listed on the x-axis, whereas the y-axis demonstrates which actor is ascribed to which role, as indicated in the data analysis. It becomes obvious that most actors not only have one role, but many. The order of roles is not structured according to their frequency of reference in the data analysis, but in alphabetical order. Which actor fulfills which role and why will be discussed in the following.

Table 4.2 Overview of multi-stakeholders' governance roles in the Kenyan digital ecosystem (own illustration)

	Advisor	Balancer	Broker	Connector	Coordinator	Driver	Enabler	Facilitator	Supporter	Manager
Bilateral Donor Org.	✓		✓	✓		✓	✓		✓	
Multilateral Donor Org.	✓	✓		✓	✓	✓	✓		✓	
Kenyan Government		✓		✓	✓	✓	✓	✓		✓
Hubs	✓						✓			
NGOs		✓	✓	✓	✓				✓	
Private Sector	✓	✓			✓	✓	✓			
Startups			✓							
ICTs			✓		✓		✓	✓		

Role Advisor

As a first role, the role of an advisor emerged from the data analysis. While this role is indirectly discussed as inherent in development cooperation roles, this role emerged specifically in the empirical data analysis in contrast to the theoretical background (Sect. 2.5.3). Advisor here is understood as an actor group giving advice to another group (here mostly to public sector organizations like the Kenyan government). Before going into detail here, it is worth discussing the need for advisement in the Kenyan digital ecosystem for a sustainable digital development.

Advisement to the Kenyan government has been briefly discussed in Sect. 4.3.4 and Sect. 4.4.3, with especially the private sector and donor organizations highlighted in the role of advisors. Hence, the directions of advisement are important to analyze. In the conducted interviews on the Kenyan digital ecosystem, various multi-stakeholder organizations take the role of an advisor to different degrees and with understandings of advisement. More importantly, the forms of advisement differ depending on which actor's role in advising is analyzed:

Firstly, I10 (multinational software company) provides further insights into the role of international tech companies in advising, which for instance entails advising customers on the product or solutions as part of the business process. The underlying business interest in this advisory process however is further emphasized in that the advisory process is accompanied by convincing the customer to buy the service or technology (I10). This statement on the other hand provides another argument in the relation between the private sector and customer (regardless whether private or public) demonstrating the role of willingness to adapt digital technologies and participate in the digital transformation as discussed in Sect. 4.4.3 and Sect. 4.4.4.

Further elaborating on the role of an advisor taken by the private sector, it is additionally important to highlight its role in advising the Kenyan government. Here it is important to note that the advisory is not only conducted externally, but the private sector also consults in-house of the government (and other public organizations):

> "[...] there are actually private sector actors who are sitting who have an official position within the government and who who provide this type of of advisory and that's something that I've also seen within the UN the understanding from the public sector that they don't know about digital and digitization is so strong that they believe it would be better to bring in the people who have that apparent knowledge and then they almost bring in their on enemy" (I2).

This form of advising conducted by the international private sector is hence criticized as it further demonstrates not only power asymmetries, but also dependencies that are factors explaining a hindrance in sustainable digital development. The argumentation provided by I2 here is that the role of advisor is taken out of business interests.

Secondly, the private sector is not the only actor group taking the role of advisors. Besides this actor group, the interview data suggests that hubs are also advising the Kenyan government and play a key role (I11), in addition to political advisory of the (mostly European) donor countries (I5).

In sum, the role of an advisor is taken by almost every stakeholder group, but in different directions. The direction of advisement towards the Kenyan government is identified as the most common one. Advisement here takes the role from consultancy (both external and in-house), to business relation advisement, further demonstrating degrees in closeness (Sect. 4.4.3). The role of an advisor in the Kenyan digital ecosystem is fulfilled by the private sector, hubs and donor organizations (bilateral and multilateral).

Role Balancer

A second role that has not been directly but indirectly derived from the theoretical background (section on a network manager, Sect. 2.5.3), is the role of a balancer. This role is understood as to level, weigh or stabilize developments in the Kenyan digital ecosystem. In the empirical interview data analysis, the role of a balancer is ascribed to NGOs, academia, the Kenyan government and (to a lesser extent) the private sector. I9) sees NGOs in the position of taking the role of a balancer for counteracting biases and self-serving interests of the actors for increasing effectiveness. This is similarly supported by I24 who likewise highlights the need to balance self-serving interests that have the potential of infringing each other, here specifically with regard to government and private sector organizations. The crucial role of the Kenyan government in balancing is emphasized by I9:

> "I think maybe what needs to happen is for the government to define the limits in terms of partnerships and to thoroughly investigate whether speci how specific policies and partnerships will affect local industries and local companies but at the end of the day alright then the they have to kind of find that sweet spot where the balance between the two aspects".

A further balancing act for the Kenyan government is seen in "[...] acting to create frameworks that allow the right amount of disruptive activity but also keep keep bounces of power someone in tech through through flows of of money through the government" (I7).

Moreover, the role of balancing is also referred to the extent of *harmonizing* interactions in the Kenyan digital ecosystem by I5, further adding that academia has the role of harmonizing the interactions (see also Sect. 4.4.3). Another shape of the balancer role identified in the data analysis is described as the role of a *moderator*. Here, one representative of a multilateral organization sees its organization as overtaking the role of moderating "[...] as a referee promoting data protection, standards" (I17). Conclusively, most expert interviewees observe that the role as a balancer in the Kenyan digital ecosystem is taken by the government and to a lesser degree by NGOs, academia and multilateral donor organizations.

Role Broker

The next governance role, the role of a broker, is identified in both the theoretical literature review (Sect. 2.5.3) and the empirical data analysis. In the latter, a broker in the Kenyan digital ecosystem is understood as initiating or functioning between entities, similar so intermediaries (see below). As illustrated in Table 4.2, various multi-stakeholder groups are ascribed with the role of a broker by interviewed experts.

Firstly, before further analyzing the role of a broker in the Kenyan digital ecosystem, it is important to highlight the expressed need for a broker (I5). Moreover, this role is understood differently by the expert interviewees. For instance, while I10 sees the role of a broker in the sense of "[...] having brokers in the middle and all that", I5 shares the perception that "the role of being the broker is also something that develops over time". Here it is important to note that the role of a broker is not entirely without criticism as emphasized by I10: "If I want something done somebody will tell me, hand me this amount of money I will quickly do it for you." This suggests that the ones taking the role of a broker might expect a compensation in return, which possibly alludes to corruption (Sect. 4.1.2).

Secondly, the question which actor group fulfills the role of a broker in the Kenyan digital ecosystem, is not widely discussed among expert interviewees. Hence there is only limited data available. Only I5 here sees both startups and civil society as well as their own organization (bilateral donor) as a broker.

Role Intermediary The role of a broker serving as an intermediary is more widely discussed in the empirical data analysis than in the theoretical background. In the theoretical background (Sect. 2.5.3), the role of an intermediary is associated with a broker. Traditionally, in social sciences intermediaries include non-governmental actors such as the media (see e.g., Sect. 2.1.4). In the conducted interviews, the role of the media is mentioned by some interviewees (I11; I15), but rather in their communication of information such as "[...] so who knows maybe maybe our media is is hyping that a little bit" (I2). In this dissertation intermediaries are approached from a governance perspective similarly to the role as a "middleman" or broker as discussed in Sect. 2.5.3. This is illustrated by the expert interviewee's statement of "I think a middleman often is providing a valuable service in a context the maybe maybe if our information isn't flowing very quickly" (I7). This perception is similar to the before mentioned results concerning the role of a broker. More importantly for this dissertation's digital development focus, (digital) platforms are ascribed this role as a middleman i.e., of a broker (I7). This is not only proposed in Sect. 2.5.4, but will also be further discussed in more detail in Sect. 4.5. With regard to multi-stakeholder actor groups in the Kenyan digital ecosystem, I6 further sees hubs and associations in the role of an intermediary, although the expert representative criticizes their performance in this role by highlighting their narrow target group that contradicts a sustainable and inclusive digital development (I6). Hence, the actor group that overtakes a certain role (in this case intermediaries) impacts the outcome and hence sustainable digital development, so that there are differences in outcome as to which actor groups fulfills this role. To conclude, the role of a broker in the Kenyan digital ecosystem is overtaken by startups, NGOs, bilateral donor organizations and digital technologies (digital platforms).

Role Connector

Similarly to the role of a broker, the role of a connector is indirectly derived from the theoretical literature review (under the role of a broker) emphasizing similarities between these roles. In the empirical analysis, the governance role of a connector in the Kenyan digital ecosystem, is understood as bringing actors in the Kenyan digital ecosystem together. Here, the data analysis suggests a distribution of the role among multi-stakeholders: Similarly to the before mentioned role of a broker Sect. 4.4.5, *digital technologies* are seen as connectors by I2, for instance in bringing the regions closer in the digital divide context: "[...] so it's just a lot easier to come together if if you have a a a digital connector inbetween". Moreover, *foundations (civil society organizations)* are mentioned by I9 (international private sector) as one actor group that in their experience serves as a connector of the private sector organization to the Kenyan government.

Expanding on the role of connectors to the public sector context, *Kenyan ministries* are argued to act as connectors across the network, i.e., ecosystem (I2). More broadly speaking, the Kenyan government including specific ministries, are attributed with the role of a connector, for instance in increasing effectiveness of (ICT) projects and partnerships (I9).

Shifting to foreign public sector actors, interviewed *development cooperation agencies* additionally see themselves in the role as a connector in the Kenyan digital ecosystem, especially in connecting the private sector and the Kenyan government. For instance, in the context of entrepreneurial subnetworks, they are acting as connectors (I4). The role of development agencies as a connector is likewise perceived with regard to multilateral donor agencies who are "bringing partners together" (I14). Here it is worth mentioning that multilateral donor organizations further see their role not only in connecting, but also in integrating. The role of an *integrator* has emerged in one interview with a multilateral donor agency, where connecting and integrating is perceived as not only synonymous, but also as a unique role (I14). Since the data analysis of this dissertation however suggests that a variety of multi-stakeholders are seen in the role as connecting and 'bringing together', the uniqueness of this role taken by multilateral donor agencies differs from their self-perception.

Conclusively, although all of the multi-stakeholder actor groups differ in their nature (see e.g., Sect. 2.1.4), almost all of them see themselves (their organizations) as connectors. The data analysis identifies the role of a connector in the Kenyan digital ecosystem as mostly fulfilled by NGOs, Kenyan government, and donor organizations (bilateral and multilateral).

Role Coordinator

Similarly to role of connector, almost every actor group see themselves as coordinators. The role of a coordinator is derived from both the theoretical (Sect. 2.5.3) and empirical analysis. Since this dissertation's research question aims at an analysis of network coordination, the role of a coordinator is most essential for this research. In the conducted interviews, the question who fulfills the role of an coordinator in the Kenyan digital ecosystem is intensely debated.

Firstly and most importantly, similar to the findings in the section on coordination (Sect. 4.1.6), I10 and I22 confirm the need for a coordinative organization for sustainable digital development, thereby underlining this dissertation's research question. Moreover, coordinative activities manifest into a role of a coordinator. Digital platforms as coordinators of activities in the Kenyan digital ecosystem are likewise discussed, which will be elaborated in more detail in Sect. 4.5 (see also Sect. 2.5.4). Hence, actor groups who are conducting coordination efforts can likewise be labeled as coordinators.

One example is the EU's D4D Hub, which coordinates the digital development activities of some EU member states (I18; I12) in like an internal political coordination role (I17). In this context, an additional role linked to the role of a coordinator emerged in the data analysis – the *role of an aggregator* (I6). However, this role has only been mentioned by one representative of a multilateral donor organization (I17) and not by other multi-stakeholders. Nonetheless, this role is highlighted for full disclosure of the empirical results. The role of an aggregator here is hence understood similarly to a coordinator and leader, which roles will be further explained in more detail in the following subsections.

With regard to an additional multilateral donor organization in the Kenyan digital ecosystem, the role of the UN as a coordinator is discussed controversially among expert interviewees. While the representative of a multilateral organization sees their role as coordinating over different governance levels in a particular country (here Kenya) (I20), I7 contests the role of the UN as a coordinator when it comes to coordinating and governing the Kenyan digital ecosystem: "[...] if you want the UN to coordinate it, it's so risk averse for one it's also slow, it's also bloated and bureaucratic". An additional negation of a single actor group overtaking the role as a coordinator in the Kenyan digital ecosystem is illustrated with regard to the international private sector, especially to US tech companies (I7). However, further research is needed here as this argument is only based on one perspective and hence poses limited generalization. These findings demonstrate a discussion among experts around one single organization in the role of coordinating versus formalized or self-governance forms, which will be discussed in more detail in Sect. 4.4.6 and Sect. 4.5

Besides these actor groups, the actor group of the Kenyan public sector is further crucial for this dissertation's research into implications of the digital transformation on governance and government action. Here, one result is that more expert representatives see the *government* in the role of a coordinator in the Kenyan digital ecosystem than not. On the one hand, I10 highlights a possible danger of the government as a coordinator since "[i]f you leave it to the government there will be some form of corruption but if it's a separate group a separate identity and ideally a private sector and not public sector". This is further supported by I3 expressing the need for a coordinative body, but not a political one: "I think I would try to keep it away from from the political sphere so rather not turning to the EU or so". Opposed to this argumentation, interviews such as I2 and I22 assign the role as coordinator to the Kenyan government. More specifically, the Kenyan Ministry of ICT is highlighted: "I mean the coordination should be driven by the ministry that has the mandate" (I6). Here, I14 further highlights that not only should the Kenyan government overtake this role, but it already attempts to play that role. The role of the Kenyan government in coordinating in the Kenyan digital ecosystem is further experienced by I1, however with mixed results.

In sum, based on this dissertation's data analysis, the role of a coordinator in the Kenyan digital ecosystem is fulfilled by various actors: NGOs, ICTs (digital platforms), multilateral donor organizations, private sector and the Kenyan government. While the interviewed expert representatives express criticism towards the discussion around multilateral organizations and (at least external) private sector organizations in the role of coordinator, the Kenyan government is argued to be the chosen coordinator in the Kenyan digital ecosystem. This discussion will continue in Sect. 4.4.6.

Role Driver of Digitalization

A role that is crucial for the ICT component of the Kenyan digital ecosystem, is the role of a driver of digitalization (as likewise highlighted in the theoretical background, Sect. 2.5.3). This role is closely linked to actors' levels of activity in the digital ecosystem (Sect. 4.4.3). The role of driving the digital transformation is hence understood as actively intervening and shaping the digital transformation. Here it is important to highlight parallels to the findings of Sect. 4.4.3, as there is a correlation between active actor groups and being a driving force, because both are characterized by higher levels of activities. Hence, the actor groups overtaking the role as driver of the digitalization in the Kenyan digital ecosystem are similar to the ones identified as active ones in the respective section. Generally, all multi-stakeholders active in the Kenyan digital ecosystem are mentioned by interviewed expert representatives as fulfilling the role of a driver since "digital transformation and everybody is in together it's a ecosystem of all players you know driving driving the same the same way really" (I21). More specifically, the data analysis identified various stakeholder groups having this role, with the private sector and the Kenyan government being the most frequently mentioned ones. However, they are discussed controversially with regard to their implications for sustainable digital development.

Beginning with findings on the *private sector* actor group, the data analysis suggests that a majority of respondents experience the private sector as the driver of digitalization. More specifically, the role of drivers taken by the private sector, including both multinationals and local private sector companies, is seen in leading (I1) and "pushing" (I10; I24). Moreover, startups in the Kenyan digital ecosystem are likewise seen in the role of driving by I5, I9 and I24. Specifically the "vivid tech scene Nairobi that is leading" (I1) and the startup ecosystems have a driving role in the Kenyan digital ecosystem (I23). The role of the private sector as driver of the digital transformation in the Kenyan digital ecosystem is even stronger if startups and multinationals are combined e.g., in a partnership format (I23). I13 (bilateral donor) and I2 (international NGO) however perceive international tech companies as being in the role of driving the digital transformation in Kenya. This is similarly perceived by I7 as "[...] a lot of the innovative tech ecosystem is foreign." Thereby the (Kenyan) society likewise indirectly overtakes the role as a driver, since their demand drives digital technologies and innovations of private sector companies (I10) and will increasingly drive the digital transformation in Kenya: "[...] my gut feeling in the next ten years I mean this transformation the young people will just be leading this transformation" (I19).

While these arguments suggest that private sector actors drive the digital transformation, *donors* are likewise seen in the role of drivers (I3), but more in the sense of "pushing it to the more marginalized groups" (I7). Some bilateral and multilateral donor organizations are mentioned by expert interviewees in the (future) role of driving the digital transformation in Kenya: GIZ (e.g., I6), EU (e.g., I6), the World Bank (e.g., I6) and USAID as the "biggest donor in in the digital ecosystem in Kenya" (I6). However, the data here is limited as it only refers to one expert's subjective assessment. The activity and impact of donors is discussed in more detail in Sect. 4.4.3.

Synthesizing these perspectives, the private sector organizations are seen as the dominant actor group that is mostly overtaking the role as driver of digital transformation. This is further confirmed by I21 (private sector): "There are the local companies with the local organizations and then when we come to and when we come to the global organizations I would probably say that you some of these big tech companies are really driving ones who have invested heavily." Here the role of a driver is also measured by investments. This is supported by this dissertation's findings emerging from the data analysis on the *role of attracting investment*. The role of attracting investment however is not discussed as an individual role for two reasons: On the one hand, it has been only referred to by a small number of interviewees and on the other hand, it is used for supporting the argument of a correlation between the role of a driver of digitalization and the role of investment in this governance role context. Attracting investment is crucial for sustainable digital development, for instance in the context of (required) financial resources (see Sect. 4.4.4). With further regard to Kenya's performance in attracting (foreign) investment, this capacity and ability has also been derived as an important determinant for sustainable digital development as for instance emphasized in Kenya's performance in the *Ease of Doing Business Index* (Sect. 4.1.1).

More specifically, the role of attracting investment in the Kenyan digital ecosystem is explicitly mentioned with regard to the *Kenyan government* in the conducted interviews (e.g., I20), where the role of attracting investment is emphasized as an instance where the government is facilitative for the digital ecosystem (see further Sect. 4.4.5). Partial success in the Kenyan government fulfilling the role of attracting investment is attested by I13 given the fostered narrative of 'Silicon Savannah' (Sect. 4.3.5, Sect. 4.1.1). On the other hand however, the role of the Kenyan government in attracting (foreign) investment is discussed critically among expert interviewees, for instance in the context of regulations and legislation posing barriers for entry (here for investment; I11, see e.g., Sect. 4.4.3, Sect. 4.3.4). In addition to the Kenyan government in the role of attracting investment, private sector actors, especially tech startups, are further highlighted with regard to Subsaharan Africa: "[...] lot of foreign direct investment so foreign capital is being attracted through tech startups in Subsaharan Africa" (I6). This is further discussed in Sect. 4.3.5.

These findings lead to the conclusion that the role of driver in the digital transformation falls either to private sector organizations or the government (donors to a lesser extent), which alludes to a setting of competition. As a first insight into this argumentation, I11 states that "[...] private entities are doing a much better job in driving digital transformation". This directly leads over to an analysis of the Kenyan government overtaking the role as a driver of digitalization, which will be discussed in more detail as follows.

The analyzed interview data shows that the *Kenyan government* is also attributed with the role of driving the digital transformation in Kenya. The Kenyan government is not only "one of the most important players when it comes to to this [driving]" (I24), but also has the ability to substantially shape the (sustainable) digital transformation: "I believe the government can literally make or break the digital transformation" (I9). This is further confirmed by I21 stating that "the government's had a big driving role". More specifically, the Ministry of ICT is repeatedly referred to as having the role of a driver (I6).

The role of driving exerted by the Kenyan government is perceived as executed firstly on a structural basis through digital transformation initiatives (I10) or also through regulatory frameworks (I10), which have been previously discussed (Sect. 4.1.2). Here however, gaps between the digital initiatives and the reality "on the ground" (I10) are argued to sometimes differ, for instance the speed in which digital transformation is driven by the Kenyan government. This is further experienced by I5 in that "generally the political side is more or is slower and not driving this change as fast as it should". Reasons for that are discussed in previous chapters such as bureaucracy (Sect. 4.4.4), digital capacities (Sect. 4.4.4) or power asymmetries (Sect. 4.3.4).

Regardless of how the Kenyan government (currently) fulfills this role, I6 highlights that the government *should* be the actor fulfilling the role of driving the digital transformation in Kenya, especially with regard to its interactions with other multi-stakeholders in the Kenyan digital ecosystem. This can for instance be supported with regard to legitimacy (Sect. 2.5.2). Driving the digital transformation is also associated with the *role as a promoter*. This role is introduced by I19, which is overtaken by the government, that can for instance take the form of diffusing technologies (I7) together with other multi-stakeholders. Although not specifically framed as promoting, it is worth mentioning that all actor groups are ultimately pursuing their own interests, which they promote (Sect. 4.3.3).

In conclusion, the role of the driver in the Kenyan digital ecosystem falls into a "mix of the private sector and the government authorities [...] (I1). The role of the government in driving the digitalization is dynamic and prone to change. I19 claims that the private sector may have overtaken the role as driver from the government recently, especially in the last five years with its young population stated as a possible reason. Balancing the interview findings leads to mix of private sector and government, which is further confirmed by I14. Bilateral and multilateral donor agencies are mentioned to a lesser extent.

Role Enabler

A further role that differs from the theoretical background (Sect. 2.5.3) is the role of an enabler. While ICTs are discussed as enabling development in Sect. 2.2, the governance role of an enabler is only briefly addressed under the role of a network broker in the theoretical background (Sect. 2.5.3). In the empirical data analysis however, the role of an enabler is extensively highlighted by expert interviewees. The role of an enabler is understood as creating possibilities through encouraging or facilitating efforts in the digital transformation (can be an actor or technical tool such as ICTs) and also includes the notion of enhancing.

Firstly, contributing to the discussion in the literature whether ICTs are actors themselves (Sect. 2.2.1, Sect. 2.5.3), this dissertation's results emphasize that expert interviewees critically discuss digital technologies and ICTs as enablers in the Kenyan digital ecosystem. More specifically, I5, I13, I19, I9, I20 and I18 assess ICTs as enablers and enhancers. According to I13, this includes infrastructure and connectivity in the role as an underlying enabler beyond a mere precondition for digital development. Moreover, digital technologies overtake the role as enablers for offering new services, improving efficiency and effectiveness of processes (I5; I18), problem solving (I21), enabling or improving the work of a range of organizations beyond tech organizations (I20) and enabling collaboration as new forms of governance (I13). Whether they are actors is however still debated since their role is acknowledged in the ecosystem, but the notion of 'a means to an end right' (I9; I17) prevails as opposed to ICTs as actors themselves (I9).

Secondly, with regard to this dissertation's focus on interactions of actor groups for a sustainable digital development, the interview data analysis demonstrates that various multi-stakeholders have the role as an enabler in the Kenyan digital ecosystem with the Kenyan government being the most central one. Beginning with the stakeholder group of hubs, I13 (bilateral donor) sees this group and associations in the role of enabling the ecosystem with regard to innovation. This external organizational perspective on hubs in the role as enablers is likewise confirmed by the interview representative of a Kenyan hub (I11). Shifting the focus to the next actor group of donor organizations, both multilateral (I19) and bilateral donor organizations (I4) see themselves as enablers in the Kenyan digital ecosystem, especially in enabling the Kenyan government and local structures. Here, their role is similar to the role of a supporter (Sect. 4.4.5) in enhancing "[...] the preconditions to to reach digital transformation of the economy and society" (I6).

The role of private sector organizations as enablers in the Kenyan digital ecosystem is seen in offering upskilling programs (I10) and also in general upskilling through the use of digital tools (I9) as discussed for instance in Sect. 4.3.2. However, the data analysis did not provide further insights into the role of the private sector in the role of an enabler. While this may lead to the conclusion that the private sector does not take the role of an enabler, enabling functions are not clear cut and can be found indirectly in the other discussed roles. Hence, the private sector is not immediately associated with enabling functions by the interviewed experts (i.e., their role lies elsewhere, see other roles).

Besides these actor groups overtaking the role as an enabler in the Kenyan digital ecosystem, the majority of expert interviewees attribute this role to the Kenyan government. The data analysis demonstrates that the role as an enabler by the Kenyan government is mostly seen in enabling the environment in which the Kenyan digital ecosystem operates (e.g., I22). Environment here is defined by infrastructure, regulation, education and polices among other factors (I14; as discussed in Sect. 4.1). Hence, this role as an enabler is closely linked to addressing the challenges of the environment as discussed in Sect. 4.1. Moreover, in reference to the Kenyan government, the role as an enabler is closely linked to the role of a facilitator (see further Sect. 4.4.5). This is further supported by I20 using the roles of enablers and facilitators interchangeably. In sum, the way in which the Kenyan government exerts this role is seen in regulations and polices that enable digital development while mitigating risks (I13). Hence, regulations and polices as enablers supplemented by legislation serve as a baseline for an enabling environment (I18). These arguments demonstrate similarities to the role of a balancer (balancing potentials and risks, see Sect. 4.4.5).

More specifically, two major instances emerged from the data analysis as to how the Kenyan government uses its enabling role: On the one hand, the government creates an enabling environment for the digital transformation (I23) and more specifically for enabling societal digital development (I7). On the other hand, the interview data analysis demonstrates a focus on enabling digital innovations on an economic level, specifically with regard to startups. The Kenyan government fulfills its role as an enabler for providing an environment for investments (I16). These investments for instance address enabling startups through polices such as the startup bill (I23), although I4 highlights that there is still room to grow for the government's role in enabling entrepreneurial development.

Conclusively, although most of the expert representatives see their own organizations in the role of an enabler, the majority of respondents agree upon the role of the Kenyan government in the governance role of enabling the Kenyan digital ecosystem. This is expressed by representatives of bilateral donor organization (interviews 4, 13 and 18), multilateral donor (interviews 20, 14), academia (interview 22), startup (interview 23), private sector (interview 16) and NGO (interview 7). To a lesser extent, the role of an enabler likewise falls to ICTs and hubs, bilateral and multilateral donor organizations and private sector.

Role Facilitator

Similar to the enabling role of the government through legal frameworks and polices, the role of a facilitator as a further coordinating role is identified. Facilitating is understood as a different angle of enabling that aims at improving an intended outcome (such as effectiveness or efficiency of the ecosystem for sustainable digital development). The role of a facilitator has likewise been derived in the theoretical background (Sect. 2.5.3).

Firstly, ICTs are discussed among interviewed expert representatives as facilitators similar to the before mentioned discussion around ICTs acting as enablers (Sect. 4.4.5), but here without further reference to the discussion around ICTs as actors. As one example, I13 suggests that ICTs are seen in the role as facilitating decentralized programs as a new governance mechanism that enable remote ICT programs. In contrast to the role of an enabler, the interview results towards the role of a facilitator demonstrate a greater focus on the governing role of principles such as the SDGs that facilitate a common understanding in handling ICTs (I5) and digital technologies (see Sect. 4.4.4). More importantly for this dissertation's research question focussing on coordination, digital technologies not only facilitate the setup and administration of networks (I5), but also digital networks or digital platforms can facilitate coordination (I5). This supports this dissertation's argument that digital technologies offer a means for coordination (see also Sect. 4.5).

Secondly, I1 for instance further mentions the actor group of associations such as KEPSA as a facilitator (and a leader), enabled by the Kenyan government. Given the limited data on associations as facilitators, this requires further research.

Thirdly, similarly to the role of an enabler, the Kenyan government is seen as the main actor group that fulfills the facilitating role by interviewed experts. This similarity between the two governance roles is for instance highlighted by I20 in using facilitating interchangeably with enabling in the context of attracting investment. One dimension where the Kenyan government exerts the role as a facilitator is in creating an environment that facilitates the digital transformation in Kenya (I14). Here the roles of facilitator and enabler are again similar. Moreover, the Kenyan government in their role as facilitator contributes to the vibrant Kenyan digital ecosystem through facilitating innovations (I5). Similar to the role as an enabler, the government facilitates action in the Kenyan digital ecosystem through legal frameworks (I22; I1) and through policies such as startup bill or ICT polices (I1). To conclude, the role of a facilitator in the Kenyan digital ecosystem is argued to be mainly overtaken by ICTs and the Kenyan government.

Role Supporter

As a further governance role that emerged in addition to the theoretically derived ones (Sect. 2.5.3), the empirical data analysis further identified the governance role as a supporter. This is indirectly expressed in development cooperation relations between donors and partner countries. Supporter here is aimed at providing support, instead of receiving support. Hence the identified actor groups in the role of a facilitator can likewise fulfill the role of a supporter. The data analysis however identified only the actor groups of bilateral and multilateral donor organizations, and NGOs (to a lesser extent) as fulfilling this role. This simultaneously illustrates the direction of support (i.e., who supports whom):

Beginning with the actor group of NGOs, their role as a supporter is for instance experienced in supporting education through digital technologies (I7). Here the direction of support is from NGOs to society. Furthermore, NGOs can likewise overtake a consulting role in offering support to the Kenyan government (I1).

The main actor group mentioned in the interviews overtaking the role as a supporter is seen in donor organizations, primarily bilateral donor organizations. The role as a supporter fulfilled by multilateral donor organizations is seen in supporting the Kenyan government, for instance in the enforcement of data protection (I17) and in reviewing funds (I20). Here the direction of support is from multilateral donor organizations to the Kenyan government. Similarly to the role of multilateral organizations, bilateral donor organizations as a supporter can be further described through implementing digital projects. As the nature of ICT(4D) projects have already been discussed in Sect. 4.3.2, this section focuses only on the role as a supporter from a governance perspective. The supporter role of bilateral donor organizations is observed in the interaction with the Kenyan government as well as with local hubs and startups.

Support by donor organizations towards the Kenyan government is offered for instance in implementing or executing agendas (I6) and also offering support in raising awareness in relation to the digital transformation (I3). One special area where support is offered refers to data markets (I6). The support here further includes building respective capacities (I6) and "[...] guidance and and and policies on data exchange and sharing and especially with the private sector" (I6). This role is similarly fulfilled with regard to the data protection act (I6). Moreover, the role as a supporter conducted by bilateral donor organizations is further directed in offering support to the local private sector, for instance in "supporting young digital entrepreneurs and new partners and stakeholders like innovation hubs or start tech startups" (I5). Additionally, it includes supporting local startups in getting access to partnerships with multinationals (I4). Bilateral donor organizations in the role of a supporter in and of the Kenyan digital ecosystem is assessed critically by I4 arguing that bilateral donors should support the ecosystem, but not be part of it. With regard to the Kenyan government in the role of a supporter, the data analysis identified supporting in the form of protecting, in this context meaning protecting the local tech economy (e.g., from external entry or activities and exploitation; e.g., I9).

In sum, these are only a few use cases where the role of a supporter is discussed. As mentioned, the other roles have a support component as well. This is only a demonstration of the direct interview results. In conclusion, the role of a supporter in the Kenyan digital ecosystem is fulfilled by NGOs and especially bilateral and multilateral donor organizations.

Role Manager

As a final role, congruent between the theoretical (Sect. 2.5.3) and empirical analysis, the interview data analysis likewise emphasizes the role of a (network) manager. This role includes the roles of a regulator, shaper, governor, interferer and operator, which all are summarized under "manager". In this dissertation, the role as a manager is twofold: one, in terms of who is managing the Kenyan digital ecosystem and two, who is managing the digital transformation in Kenya.

Firstly, the role of a manager is discussed among interviewed expert representatives with regard to managing the digital transformation in Kenya. Here, the managing role is further described as a "governor" and "regulator", highlighting the governance role. The role of a governor of the digital transformation in Kenya is argued to be overtaken by the Kenyan government, especially in form of "legislation governing" (I18) and "putting up laws that govern" (I11). Hence the managing role of the Kenyan government is seen in their unique ability to legislate and also to regulate (I15). Here it is worth mentioning however that it has been previously established that the private sector (and other multi-stakeholders) are actively involved in policy processes, so this is to be further discussed (Sect. 4.3.4).

Secondly, the role of a manager is discussed among interviewed expert representatives with regard to managing the Kenyan digital ecosystem. This is closely linked to the role of coordination of the Kenyan digital ecosystem as discussed in Sect. 4.4.5

and Sect. 4.1.6. This section however only concentrates on the governance role of the manager. Here, the role of the manager falls to a variety of multi-stakeholders instead of just one. Multi-stakeholders are identified as 'strongholders' of the Kenyan digital ecosystem: "I think it should be a mix because we cannot have just government and being part of it but I mean it it it's always good to have a mix of different stakeholders who are holding this ecosystem." (I10) While the private sector is seen as a 'network interferer' (I2) due to their power relation (see Sect. 4.3.4), the actual 'shaper' of the ecosystem is the Kenyan government (I5). In this the Kenyan government is conclusively perceived as the leading manager, especially in setting agendas and attracting or mobilizing investment (I19) as discussed in Sect. 4.4.5. Similarly, the government is seen as the "network operator" (I2). To conclude, the role of a manager of and in the Kenyan digital ecosystem is overtaken by the Kenyan government.

In conclusion, the interview data analysis delivered further insights into coordination of and in the Kenyan digital ecosystem through various governance roles that are overtaken by multi-stakeholders as summarized in Table 4.2. Besides governance roles, this dissertation further proposes an additional 'solution' for coordination in form of governance forms, which will be discussed in the next subsection.

4.4.6 Determinant 4: Governance Forms in the Kenyan Digital Ecosystem

Before going into detail of possible governance forms for coordination, expert representatives assess that there is a need for governance in the digital transformation for sustainable digital development exemplified by the Kenyan digital ecosystem. Firstly, a need for governance in the sense of regulating the digital transformation is discussed (e.g., I19) with both terms mostly used interchangeably. Here it is worth referring back to Sect. 4.1.2 where the implications of (a lack of) regulation for sustainable digital development in Kenya are discussed. More specifically, the need for governance is further expressed with regard to the Kenyan ICT industry (I15), governance of data (I5), the internet (I11) and for governing interactions in the Kenyan digital ecosystem (I15). This is especially relevant for this dissertation, as it emphasizes the relevance of both analyzing interactions in the (Kenyan) digital ecosystem, as well as analyzing possible governance mechanisms, which is the main focus of this section.

Secondly, the need for governance structures for achieving sustainable digital development in the Kenyan digital ecosystem i.e., governing the ecosystem is discussed. Here the two angles of governing the ecosystem (need for governance structure to align the ecosystem I5) and digital technologies (I5) are simultaneously mirrored. Moreover, a need for a concept or framework for multi-stakeholder collaboration in the Kenyan digital ecosystem is expressed, for instance by I21. While possible governance forms have been derived in the theoretical background in Sect. 2.5.4, this section will further provide empirical evidence into these possible governance forms for sustainable digital development in Kenya. Therefore, this section is closely linked to the before mentioned governance roles (Sect. 4.4.5) in that these roles cannot only be fulfilled by multi-stakeholder actor groups in the Kenyan digital ecosystem, but also by other institutionalized governance forms such as PPPs or MSPs. More specifically, different degrees of formalization or institutionalization (less formalized vs. formalized) as discussed in the theoretical background (Sect. 2.5.4) and their impact on coordination and sustainable digital development are distinguished here. Less formalized governance forms in this research project are self-governance forms in contrast to formalized forms such as associations and alliances, standards, working groups, committees, PPPs and MSPs. These will be discussed in more detail with regard to the Kenyan digital ecosystem in the following.

Less Formalized Governance Forms

Beginning with the analysis of less formalized governance forms, in addition to the theoretical discussion in Sect. 2.5.4, the empirical data analysis further identified governance forms of events and workshops as less formalized coordination mechanisms of multi-stakeholder interactions in the Kenyan digital ecosystem.

Events and Workshops As one less formalized governance form, *events* such as conferences are mentioned by expert interviewees as governance formats that bring multi-stakeholders together (I10). Local tech events are perceived as collaboration means (I16). Moreover, multi-stakeholder events are also seen as one determinant that characterizes the Kenyan (Nairobian) digital ecosystem as vivid and open (I1). Openness also in the sense of inclusiveness, e.g., invitations of all multi-stakeholders, however is discussed controversially in Sect. 4.4.3. Inclusiveness in events is similarly discussed in the context of efficiency (Sect. 4.4.3) with I12 sharing the experience that it is more constructive to work with a smaller group of actors which is why (their) events are often divided into sessions.

Another less formalized governance form mentioned by the interviewed expert representatives are *workshops*. Workshops here are understood as less formalized because they are limited in time or just on temporary basis. Workshops are mentioned as a format for "bringing in people together" (I14), more specifically including (in the best case) a variety of stakeholders from various stakeholder groups in workshops (I20). In the Kenyan digital ecosystem, workshops between the EU and private sector actors are named as one example (I12). Moreover, workshops are also a format used by the private sector in their interaction with stakeholders, also as part of their selling process (I10).

Self-Governed Structures A further governance form that is understood as less formalized or institutionalized are self-governed structures in line with the respective theoretical section (Sect. 2.5.4). When discussing governance mechanisms for coordinating digital ecosystem activity in Kenya, an alternative to active governance through e.g., a single institution is identified in form of self-governance. Similarly to the blurring understandings and usage of the terms regulation, governance and coordination (see beginning of Sect. 4.4.6), self-governance is likewise used interchangeably with other terms such as self-regulation, self-governance and self-coordination. This governance form is here discussed under less formalized forms since it is understood as governance without the active engagement of any governance form(at) or institution.

Firstly, the value of a "self geared towards self coordination in the sense of self govern system" (I1) for the purpose of information sharing, connecting actors and making actors' actions transparent is emphasized. This is a further instance where coordination and governance are discussed interchangeably. Moreover, self-governance is not only expressed as needed, but it is even seen as a goal to be achieved (I5). Expanding this goal perspective of governance formats, self-governance is further stated as an alternative to hierarchical coordination of the Kenyan digital ecosystem for achieving "a flourishing ecosystem where innovation can happen" (I13). This argument demonstrates a counterargument against institutionalized governance or coordination, which is further supported by I1. More concretely, this opposition to coordination and instead a preference of self-regulation or self-governance hence specifically applies to the setting of (digital) innovations. In innovation for development, a certain amount of space is argued to be needed that is enabled through self-governance (I16). As previously discussed in Sect. 4.4.3, Sect. 4.3.4 and Sect. 4.1.6, attention here is drawn to avoiding stifling of innovation (e.g., I24). Hence coordination through the governance mode of hierarchy is contested (see Sect. 2.5.4) and a more passive or laissez-faire approach preferred with not much regulation or restriction (I1; I24), leaving room for innovation (I3; I12) without stifling innovation. Challenges of this approach for sustainable digital development in Kenya however have been discussed in Sect. 4.3.4 and Sect. 4.1.2. Here it is worth mentioning that in the expressed statements towards a liberal governance approach, possible biases of the respondents in terms of political ideology or paradigm cannot be ruled out completely.

Secondly, some expert interviewees expressed mixed perceptions towards the impact of self-governance in the Kenyan digital ecosystem. I15 offers a diplomatic approach towards the role of self-governance in the Kenyan digital ecosystem as the expert argues that it depends on the targeted area. While I13 takes the stance of "the middle way is is kind of the highway", I11 was the only interview that directly proposes an argument against self-regulation given that individual actors' interests are favored against interests of society. However, this limited availability of insights can possibly be referred back to the research project's sample size and the fact that not all of the interviewed actor groups delivered insights on self-governance.

Formalized (Multi-Stakeholder) Governance Forms

After discussing possible less formalized governance forms for a sustainable digital development in the Kenyan digital ecosystem, this section will deliver insights into more formalized governance forms. In line with the theoretical background (Sect. 2.5.4), the empirical data analysis has identified the following formalized governance forms in the Kenyan digital ecosystem: standards, alliances, associations, committees, working groups, PPPs and MSPs and thereby almost identically following the theoretical approaches. With regard to Kenya, Lundsgaarde and Keijzer (2019) for instance argue that more formalized mechanisms for coordination have in recent years included various types of country-level dialogue forums or working groups, which guided by various forms of 'joint assistance strategies' (Lundsgaarde and Keijzer 2019:223). This section will deliver further empirical insights from the analyzed expert interviews.

Standards for Coordination

Beginning with a formalized governance form, standards for coordination in the sense of governance have been identified in this dissertation's theoretical (only to a minor extent in Sect. 2.3.6) and empirical analysis. Standards are understood as forms of standardized regulations that can be of both technical and regulatory nature. Standards are further identified in regulating stakeholders' actions in digital development, especially with regard to the ICT industry, that can be implemented both on the global and the local level. Before analyzing these in more detail, it is important to note that expert interviewees share experiencing different understandings of standards (e.g., I20; similar to other topics in the digital transformation, see Sect. 4.4.4). This is another instance that complicates sustainable digital development and coordination, here through governance form of standards. With regard to this dissertation's research question, I7 assesses the potential of standards for achieving coordination transparently.

Expanding on transparency through *technical standards* in the digital transformation, open standards are argued to be crucial. One example that experts refer to is 'https' which "[...] creates a type of coordination through standards that can be very effective and low cost." (I7) Open standards as well as open source can potentially increase transparency (also with regard to open data or open source, see Sect. 4.2.2, Sect. 4.4.3), and likewise serve the purpose of preventing vendor lock-ins as discussed in Sect. 4.4.4.

Moreover, and closely linked, I10 calls for (more) *standards to regulate the IT industry* as "[...] there should be also standards to regulate the IT industry like what players are there, how do we support emerging players in the market and all that." I9 highlights that IT industry standards should likewise apply to IT companies, especially to multinationals. Hence, standards for regulating the IT industry in Kenya needs to be further explored in future research as they have an impact on the sustainable digital development of the ecosystem in their regulating function (Sect. 4.1.2).

In addition, the expert interviewees emphasize the value of *global and local standards* for collaboration and coordination in sustainable digital development. I20 for instance refers to the cross-cutting nature of standards that can be applied to a variety of countries.

With regard to the role of local standards for sustainable digital development, I11 for instance argues that standards should be decided locally. Similarly, I7 claims that one value lies in implementing standards for protecting local innovation, especially with regard to foreign tech companies in the Kenyan digital ecosystem. Standards are also mentioned as one determinant for Kenya being a rolemodel that influences other countries as well (I17). Another use case for local standards is seen in allowing harmonization through standardization (I20) and the potential of standards to allow alignment (see further Sect. 4.4.3).

Another area for applying standards as a governance form in digital development of Kenya, *standards for ethical and normative purposes* are identified by the data analysis. Standards are highlighted as an important factor in explaining the performance of countries in equal and fair regulations of the digital transformation (I7) as opposed to countries perceived as lacking ethical standards (I3). This for instance relates to standards for digital identity (I7). One standardized way of governing is for instance seen in principles such as the *Digital Principles* in development cooperation that serve as a guideline for action (I5, see further Sect. 2.3.2).

For this dissertation, governance of digital ecosystems such as the Kenyan one comprising of multi-stakeholders is crucial. It is important for this research project to elaborate on this actor perspective, for instance with regard to governance roles. This is likewise important in the discussion around *actors' roles in standards* for digital development. Similarly to the governance roles discussed in Sect. 4.4.5, a representative of a multilateral organization assesses its organization in the role of promoting standards (I17), while likewise mentioning another multilateral organization (ITU) as the multilateral coordinator for (technical or digital) standards. The role of the Kenyan government in setting standards is similarly discussed by I11 and in the previous passages of this subsection. In sum, standards are highlighted especially in the empirical data analysis as possible governance forms of coordination for sustainable digital development.

Alliances and Associations
Additionally, alliances and associations emerged in the data analysis of this research project as further governance forms, both being often used interchangeably by interviewed experts. Alliances are understood as an institutionalized form of actor groups being united under an alliance umbrella. While alliances and associations have been subsumed under self-governed forms in the theoretical background (Sect. 2.5.4), the empirical data analysis suggests that they are more formalized since they can also be seen as individual actors or organizations representing their members.

On the one hand, alliances are seen as a trend (see Sect. 4.4.1, Sect. 4.2), similarly to the emergence of hubs expressing an organization form of the innovator scene with increased professionalism (I5, see Sect. 2.1.4). On the other hand, the value of alliances beyond a mere trend is experienced as a well-working or even best-practice format: "I think what we see across Africa what tends to work quite well are alliances alliances" (I2). One explanation is the impact of alliances on increasing growth, with KEPSA named as one major example in the private sector (I24), which is also a (coordinative) format for collaboration (I16). The governance format of alliances is seen as predominantly used for following strategic interests (I2, see further Sect. 4.3.3). Through the formation of a multitude of (similar) actors under the umbrella of alliances, they bestow the individual actors with a greater voice, especially in the interaction with the multi-stakeholders in the Kenyan digital ecosystem (I5). The importance of voice for sustainable digital development has been discussed in more detail in Sect. 4.4.3. In sum, alliances are formed for strategic interest purposes in a less formal way than associations, which are understood as a separate actor group.

Similarly to alliances, associations are understood as a group of organizations that align and together form an association for a joint goal. According to I11, associations are further understood as "a consortium of active players in tech innovation in Africa" (I13) and also used to define a network as discussed in Sect. 4.4.1. With regard to the Kenyan digital ecosystem, the governance form of association is mentioned with regard to the organizational body around tech hubs that plays a role in achieving coordination and sustainable digital development with individual tech hubs uniting under the association umbrella (I11). In the Kenyan digital ecosystem, associations not only fulfill the role as an intermediary as discussed in Sect. 4.4.5, but also function as a channel for individual actors in their interaction with different multi-stakeholders such as bilateral donor organizations (I18) and multilateral donor organization (I12). Hence, the benefit or value of association as a governance form for sustainable digital development is seen in the potential of alignment of individual actors under the umbrella of an association for achieving coordination (I11). With regard to their performance in the Kenyan digital ecosystem, associations are assessed by I11 as "doing well", while other expert interviewees did not comment on their impact or performance.

Committee

A further governance form are committees, which are subsumed under associations and alliances in the respective theoretical section (Sect. 2.5.4). Committees in the empirical analysis are defined as a group within an organization or from different organizations that have chosen to align. With regard to committees as a governance form in the Kenyan digital ecosystem, they are experienced as quite established by I10. More specifically, the insight is shared that committees are created for specific digital (development) topics, such as cybersecurity or data protection (I15, see e.g., Sect. 4.4.4). Supporting this dissertation's angle on committees as a governance form for sustainable digital development, I15 confirms the setup of committees for coordination, but rather in the sense of representation: "[...] what they do is they have representative committees of the stakeholders." Another example for this representation are 'steering committees', for instance within multilateral organizations (I19).

The goal of creating committees for governance of the digital transformation in Kenya is, among others, to have "an inclusive committee of the different stakeholders" (I15). However, the role of committees for inclusion and sustainability is discussed controversially among expert interviewees. I10 criticizes that, in their impression, committees sometimes lack experts with digital know-how (see further Sect. 4.4.4). Moreover, I7 even questions the outcome or impact of such committees: "[...] then you form a committee with multiple people, multiple interests who are already fairly risk averse you know what do I expect that group to come up with an innovative bold and transparent standard that really takes a thesis that's gonna move the future forward?".

Working Group

Closely related to the governance form of committees is the format of working groups. In the theoretical background, working groups are discussed as part of alliances and associations (Sect. 2.5.4). Working groups in the empirical analysis are understood as a group of actors (or people from several organizations) coming together for discussing and solving a specific problem.

With regard to the Kenyan digital ecosystem, "Kenya is an example of good practice, with the government using multi-stakeholder sector working groups to spearhead development planning and support budget allocation decisions" (OECD/UNDP 2019:65). Working groups usually comprise of members from ministries, non-governmental organizations, the private sector, UN bodies and other relevant groups (Ceelen, Wood and Huesken 2019). The following passages will deliver further empirical insights into the impact of working groups on sustainable digital development assessed by interviewed experts.

Firstly, various forms of working groups in the Kenyan digital ecosystem are highlighted by the interviewed experts. What these have in common is "bringing actors together" (e.g., I13). One example are donor working groups on private sector development where for instance entrepreneurship builds a subgroup (I4). Additionally, working groups are established as formats of advisory with a possible duration of weeks or months (I12; see also Sect. 2.5.1, Sect. 4.4.3). More specifically with regard to donor working groups, I4 (bilateral donor organization) criticizes that the goal of donor working groups is not (always) alignment, but merely informing each other (see Sect. 4.4.3).

Secondly, the data analysis highlights that working groups differ in involvement or participation of multi-stakeholder actor groups. With regard to donor working groups, I17 shares the experience of participating in three donor sub-working groups. In addition, the amount of donor working groups (for the purpose of coordination) is further explained as evolving over time and as a source of confusion due to a lack of awareness of involved stakeholders (I13, Sect. 4.4.3). One expert representative of a multinational tech company expresses no awareness of multi-stakeholder working groups and hence no participation of their company in working groups (I10). This is contrary to the perception of I19 that (other) private sector actors are involved in multilateral Smart Africa working groups, especially if they are members. This similarly applies to ICT donor coordination working groups, which are mostly comprised of Western donor agencies (I13), although hubs are claimed to be a part of these ICT donor working groups as well (I11). Besides the mere participation of multi-stakeholders in ICT (coordination) groups, different forms of leadership of these working groups are illustrated (I6).

In addition to the mentioned criticism towards alignment, one representative of a Kenyan NGO criticizes the impact of such working groups by sharing the experience of participating in multi-stakeholder formats such as working groups as wasting time (I7, see Sect. 4.4.3). One host of such an ICT donor working group however shares the opposing (self-)perception of successful working groups:

> "[...] in other sectors we've been quite successful in Kenya already with setting up technical working groups or sub sector according nation groups always under the leadership of the respective authority or ministry of the Kenyan side so like shared by the implementing partner by the political partner or line ministry on the Kenyan side and I think that is crucial and key but I feel like specifically for the digitalization it can and will lower dependency." (I6)

In sum, expert representatives in the Kenyan digital ecosystem acknowledge working groups as a format for coordination in terms of coming together to attempt alignment, however their impact on sustainable digital development is debated.

Public-Private-Partnerships

As discussed in the theoretical background (Sect. 2.5.4), and most importantly contributing to this dissertation's research focus on multi-stakeholder activities in the Kenyan digital ecosystem, PPPs and MSPs are important governance forms in the context of sustainable development. PPPs are subsumed under formalized governance forms due to their contractual nature between public and private partners and thereby demonstrate a formalized closeness in a contractual setting (Sect. 4.4.3). Given their importance in sustainable digital development and for this dissertation's focus on network governance as coordination mechanisms, both formats are often used interchangeably with alliances and cooperations by the interviewed expert representatives (e.g., I2). While MSPs are discussed in more detail in the following subsection, this section will deliver empirical insights into PPPs using the Kenyan digital ecosystem. PPPs are (only) formed between public and private actors, here especially the Kenyan public sector and (foreign) tech companies. Similarly to the preconditions derived in this dissertation's developed framework (Sect. 2.5.1, Sect. 4.4.3), these likewise apply to the governance format of PPPs. In addition to an "appropriate partner selection" (I18), these preconditions are for instance highlighted by I5 as "[...] before such a partnership is formed usually you ask for a mutual interest and mutual objec objectives and then see how the capacities and capabilities are complementary to each other [...]". Here especially the preconditions of shared interests and objectives (Sect. 2.5.1) are highlighted.

Firstly, with regard to PPPs for sustainable digital development in the Kenyan digital ecosystem, PPPs are highlighted by I6 as a network format that is an important coordination mechanism for sustainable digital development in Kenya (I4). PPPs and MSPs are important formats for sustainable digital development as they can be helpful mechanisms to leverage and create societal impact (I5). More specifically, I20 sees the potential of PPPs for sustainable digital development in their longer-term learning opportunities, especially for public sector actors. Specifically relevant for this dissertation, PPPs are one established collaboration format of multi-stakeholders in the Kenyan digital ecosystem. One expert representative of a multilateral donor agency (I17) shares experiences of collaboration with the private sector in PPPs, while highlighting "we have very limited tools". Through its Public Private Partnership Act (2013), Kenya has set up a specialized PPP unit within the treasury to promote and to oversee Kenya's PPP program. Here ICTs are one of the seven priority sectors for PPP project implementation in the country.[12] The empirical data analysis highlights a special use case for PPPs in the Kenyan digital ecosystem that is mostly seen in (digital) infrastructure developments (I20).

Secondly, however, the impact of PPPs for sustainable digital development in the Kenyan digital ecosystem as a format for coordinating multi-stakeholders activities, is discussed controversially among expert representatives. This dissertation analyses PPPs as a format for not only coordinating multi-stakeholder activities for sustainable digital development, but also as a format for addressing asymmetries between multi-stakeholders by being 'partners'. With regard to the discussed power asymmetries in the Kenyan digital ecosystem (see Sect. 4.3.4), especially between public and private actors in the digital transformation, interviewed expert representatives however assess the potential of PPPs for solving power asymmetries as limited. I2 for instance shares the experience of PPPs as "[...] that's been tried and tested in what in multiple ways a a private public partnership has proven to be a very dangerous construct because often the public side loses out against the private side who knows more and has better arguing in power is like for the digital space that that certainly holds true". Hence PPP's potential to balance power asymmetries is questioned. Instead, they potentially even increase asymmetries in the digital transformation. This experience is shared and underlined by I3 arguing that PPPs further strengthen the power of the private sector, which is counterproductive to sustainable digital development. The impact of PPPs on sustainable digital development is further questioned by I2 sharing the experience that PPPs "[...] don't necessarily last very long because they figure out that they've basically not been not been achieving

[12] More information available at https://www.pppunit.go.ke.

the right results". This is mirrored and further manifested in the discussed failure of ICT projects as discussed in Sect. 4.3.2

One reason for the failure of PPPs, failure here in the sense of failing to meet the goal of fostering sustainable digital development, is identified in the data analysis with regard to the *hype around PPP* (similarly to the hype around digital transformation, Sect. 4.3.1). For instance, a hype around PPPs and the role of private sector actors in the partnership formats are emphasized by I17. More specifically, both the hype around PPPs and the power asymmetry in the partnership format are mentioned as a potential cause of unsustainable digital development, for instance in terms of drained financial resources of the public sectors:

> "Aaand I think the reality is quite different in the sense that PPPs have a contentious history in the last few decades many organizations have been promoting PPPs as the best in the world and in the majority of the cases they have ended up costing a lot of money to the public sector 'cause there is an asymmetry or the private sector has huge incentives of making profits and pro public sector doesn't have so much incentive to protect itself from losses and that has generated some good projects but a lot of depth[debt]." (I17).

The hype around PPPs and also around MSPs is specifically ascribed to public sector actors (I17), especially with regard to the role of the government in ensuring sustainability and durability of PPPs (I20). A lack of alternative governance tools is mentioned as a further cause by I17.

Multi-Stakeholder Partnerships

Closely related to the governance format of PPPs, this dissertation proposes MSPs as a further means to address the discussed dependencies and power asymmetries in the Kenyan digital ecosystem for sustainable development (Sect. 2.5.4). Similarly to PPPs, MSPs are here understood as a further formalized governance form that is not necessarily structured via contracts, but manifested through international principles such as the SDGs, especially SDG 17 (Global Partnerships, see Sect. 2.1.3). Moreover, similar preconditions for the performance of and governance through PPPs, likewise apply to MSPs such as willingness, capability, and resources of partners (Pattberg and Widerberg 2016:46, Sect. 2.5.1).

Firstly, the importance of MSPs for sustainable digital development in the Kenyan digital ecosystem is highlighted among expert interviewees ranging from "quite important" (I24), "very very important" (I14), to "it's key" (I21). The importance and value is for instance seen in MSPs as a format for "bringing together" multi-stakeholders across different sectors in the Kenyan digital ecosystem (I6). Potentials of MSPs according to expert representatives lie in trade and market opportunities

(I14), economic development (I14), opportunities for funding (I7) as well as in information sharing and awareness raising around digital development or digital technologies (I3). From a development cooperation perspective, the SDGs highlight multi-stakeholder partnerships for achieving SDG 17 (Global Partnerships) as discussed in Sect. 2.1.3. More importantly for this dissertation's research focussing on coordination, I6 refers to MSPs as "classic tools" for coordination and "an advantage to have more of them".

Secondly, however, the data analysis of this dissertation highlights challenges of MSPs as coordination mechanisms that limit their impact on sustainable digital development. While the awareness around multi-stakeholder dialogues or partnerships are recognized as being important (e.g., I13), the interviewed experts nonetheless highlight challenges of multi-stakeholder processes. Here the experience is shared that

> "[...] because multi stakeholder often means that process is take long and if you want to have a certain political agenda achieved in a certain amount of time that impacts that. So I think that is something that that has clearly also an impact on how this coordination between actors is working". (I13)

Hence a delay of processes due to multi-stakeholder involvement in a given political time frame is one challenge for the performance of MSPs and hence sustainable digital development. More specifically, the multi-stakeholder rhythm of MSPs is perceived as time and resource intensive (I13). This is especially challenging for NGOs since "[...] and civil society always is in a way looking out for for for funding supports so being able to for instance participate in a a law making process that takes three-four years that's not something that anybody can do." (ibid.). Besides the durability of MSP processes (Sect. 2.5.1, Sect. 4.4.3), the goal of MSPs to include a variety of multi-stakeholders poses challenges to sustainable digital development due to the number of participants in MSPs (as discussed in Sect. 4.4.3) and transaction costs (I18). A further challenge is seen in the tender processes of MSPs (I21, see also Sect. 4.3.4). The impact of MSPs on sustainable digital development is questioned by expert interviewees such as I1, experiencing "not so much" sustainable impact.

Most importantly for this dissertation, the potential of MSPs as a coordination mechanism for a sustainable digital development, I5 and I7 argue that MSPs are complex to manage due to the specific interests, expectations and requirements of multi-stakeholders (as further discussed in e.g., Sect. 4.3.3, Sect. 4.4.3, Sect. 4.4.4, Sect. 2.5.4). From these challenges, preconditions for the impact of MSPs on sustainable digital development can be derived. Complementary to this dissertation's identified preconditions for governance (Sect. 2.5.1), these likewise apply to MSPs as a governance format. This is for instance highlighted with regard to the transparency and awareness of multi-stakeholders around requirements and expectations of multi-stakeholders (I15; I3, see also Sect. 4.4.3). Moreover, the performance of MSPs further depends on the MSP process and design, including common goals, communication and management (I18). Hence common goals are further crucial for MSPs to develop sustainably (I21). This underlines the importance of the identified precondition of goal consensus for a sustainable digital development and digital ecosystem governance (Sect. 2.5.1; Sect. 4.4.3).

With regard to the coordination of multi-stakeholder activities in the Kenyan digital ecosystem, I15 highlights the need for governance of MSPs, which would provide a framework around how to conduct MSPs. Similar to the discussed governance roles (Sect. 4.4.5), the Kenyan government here is seen as "key in this" (I24) for fostering collaboration between multi-stakeholders, including in MSPs.

4.5 Outcome: Coordinative Digital Platform and Meta Organization as New Governance Structures in the Kenyan Digital Ecosystem

As a last section of the empirical analysis of the Kenyan digital ecosystem, the outcome of the empirical analysis will be discussed. The data analysis in synthesis with the theoretical background (Sect. 2.5.5) identified a coordinative digital platform and a meta organization as possible ways to achieve sustainable digital development in the Kenyan digital ecosystem. More concretely, both the findings of the theoretical background and of the empirical analysis, especially in reference to the developed framework, culminate in this outcome section. Here, firstly, the need for new governance formats and governance structures for a sustainable digital development will be discussed based on the expert interviews. Specific reference to the need for an overview and mapping of multi-stakeholders and their activities in the Kenyan digital ecosystem will be provided. Secondly, the two developed proposals as to how ICT, development and governance can be combined for achieving and fostering sustainable digital development will be discussed. Proposal 1 refers to the governance form of a meta organization and proposal 2 is a coordinative digital platform for governing the Kenyan digital ecosystem's activities.

While some discussed forms of coordination would classify as "new" due to their simultaneous emergence with the digital transformation (Sect. 4.4.1, Sect. 2.4.3) such accelerators, lab networks (I14), and hubs that "revolutionize provision of the governance so different solutions" (I18), they are now quite established. Their role and impact in the Kenyan digital ecosystem has further been discussed in e.g., Sect. 4.4.5. Moreover, a discussion in the academic literature and among expert interviewees is introduced in terms of centralized (i.e., brokered) vs. decentralized (i.e., non-brokered) (self-)governance forms (see e.g., Sect. 4.4.6, Sect. 2.5.4). For instance, I13 calls for "new governance mechanisms, a new de decentralized governance mechanisms". The exact form however is debated. I6 for instance emphasizes a need for subgroups on digital transformation topics (e.g., for data protection, digital skills) on a technical instead of a macro level of coordination for alignment and knowledge sharing. Nonetheless, experts highlight this dissertation's relevance in asserting a lack of tools to collaborate with multi-stakeholders and a difficulty "to see many examples of effective collaboration" (I17).

After an introduction into the need for alternative governance forms in the Kenyan digital ecosystems, coordination through approaches of mapping for providing an overview are discussed among expert interviewees. This will be analyzed in the following before discussing the two coordination 'solutions' proposed by this dissertation.

4.5.1 Coordination through Overview and Mapping

Firstly, expert interviewees emphasize the importance of and a need for an overview in order to achieve a sustainable digital development. This includes the form of a mapping of multi-stakeholder activities in the Kenyan digital ecosystem. In contrast to the literature review in the theoretical background (Chap. 2), gathering an overview of multi-stakeholder activities has only been derived from the empirical data analysis. Moreover, this rather falls into the scope of ANT[13] than this dissertation's network governance focus. Nonetheless, the empirical data analysis suggests a relevance of actor mapping for coordination that needs further research. While I6 assesses coordination and having an overview as "[...] the way forward", the need to identify the key stakeholders in the ecosystem is further highlighted by I12. More specifically, not only an overview of who are the (key) multi-stakeholders in the Kenyan digital ecosystem is argued as crucial for sustainable digital development, but also an overview of power relations (as discussed in Sect. 4.3.4):

> "But having a clearer overview of who has to be involved, what are the power relations between the actors because I think that's also something that is not taken into account

[13] For instance The World Bank uses a tool 'Net-Map' for stakeholder mappings (The World Bank 2016b), see also Bella, Grant, Kindornay and Tissot (2013).

> enough, what are the power relations of of the actors between themselves when we have multi stakeholder mechanisms." (I13)

More importantly, this underlines that coordination is argued to be beneficial for achieving this overview (I18).

Secondly, having an overview and a mapping is argued to be a precondition for coordination of multi-stakeholder digital activities. More specifically, coordination here is valuable for managing contacts, categorizing fitting counterparts and inclusively addressing each stakeholder without leaving anyone behind (I12). This emphasizes coordination for providing an inclusive overview of multi-stakeholders and a tool for 'managing' these interactions. One indicator as discussed in Sect. 2.5.1 and Sect. 4.4.3 is visibility, more specifically visibility of stakeholders that have participated in a policy making process (e.g., I13). However, this approach has been identified as not being holistic and inclusive since not all multi-stakeholders were identified as equally involved or recognized (see Sect. 4.4.3 and Sect. 4.3.4).

Thirdly, this leads to an impact of coordination for achieving an overview of multi-stakeholder actions in the Kenyan digital ecosystem. The value of coordination is for instance perceived in not only mobilizing resources (see further Sect. 4.1.6, Sect. 4.4.4), but also in having a (visual) overview about the (equal) distribution of resources (I14). Similarly, coordination for achieving an overview would also address the closely linked challenge of fragmentation, both in terms of aid fragmentation, but also in terms of fragmentation of knowledge, power and resources (Sect. 4.1.5, Sect. 4.3.4). I14 here highlights that fragmentation in the ecosystem can be reduced if there is a "[...] visual on what is being done and what is not being done." Another dimension of coordination for sustainable digital development has been identified with regard to alignment of actors (Sect. 2.5.2, Sect. 2.5.1, Sect. 4.4.3). An overview or a mapping is assessed as crucial for facilitating further alignment of actors: "[...] have a mapping of who works in which field and what intervention, how long, which is already a big improvement because that helps obviously to align things into future." (I13) This is also argued to potentially decrease transaction costs (Sect. 4.1.6).

Fourthly, and most importantly for this dissertation's analysis of multi-stakeholder interactions in the Kenyan digital ecosystem, it is further crucial to analyze which stakeholder group claims to have an overview and which does not. Here, the data analysis identified a fragmentation, since some actors do claim to have an overview and many do not, which likewise feeds into parallel working and doubling of activities (Sect. 2.5.2, Sect. 4.4.3). This dissertation argues that a coordination of multi-stakeholder actions would decrease parallel working and doubling of activities for a sustainable digital development.

On the one hand, the following actor groups stated having an overview and a mapping of the multi-stakeholders (and digital projects):

- I11
- I12, specifically of "[...] digital projects in different African partner countries by from the European side"
- I17
- I20, states having several stakeholder mappings in Kenya

On the other hand however, the following actor groups stated having no overview of the multi-stakeholders (and digital projects):

- I13, specifically as "[...] this is always a little bit confusing when suddenly there's a new actor on the stage and 'oh oh okay you also work in ICT, good'."
- I9, specifically with regard to awareness and visibility as "sometimes it's not always clear what exactly is going on and who's doing what in the ecosystem"
- I14, specifically as "[...] sometimes very difficult to know who's doing what where they're doing it where where or what purpose they're doing it."

Fifthly, these findings highlight the need for a holistic and inclusive overview in form of a stakeholder and activity mapping of the Kenyan digital ecosystem for a sustainable digital development enabled through coordination. The need for a mapping is further emphasized by I13 calling for a "whole stakeholder map of the digital economy in Kenya" and by I20 highlighting the need for a holistic mapping. This is why this dissertation has derived two proposals as to how this can be achieved.

4.5.2 Proposal 1: Meta Organization for Coordination

Beginning with the first proposal, this dissertation analyses a potential meta organization governing and coordinating the Kenyan digital ecosystem's activities for a sustainable digital development. A meta organization is understood as a central coordinative body or authority that can emerge from the ecosystem, but can also be an external actor. Moreover, a meta organization is seen as a formalized or institutionalized form of meta governance as proposed in the respective theoretical section (Sect. 2.5.4). Hence, it stands both in contrast to discussed forms of self-governance (Sect. 2.5.4, Sect. 4.4.6) and to the second proposal of a digital coordinative platform (Sect. 4.5.3) that is not institutionalized. However, the suitability of a meta organization, especially the creation of a new (additional) organization, is discussed controversially among expert interviewees. In this, experts did not always distinguish or indicate whether they are referring to an existing or new organization/governance body.

On the one hand, expert representatives argue that an organization is needed for coordinating multi-stakeholders and their actions in the Kenyan digital ecosystem. Experts explicitly highlight the need for a coordinative instance (I3), for a coordinative body (I10) or "a need for a regulated body for all that" (I10). More specifically, a meta organization for coordination is especially seen as relevant for kick-starting sustainable interactions "to govern all that" (I10). This is similarly supported by I5. Furthermore, I5 argues that after a coordinative body sets up new structures for coordinating multi-stakeholders' digital activities, the entire ecosystem then focuses on this organization with some of the multi-stakeholders taking over the coordinative role. However, this is criticized due to the level of experience and capability of existing multi-stakeholders taking the role as a coordinator as discussed in Sect. 4.4.5. Another benefit of a meta organization in contrast to multi-stakeholder self-coordination is seen with regard to the issue of political stability and corruption: "[...] it should be a separate body actually so that you can not really influence it because you see corruption is huge here." (I10, Sect. 4.1.2) Further use cases where a meta organization is assessed as needed, are seen by experts with regard to ensuring equal representation, especially of smaller players (I2) and for deriving economies of scale from a central coordinative body (I2).

On the other hand, expert representatives however also oppose the need for a meta organization coordinating the Kenyan digital ecosystem. For instance, I4 is "[...] finding it very hard to imagine anyone in particular managing such network on their own". Similarly, I1 counters the argument for a coordinative meta organization due to "[...] simply too much overhead too much work and friction and cost and everything" and the need for "super super power and super for [unk] super information" (I1). These arguments highlight a possible increase instead of a decrease in transaction costs (Sect. 4.1.6). This is likewise mirrored in the discussion around benefits and challenges of self-governing forms (Sect. 4.4.6).

While I5 questions the need for "a new organization for the coordination", I19 further argues that not a new organization is needed, but the support and reinforcement of existing ones. This is similarly perceived by I5 arguing in favor of multi-stakeholders leading coordination instead of a meta organization, as "[...] if you then set up a new organization this may irritate or be contraproductive actually". Here, empowering multi-stakeholders' roles (Sect. 4.4.5) in coordination is seen as less complicated and less counterproductive to sustainable digital development than a coordinative meta organization. Another argument that feeds into this perspective is the question of legitimacy of multi-stakeholders and of a (new) meta organization since "[...] I would fear that a meta organization to to manage managing such a network would just complicate things who would you know what what would their legitimacy to do so be" (I4). This can be interpreted as an argument in favor of the Kenyan government as a coordinator considering their legitimately elected role, which however is discussed critically in Sect. 2.5.2 and Sect. 4.4.5. Moreover, in addition to the before mentioned argument of a new organization for tackling corruption, corruption is a further factor to consider in this discussion around a coordinative meta organization (Sect. 2.1.2, Sect. 4.1.2). I19 asserts that a meta organization would possibly lead to more corruption instead of less by introducing a meta organization.

Factoring in the digital (transformation) component of this dissertation, a coordinative meta organization is assessed as counterproductive and is argued to not be compatible with the innovator scene (I5) and digital spirits as "they don't want to be coordinated" (I1).

Besides the discussion around the impact of a meta organization as a coordinative governance mechanism for sustainable digital development, this dissertation highlights the introduction of such a meta organization as an example of governance through hierarchy. Building on the contra-arguments of expert interviewees, I4 further argues that coordinating (in) the Kenyan digital ecosystem through an top-down operating organization is challenging. More importantly for this dissertation's governance perspective, I7 further highlights network governance approaches instead of hierarchy mechanisms of governance: "[...] kind of decentralized coordination and and networks to form and learn from different nodes rather than trying to do a top down on co co coordination that ends up being inefficient [...]". This however is contested by I2 arguing that some sort of hierarchy is always needed, only the form of executing hierarchy for sustainable digital development is left open and rather seen in regulation and legislation.

In sum, considering the arguments, the data of this analysis suggests that a coordinative meta organization for coordinating the Kenyan digital ecosystem for sustainable digital development is not universally accepted as the most suitable solution. Hence, a second coordination form is proposed by this dissertation that incorporates the digital angle on governance in this digital development research project.

4.5.3 Proposal 2: Coordinative Digital Ecosystem Platform

Synthesizing this dissertation's findings from the previous empirical subsections, this research project proposes a second alternative solution to coordinating the Kenyan digital ecosystem for sustainable development: a coordinative digital ecosystem platform (see Sect. 2.5.4). The rationale here follows the question of why not use digital technologies such as platforms as a coordination mechanism? At a first glance, this may not be particularly new as there are already a lot of platforms trying this. Kenya for example is part of a SDG platform (I20) and has a 'White Box Platform', an online tool that allows a visual around innovations, funding and support (I14).[14] Furthermore, and expanding this to the development perspective, the D4D Hub recently initiated a "[...] multistakeholder platform to drive forward digital partnership and cooperation and an an enhance digital partnership and cooperation between European partners and global partners" (I12). This platform aims at coordinating and increasing transparency (I12) at least of (some) European donors activities. Notwithstanding this enthusiasm, this platform is (as of interview date) not digitally operated but a "[p]latform at the moment in literal sense" (I12), i.e., an analogue platform. This emphasizes that platforms are used, but again in a fragmented way as multi-stakeholders create their own platforms rather than contributing to an holistic platform (also due to the hype around platforms as discussed in Sect. 2.2.3, Sect. 2.2.5).

This dissertation hence delivers a novel insight into a digital platform incorporating multi-stakeholder activities in the Kenyan digital ecosystem and also in a holistic approach where every multi-stakeholder can participate, thereby enabling coordination (see also Sect. 4.1.6). Moreover, this is underlined by I22 arguing that "I think is critical for a country like this that all this stakeholders find a collaborative platform much these things can come together". This dissertation argues that such a digital platform would also decrease parallel working, intergovernmental doubling of activities and cross-sectoral issues as discussed in Sect. 4.4.3. Hence, this dissertation proposes a platform as a mechanism how ecosystem can be coordinated for sustainable digital development as per its research question. In the reviewed literature the question of how to govern and by whom, also in terms of platform ownership, is debated (e.g., Bonina et al. 2021; Hein et al. 2020), for instance with regard to gate-keeping, platform evolution and decision rights (e.g., The World Bank 2014; Bonina et al. 2021). This however is not the direct focus of this dissertation, which instead focuses on the previous step as to how to implement or govern such a

[14] In academic literature, digital platforms for development are for instance discussed by Bonina et al. (2021) as further analyzed in Sect. 2.2.5.

platform, namely a) is it beneficial for sustainable digital development of the Kenyan digital ecosystem and b) where do benefits and values lie.

More specifically, the data analysis of this research project has identified the importance of mapping and an overview of multi-stakeholders and their activities in the Kenyan digital ecosystem as part of coordination mechanisms (Sect. 4.1.6, (Sect. 4.5.1). As a governance tool, the interview data further suggests digital platforms in the governance role of not only providing a mapping as highlighted by I1, but also ascribe platforms with the roles of a broker (I7) and a facilitator or admin of networks (I5, see also Sect. 4.4.5). Hence, this platform would be a solution in that it would allow a mapping for gathering an overview of the Kenyan digital ecosystem. Moreover, the digital platform can make use of ICTs as connectors (I2). More specifically and especially relevant for this dissertation, digital platforms are further seen as facilitators of coordination (I5), whereby the platform would likewise enhance transparency (I2, see also Sect. 4.4.3). Most importantly for this dissertation's research question, the data analysis identified digital platforms in the role of coordinators in the Kenyan digital ecosystem. Here, the findings are additional to the findings of Sect. 4.4.5 discussing the role ICTs in general. I3 emphasizes digital platforms as a potential solution for coordination while highlighting the relevance of multi-stakeholder involvement and their willingness. This is further confirmed by platforms' potential of bringing multi-stakeholders together, coordinating resources and enabling capacity building while identifying gaps. Only the platform's operation in a self-governed manner is contested since the Kenyan government is argued to still have (and should have) a governing role (I14). However, the role of the Kenyan government is not universally agreed upon by expert representatives as discussed in Sect. 4.4.5.

Lastly, these findings demonstrate the challenges of platforms as coordinators in the Kenyan digital ecosystem. What the data suggests is that the here identified preconditions, such as willingness, transparency, etc. (Sect. 2.5.1, Sect. 4.4.3), likewise apply to the performance of platforms. In how far a really holistic and inclusive platform can be created for coordination is questioned due to the amount of actors (Sect. 4.4.3) and the diversity of actors (Sect. 4.4.3). Experts I10 and I9 assess this as difficult, especially considering the precondition of how 'easy' it is to use the platform. Also the local context of Kenya is highlighted where political stability and political dynamics need to be considered in the context of a digital platform (I19).

In sum, a digital platform as a coordinative mechanism for achieving sustainable digital development is discussed controversially among interviewed expert representatives, but is highlighted as having potential benefits for coordination. This includes providing a framework for coordinating a multitude and complexity of multi-stakeholder interactions in the Kenyan digital ecosystem. While the design of such a digital development platform exceeds the scope of this dissertation, its proposed framework can deliver valuable insights for securing a sustainable digital development.

5 Discussion

This dissertation has dedicated itself to the highly dynamic and complex research field of (sustainable) digital development / ICT4D with multidisciplinary theoretical and empirical approaches. Taking up the introduced model of micro, meso and macro levels (Fig. 1.1), this dissertation has offered a novel approach to analyzing implications of the digital transformation on sustainable development using digital ecosystems. More specifically, it has contributed to both theoretical and empirical research gaps through synthesizing network governance and development theory and has delivered empirical insights from Kenya. Given the "plethora of theories" (Sein et al. 2019:8) and no coherent narrative in ICT4D, through its synthesis of development, sustainability and governance research, this dissertation offered a more holistic approach as is called for in the literature (e.g., Avgerou 2017; Walsham 2017; Sein et al. 2019). Especially the research gap around 'digital ecosystems' has further been emphasized (e.g., Sect. 2.2.5, Sect. 4.4.1) to which this research project contributes in providing a further understanding.

Against this background, coordination has been identified as a central pillar of both sustainable (digital) development, but also of governance (e.g., Sect. 2.1.7, Sect. 4.1.6). For this purpose, this dissertation has developed a novel framework for governing (in) digital ecosystems (Fig. 2.6). Answering the research question "how can digital ecosystems in developing countries be coordinated in such a way that they contribute to a sustainable development?" was approached theoretically via network governance and ICT4D research, methodologically through qualitative guideline-based expert interviews and empirically using the case of Kenya.

Hence, this discussion chapter is inspired by answering the research question and five guiding questions (Chap. 1). It is thus structured around a further understanding of digital ecosystems, factors that influence sustainable development of digital ecosystems as well as around coordination in the Kenyan digital ecosystem through governance roles, governance forms and alternative forms.

C. Stötzel, *The Digital Ecosystem of Kenya*,
https://doi.org/10.1007/978-3-658-45553-8_5

5.1 Characteristics of Digital Ecosystems Using the Case of Kenya

Firstly, digital ecosystems are characterized by the context in which they operate (macro level), especially with regard to the political, economic, social and infrastructural context (see Sect. 2.1). These factors influence the inner workings of the ecosystem from the outside. Approaching the 'D' in ICT4D representing contextual factors, the nature of 'development' is discussed in the literature (e.g., Sein et al. 2019; Zheng et al. 2018; Avgerou 2017; Chipidza and Leidner 2019; Walsham 2017) and also the importance of "[w]hose development do we study?" (Sein et al. 2019:21). Moreover, "[e]ven as scholars ascribe different meanings to development in the context of ICT4D, so too do the various stakeholders involved" (Chipidza and Leidner 2019:145). This dissertation contributes to this by delivering insights into the importance of the context in which digital ecosystems operate (macro level) using the case of Kenya (Sect. 4.1). It shows that these factors have direct impact on the digital transformation developments (meso level) and also inner workings of the ecosystem (micro level). Thus, the respective contexts likewise have an influence on the sustainable development of not only the country, but also its digital ecosystem.

Secondly, this dissertation has elaborated that the nature of the digital ecosystem can be further explained by preconditions that need to be considered for coordination purposes. These preconditions are summarized under determinant 1 of this dissertation's proposed framework (Sect. 2.5.1, Fig. 2.6) that emerged from the literature review on network governance and ICT4D. In addition to these theoretical findings, further characteristics defining the ecosystem have been derived from the empirical data analysis in Sect. 4.4.1. With regard to the empirical findings on the nature of the Kenyan digital ecosystem, the following determinants were derived and can be summarized as follows (see also Sect. 4.4.3). Theoretical findings (listed first) stem from the literature review. Empirical findings (second item) stem from empirical expert interview analysis. Note that the following list is only a summary, a detailed discussion is conducted in the respective sections.

1. **Network size** (e.g., Hollstein, Matiaske and Schnapp 2017; Provan and Kenis 2008; Molin and Masella 2016a)

   a) Describes number of stakeholders in the ecosystem; indicator for interest and investment in the Kenyan digital ecosystem; the higher the number of stakeholders, the more danger of fragmentation prevails; existence of sub-ecosystems (Sect. 2.5.1)
   b) Kenya: High number of stakeholders involved (Sect. 4.4.3)

2. **Level of activity** (e.g., Klijn and Koppenjan 2016)

   a) Metric for degree of participation (and visibility) (Sect. 2.5.1)
   b) Kenya: Generally described as a vibrant ecosystem; active actors: private sector and hubs; less active actors: NGOs and Kenyan government (Sect. 4.4.3)

3. **Visibility** (e.g., Jansen 2017; Lundsgaarde and Keijzer 2019)

   a) Correlates with awareness, activity and actors' interests (goal to be visible, actor proliferation); indicated through participation in policy processes and physical presence in the ecosystem (Sect. 2.5.1)
   b) Kenya: Actors perceived as visible are foreign tech sector (donors to a lesser extent); correlation between visibility and impact is questioned (Sect. 4.4.3)

4. **Transparency** (e.g., Pattberg and Widerberg 2016; Klijn and Koppenjan 2016)

   a) Crucial for achieving sustainable digital development, e.g., in case of a lack of transparency in stakeholder actions such as in ICT4D projects; needed for achieving synergies through coordination; can be further achieved through (technical) standards (Sect. 2.5.1)
   b) Kenya: Lack of transparency in multi-stakeholders' activities, e.g., due to actors' interests or hidden agendas; also problematic in terms of corruption; can potentially be enhanced through digital technologies (Sect. 4.4.3)

5. **Voice** (e.g., Klijn and Koppenjan 2016)

   a) Correlates with activity and visibility as well as power of actors: those who are 'loud' get recognized more in policy processes (Sect. 2.5.1)
   b) Kenya: Inclusion of multi-stakeholders and recognition of their voices often flawed due to power asymmetries, especially with regard to NGOs and also Kenyan startups (Sect. 4.4.3)

6. **Type of actors** (e.g., Span et al. 2012; Molin and Masella 2016a, b; Raeymaeckers et al. 2020)

   a) Variety of diverse multi-stakeholders that ranges from public to private actors; includes 'new' actors of tech companies (Sect. 2.1.4, Sect. 2.5.1)

   b) Kenya: Variety of diverse multi-stakeholders in the Kenyan digital ecosystem (both local and foreign, see Sect. 4.4.3)

7. **Actors' interests** (e.g., McGuire and Agranoff 2011; Klijn and Koppenjan 2016; Molin and Masella 2016b; Sørensen and Torfing 2017)

   a) Often diverging interests between different actor groups (especially between public and private); poses challenges to coordination
   b) Kenya: Diverging interests of multi-stakeholders challenge sustainable development, e.g., in performance of multi-stakeholder projects (Sect. 4.3.3)

8. **Willingness** (e.g., Klijn and Koppenjan 2016; Pattberg and Widerberg 2016; Molin and Masella 2016b)

   a) Willingness to collaborate and to commit to common goal (Sect. 2.5.1)
   b) Kenya: Generally high level of willingness of multi-stakeholders to collaborate; lack of willingness (also linked with capacities) in traditional structures towards implementation of digital technologies (Sect. 4.4.3)

9. **Goal consensus** (e.g., Molin and Masella 2016a; Provan and Kenis 2008; Raeymaeckers et al. 2020)

   a) Precondition for collaboration, especially in partnerships (Sect. 2.5.1)
   b) Kenya: Finding common goal or speaking with joint voice is assessed as hard to reach, e.g., due to power imbalances and diverging interests (Sect. 4.4.3, Sect. 4.4.4)

10. **Perceptions: Awareness**

   a) Awareness (of each other's activities) as starting point for collaboration in ecosystem (Sect. 2.5.1)
   b) Kenya: Lack of awareness leads to silos and duplicate workings; awareness around ICTs and possible dangers in usage differs between multi-stakeholders e.g., in terms of data protection (Sect. 4.4.3)

11. **Trust** (e.g., Provan and Kenis 2008; Molin and Masella 2016a; Klijn and Koppenjan 2016)

a) Coordination dependent on trustworthy actors and precondition for performance of multi-stakeholder activities such as partnerships or ICT projects (Sect. 2.5.1)
b) Kenya: Trust especially highlighted in relationship between Kenyan government and private sector (Sect. 4.4.3)

12. **Network rationale** (e.g., McGuire and Agranoff 2011; Sørensen and Torfing 2009; Torfing 2005)

a) Seen in sustainable digital development, more precisely sustainability, fairness and equality as well as inclusion
b) Kenya: Network rationale directed towards sustainability, fairness, equality and inclusion (Sect. 4.4.3) aiming at need-based solutions for a sustainable digital development (with multi-stakeholders balancing their own interests)

13. **Duration and frequency of interactions** (e.g., Klijn and Koppenjan 2016; Sørensen and Torfing 2009; Isett et al. 2011; Molin and Masella 2016b)

a) Ecosystem characterized by frequent interactions between stakeholders such as regular multi-stakeholder dialogues (Sect. 2.5.1); duration of multi-stakeholder processes
b) Kenya: Activities rather on short-term basis; duration further characterized by political rhythms or cycles; short-term based ICT projects of multi-stakeholders (Sect. 4.4.3, Sect. 4.4.3)

14. **Ease of entry (degree of inclusivity or exclusivity**, (e.g., Klijn and Koppenjan 2016; Provan and Kenis 2008; Molin and Masella 2016a) and **Degree of Centralization** (e.g., Provan and Milward 2001; Sørensen and Torfing 2009; Klijn and Koppenjan 2016)

a) Determined by resources such as digital skills, access to hardware, software and infrastructure as well as by macro level factors such as business climate and regulations for businesses (Sect. 2.5.1)
b) Kenya: Generally perceived as open towards new actors and also for foreign actors, but majority of interviewees perceive it to be exclusive due to structural barriers (e.g., required resources, elitism, digital skills), informal networks and degree of centrality (focus on Nairobi); ease of entry into ICT projects easier for already known actors (Sect. 4.4.3, Sect. 4.4.3)

15. **Intensity (degree closeness or distance)** (e.g., Raeymaeckers and Kenis 2016; Klijn and Koppenjan 2016)

    a) Expressed need for alignment, streamlining and harmonization of multi-stakeholder activities (Sect. 2.5.1)
    b) Kenya: Alignment mostly on level of multilateral agendas and joint programs; impeded by actors' interests and different understandings; degrees of distance expressed in relations based on contracts, notions of parallel workings instead of collaboration or cooperation to the extent of silos; these degrees of distance are likewise hindering factors to coordination and sustainable development (Sect. 4.4.3)

16. **Expectations** (e.g., Klijn and Koppenjan 2016; Molin and Masella 2016b)

    a) Expectations and especially management of expectations are crucial for sustainable digital development and coordination of multi-stakeholder activities; especially expectation management around the value and capacity of ICTs ('overexpectiations'): gap between expectations of actors and reality (Sect. 2.5.1)
    b) Kenya: Expectations concerning digital technologies are argued to be accompanied by notions of 'overexpectations' (Sect. 4.3.1); need for expectation management, here serving the purpose of coordinating, is expressed, especially for the success of MSPs in digital development (Sect. 4.4.3)

17. **Network effectiveness and efficiency** (e.g., Provan and Kenis 2008; Choi and Chandler 2020)

    a) Both can be achieved and enhanced through using digital technologies and services; inefficiencies occur due to fragmentation of action and especially aid fragmentation that can be addressed through ICTs; coordination is emphasized as enhancing network effectiveness and efficiency (Sect. 2.5.1)
    b) Kenya: Ineffectiveness of multi-stakeholders' activities such as failures in ICT4D projects and misdirected focus of efforts (e.g., Sect. 4.4.3)

In contrast to the theoretical preconditions, interdependence (e.g., Molin and Masella 2016b; Klijn and Koppenjan 2016; Monkelbaan 2019; Provan and Kenis 2008; Sørensen and Torfing 2009) and perceptions: common language and understanding (e.g., Klijn and Koppenjan 2016), have been identified in the empirical evidence as a crucial challenge to sustainable digital development and hence have been discussed under Sect. 4.3 and 4.4.4 in the empirical analysis.

5.2 Factors Influencing Sustainable Digital Development

Factors influencing sustainable digital development (meso level) on the one hand include forces that drive development, such as in terms of leapfrogging, as discussed in both the theoretical background (Sect. 2.2) and empirical analysis (Sect. 4.2). On the other hand, the digital transformation influences development in form of posing challenges. Challenges emerged from the development context in which the ecosystem operates (macro), but also specifically from the digital transformation influences (meso) and from multi-stakeholders interaction (micro). These challenging factors were derived from the synthesis of ICT4D research, development research and (network) governance (see Sect. 2.5.2, Sect. 4.4.4). These challenges will be discussed in more detail in the following as a summary of the respective sections in the theoretical and empirical analysis.

Digital Complexities Requiring Digital Literacy and Skills The complexity associated with digital technologies and services requires literacy and skills as well as resources, which are often lacking. A lack of digital literacy and skills e.g., of staff and decision makers is one challenge for ICT projects' success, as highlighted for instance by King'ori Ndegwa et al. (2016), Chipidza and Leidner (2019) or the The World Bank (2016a) (Sect. 2.5.2). The interviewed experts in the Kenyan digital ecosystem likewise highlight the importance of digital literacy and skills of *all* actors involved (Sect. 4.4.4). It is emphasized that digital literacy is a precondition for (sustainable) digital development similar to 'hard' factors such as hardware, software or electricity (Sect. 4.4.4). This is further closely linked to the digital divide (Sect. 2.3.4, Sect. 4.3.6). The data analysis shows that digital complexities and corresponding digital literacy and skills are especially challenging for public sector organizations operating in the Kenyan digital ecosystem. Here, a lack of digital skills and capacities among governmental staff is attested not only due to competition with the private sector around employees, but also with regard to the issue of brain-drain (Sect. 4.3.4). More importantly, a lack of digital skills is not only associated with the bureaucratic nature of governmental institutions (Sect. 4.4.4), but also includes (public) donor organizations. The expert interviewees of both bilateral and multilateral donor organizations state that in digital development activities (in Kenya) they often "start from scratch" (e.g., I12). Given their role as 'advisors' to the Kenyan government (Sect. 4.4.5), this poses further challenges to sustainable digital development.

Bureaucracies and Digital Capacities Closely linked to digital capacities of public sector actors, both in the reviewed literature and in the empirical analysis, bureaucracies and traditional structures are discussed in the context of lacking flexibility to adapt to and to adopt digital technologies, while also highlighting the importance of underlying willingness (reasons further discussed in Sect. 4.4.4). One challenge can be described by the 'Innovation Paradox' (Cirera and Maloney 2017). Here, it is expressed that policies are often not coordinated with other ministries so that different agencies are responsible for implementing innovation policies and instruments, which simultaneously leads to internal and external confusion (e.g., by the private sector in terms of responsibility; see also Sect. 2.5.2). This challenge of intergovernmental coordination in digital transformation is likewise experienced by interviewed experts in the Kenyan digital ecosystem, (Sect. 4.4.3, Sect. 4.1.6) which is argued to lead to duplicative work i.e., unsustainable developments.

Similarly, the political context further influences the capacities and capabilities of governments such as the Kenyan government. Regarding the development context, the development landscape is still described as an "inter-related but uncoordinated environment" (Davies and Klasen 2019:244). Against the background of the much discussed aid fragmentation (Sect. 2.1.5, Sect. 4.1.5), governments' capacities are argued to be overstretched by the fragmentation of aid, for instance in terms of the number of donors leading to increased transaction costs for the Kenyan government (Sect. 4.1.5, Sect. 4.4.3).

Dependencies Dependencies are identified as important challenges to sustainable digital development by both the reviewed literature and interviewed experts. More specifically, identified dependencies refer to dependence on actors (e.g., Wakunuma 2019; especially on the private tech sector), dependence on technology (e.g., Heeks 2020a; Pistor 2020), dependence on resources (e.g., Chipidza and Leidner 2019; Bailur 2007; Ismail et al. 2018), interdependence, dependencies on (donor) protocols as well as existence of lock-ins (Sect. 2.3.6).

Firstly, with regard to donor protocols, dependencies and lock-ins for partner governments are highlighted for instance by Lundsgaarde and Keijzer (2019) as one challenge emerging from the fragmented character of individual donor systems with different organizational approaches. This emphasizes room for aid and donor coordination (Sect. 2.1.6). In the empirical interview analysis, the impact of protocol dependencies are discussed with regard to both donor organizations and private sector companies Sect. 4.4.4. Their requirements are criticized as not matching innovative target groups (I4) or protocols leading to exclusion when actors fail to meet or comply with the requirements (Sect. 4.4.3). With regard to dependencies on donor protocols in the Kenyan digital ecosystem, it is highlighted that the Ministry of ICT for instance has to manage all donor relations with their different requirements and regulations, which binds its capacities (e.g., I13; I12). Additionally, dependencies on donor protocols are likewise experienced with regard to Kenyan tech startups, especially the ones that seek donor support for instance in acceleration programs (as discussed in Sect. 4.3.2). Expert interviewees assert that new startups or the ones inexperienced with donor regulations and processes are disadvantaged in receiving donor support (e.g., I4).

Secondly, dependence on the private sector is further crucial in the digital development setting. Here, expert interviewees highlight a dependency on foreign private tech organizations (e.g., I20; I1), their (foreign) technologies (I17) and on the provision of digital infrastructure (I5). In addition to dependencies on the private sector as actors in the digital ecosystem, which will be discussed in more detail in the next challenge (power asymmetries), these are further associated with a technological dependency. Focussing on technological dependencies, these can be described by lock-ins. From a theoretical perspective (Sect. 2.3.6), lock-ins can occur with regard to whole technological systems due to economies of scale and influence (Ayres 1991:255) or a certain technology can have an initial advantage over competitors due to switching costs that hence lock out alternative solutions (Cecere et al. 2014). These are also described as 'network externalities' (e.g., Katz and Shapiro 1986; Shapiro and Varian 1999). With regard to the Kenyan digital ecosystem, technological lock-ins are experienced by interviewed expert representatives (e.g., I11), for instance with regard to hardware (e.g., I23, see Sect. 4.2.2), infrastructure and software (I5), or lock-ins through contracts (e.g., I10, Sect. 4.4.3). Lock-ins on an actor level, especially with regard to the private sector, are experienced in the interaction between multi-stakeholders that hinder collaboration in the Kenyan digital ecosystem. This can for instance be explained by an interdependence of multi-stakeholders in the digital transformation (Sect. 2.5.1, Sect. 4.4.4).

In network governance, interdependence or mutual dependency is a) a starting point for any networked governance process (Monkelbaan 2019:137), b) a measure to assess network governance effectivity (Sørensen and Torfing 2009), but also c) an incentive for collaboration (Monkelbaan 2019:137). As argued by Klijn and Koppenjan (2016), the greater the importance of the resource and the lower the substitutability of the resource, the higher is the mutual dependence between actors. The cost of software is especially highlighted with regard to the Kenyan digital ecosystem by the interviewed experts (Sect. 4.4.4, Sect. 4.2.2). Similarly, many ICT4D projects are further dependent on bilateral and multilateral donors and private sector organizations for financial resources and technology expertise (Delponte et al. 2015; Chipidza and Leidner 2019). Resources are hence a crucial factor for dependency and their effect on sustainable digital development, which will be discussed in more detail in the following. Technological means for counteracting the technological dependence, e.g., on software and software providers, are frequently discussed with regard to open source. Their impact however is discussed controversially in the literature (Sect. 2.2.4) and among expert interviewees with regard to counteracting dependence in the Kenyan digital eocosytsem (Sect. 4.4.4, Sect. 4.2.2).

Power Asymmetries As noted by Chipidza and Leidner (2019), ICT4D intergroup interactions occur in settings of underlying power imbalances, although they are not debated in the literature to a great extent. Similarly, in the ecosystem literature, Vallet et al. (2020) for instance demonstrate that power asymmetry shapes governance due to their influence on the access, use, benefit and participation in the management of ecosystem services (here digital services). Power exerted in IT endeavors with different and often conflicting interests, offers conflict potential and involves negotiation processes based on control of resources and information (Medaglia et al. 2018). In the reviewed literature, it is discussed further whether tech companies possess the greatest influence in terms of power and a dependency, especially in terms of power over their users (Moore 2016; Wakunuma 2019). Power asymmetries are both noted as a challenge to the performance of ICT projects (Lock and Seele 2017:238) as well as to sustainable development (Monkelbaan 2019). Moreover, combing development and ICT research, power asymmetries between foreign and local actors in the Kenyan digital ecosystem are critically discussed e.g., in terms of digital colonialism of both public and private actors (Sect. 2.3.6).

The empirical interview data analysis highlighted a regional division of dominance in the digital transformation, especially with regard to the US and China (e.g., I17; I5), i.e., between 'the West' and 'East', but also in comparison to 'the South'. Secondly, power imbalances in the context of the Kenyan digital ecosystem are expressed by expert interviewees with regard to local (Kenyan) tech companies and the international ones (e.g., I2; I19; I14; I10 and I21). Thirdly, the expert interviewees highlight power asymmetries between the private sector and the Kenyan government (e.g., I19), for instance in terms of information asymmetries (I3) and a knowledge vacuum between the private sector and the Kenyan government (I22). Additionally, power asymmetries are highlighted by expert interviewees with regard to resource asymmetries where the private sector dominates, especially in terms of financial resources in the digital transformation (I19).

These power asymmetries between the private tech sector and the Kenyan government are experienced by expert interviewees even to the extent of private sector actors overtaking government functions (e.g., I2), especially in policy making (e.g., I19; I9). This dominating influence is even assessed to the extent of 'dictating' (I4). This dissertation argues that these power asymmetries not only challenge legitimately elected governments (e.g., I3), but also sustainable and inclusive digital development. Expanding on the power asymmetry between the private tech sector and the Kenyan government, the dominance of the private sector over parts of the Kenyan government in the digital transformation is further manifested in their role as advisors to the government. This offers the opportunity to take influence 'from within' (e.g., I2; I19).

Fourthly, power asymmetries between donor organizations and the Kenyan government are highlighted among experts that are already inherent in the development status of a country and in development cooperation activities (e.g., I2; I4). Fifthly, power asymmetry between the Kenyan government and NGOs are emphasized (e.g., I15) with NGOs given less priority in policy consultations than other actors by the Kenyan government (e.g., I15; I15). The discussed 'digital colonialism' in the literature (Sect. 2.3.6) is experienced differently among interviewed experts with regard to the Kenyan digital ecosystem, although the majority argues to experience at least some notions of digital colonialism (e.g., I11; I5; I22; I13 and I10).

Common Understanding Most importantly, this dissertation emphasizes that sustainable digital development requires a common understanding of the digital transformation and its components. From a theoretical network governance perspective, networks (here ecosystems) can be characterized by the degree of shared perceptions between actors (Klijn and Koppenjan 2016:104). Outcomes of network governance further imply a certain degree of joint perceptions and a shared language among others (Klijn and Koppenjan 2016:95). Additionally, with regard to sustainable development, a lack of common understanding can be a barrier to sustainable digital development (see e.g., Monkelbaan 2019:132) and also for coordination (Sect. 2.5.1). Given the range and diversity of multi-stakeholders in the Kenyan digital ecosystem (donors, private tech companies, NGOs etc.), a common understanding further draws on different bases of knowledge (Mulder 2020) and hence is linked to digital literacy and skills of multi-stakeholders. More importantly for this dissertation's development angle, different understandings in terms of (ICT for) development, are highlighted in the literature, for instance by Sein et al. (2019), Zheng et al. (2018), Avgerou (2017), Chipidza and Leidner (2019).

Empirically, expert interviewees highlight the importance of common understanding as a necessity for cooperation (I18) and for communication among multi-stakeholders (I12). Most importantly for this dissertation, the data analysis further emphasizes a common language and understanding for coordination (e.g., I2) and for sustainable digital development (e.g., I11). A lack in common language and understanding in the Kenyan digital ecosystem identified by the data analysis is expressed with regard to sustainability (e.g., I1), digital transformation and ICTs in general (e.g., I18; I17; I22, see Sect. 4.2), *both* digital transformation and development cooperation (e.g., I10; I11; I14), human-centered digital transformation, network and ecosystem, governance as well as harmonization (Sect. 4.4.4). Hence it is crucial for digital development that all multi-stakeholders share a common understanding of what the digital transformation is, what sustainable digital development entails as well as what an ecosystem is and how it functions.

Hype This is further mirrored in an associated hype or 'dangerous enthusiasms' (Gauld and Goldfinch 2006) of technological infatuation (Sect. 2.2.3). Both the literature review (e.g., Heeks 2010; Ansell and Miura 2020) as well as the empirical interview analysis suggests notions of hype in the context of digital technologies. Firstly, this includes a hype around the impact of the digital transformation (Sect. 2.2.3, Sect. 4.3.1). More specifically, the impact of digital technologies and services on development is discussed as being overestimated by multi-stakeholders (Sect. 4.3.1). Secondly, notions of hype are discussed with regard to overestimating certain digital technologies (Sect. 2.2.3), such as AI (Sect. 4.2.2) or platforms as indicated for instance by the Gartner's hype cycle ((Gartner 2021)). The empirical analysis likewise identified a dualism between 'basic' and 'innovative' digital technologies (see Sect. 4.2.2). Thirdly, a hype in the sense of actor proliferation in 'shiny topics' is highlighted. From a theoretical perspective, a (general) global proliferation of bilateral and multilateral agencies focussing more on global presence than aid efficiency is a discussed challenge in the reviewed literature (e.g., Lundsgaarde and Keijzer 2019; Bigsten 2006; Aldasoro et al. 2010). Expert interviewees deliver digital transformation topics as one example for actor proliferation. The data suggests that actors' motivations to proliferate themselves through engagements in "nice and shiny" (e.g., I3) topics of digital transformation hinder not only coordination, but also sustainable digital development (see e.g., Sect. 4.3.1, Sect. 4.3.1).

Project Failures Closely related to these challenges, the fragmentation and failure of ICT(4)D projects due to failed suitability (misled focus of action) is identified as a further challenge to sustainable digital development. Failing ICT projects are accompanied by negative dynamics that need to be analyzed and addressed, which is where a research gap prevails (Choi and Chandler 2020). This dissertation has contributed to the research here in that its empirical analysis attests a misregard of the importance of the infrastructure context in the Kenyan digital ecosystem, especially of donor organizations, which instead focus more on digital skills and entrepreneurship projects (Sect. 4.3.2). A lack of proper (IT) infrastructure however is a challenge to sustainable digital development (e.g., King'ori Ndegwa et al. 2016; Chipidza and Leidner 2019, Sect. 2.3). The empirical data analysis of this dissertation likewise confirms the importance of infrastructure in the context of ICT4D projects (e.g., I10; I20). These findings are further in line with Delponte et al. (2015) and Chipidza and Leidner (2019) questioning the suitability and sustainability of (Western donors') ICT projects. Expert interviewees for instance see one reason behind this misdirected focus of ICT4D projects in the discussed hype and actor proliferation (Sect. 4.3.1) and multi-stakeholders' duplicative activities (Sect. 4.4.3, see e.g., I13). Similarly, multi-stakeholders involved in ICT projects are dependent on each other and also the performance of the projects dependents on the actors' activities, for instance if one actor withdraws its resources, the ICT4D project might fail (e.g., Chipidza and Leidner 2019:153). All in all, these are mentioned in the conducted interviews as factors leading to failed sustainability of ICT4D projects (e.g., I4 and I4), that contribute to a discussion around (aid or project) efficiency.

In sum, the theoretically developed framework (Fig. 2.6), especially determinant 2, can be modified based on the empirical analysis as illustrated in Fig. 5.1.

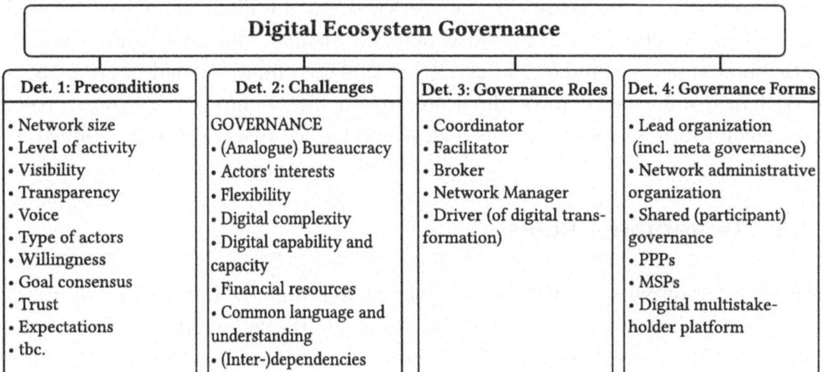

Figure 5.1 Modified theoretical framework: Digital ecosystem governance for sustainable development (own illustration)

As the empirical results suggest (Sect. 4.3; Sect. 4.4.4), the theoretical assumptions for the framework differ from the empirical results, so that 'Common language and understanding', 'actors' differing interests' and '(inter-)dependencies' are to be identified as challenges for digital ecosystem governance (as opposed to preconditions). Hence, the framework has been modified to these results (Fig. 5.1).

Ultimately, if these listed challenges are not reflected and tackled by multi-stakeholders, these can lead to further unsustainable development as they amplify the problem. Moreover, similar to ecological ecosystems (Chap. 1), it is emphasized that species (here actors) need each other to survive, or else, if one species is diminished or vanished, it effects all others. Hence, addressing asymmetries not only in terms of a digital divide, but also in terms of power and knowledge asymmetries need to be addressed for achieving inclusive sustainable digital development. This dissertation argues that coordination of multi-stakeholders' efforts can address these challenges and delivers insights into possible forms.

5.3 Coordination in and of the Kenyan Digital Ecosystem

A crucial part of this dissertation's research question is coordination for sustainable digital development that is argued to address (and potentially overcome) these challenges. This dissertation has provided detailed insights into coordination in both the theoretical background (e.g., Sect. 2.1.6) and the empirical analysis (e.g., Sect. 4.1.6). It has established that coordination can be achieved through governance roles, governance forms and alternative forms, which will be discussed in the following.

5.3.1 Governance Roles

From the literature review in the theoretical background, the (network) governance roles of a coordinator (e.g., Raeymaeckers et al. 2020), facilitator (e.g., Provan and Kenis 2008; Raeymaeckers et al. 2020), broker (e.g., Börzel and Risse 2010; Klijn and Koppenjan 2016; Kenis and Provan 2009; Jansen and Wald 2007), network manager (e.g., Klijn and Koppenjan 2016) and driver of digital transformation were derived (Sect. 2.5.3). With regard to governance roles in the Kenyan digital ecosystem (see Sect. 4.4.5), these theoretical roles were added by roles of an advisor, balancer, connector, enabler and supporter that were derived from the empirical data analysis (in addition to the theoretically derived ones).

These roles were distributed among multi-stakeholder groups in the Kenyan digital ecosystem as follows: role of advisor (multi-stakeholders), role of balancer (academia, NGOs and Kenyan government), role of broker (multi-stakeholders), role of connector (multi-stakeholders), role coordinator (multi-stakeholders), role of driver of digitalization (private sector and Kenyan government), role of an enabler (ICTs and multi-stakeholders), role of facilitator (Kenyan government), role of a supporter (donors and NGOs) and role of a manger (Kenyan government).

The distribution of actor groups fulfilling governance roles in the Kenyan digital ecosystem has further been illustrated in Table 4.2.

While the literature review has established that there is now broad consensus in ICT4D literature "that ICTs are no silver bullets providing simple and linear solutions to complex and wicked social problems" (Zheng et al. 2018:2), a focus on the ICT tool rather than on its implications and its role are critically discussed. In terms of a governance role of ICTs as actors themselves, ICTs' abilities to connect are highlighted, while the nature and actual role are still debated in both the reviewed literature (Sect. 2.2.1, Sect. 2.5.3) and among expert interviewees. The interviewed experts in this dissertation critically discuss the role of digital technologies and ICTs in the Kenyan digital ecosystem: The majority of interviewees assess ICTs in the role of enablers for development (e.g., I5; I13; I19; I9; I20 and I18). Whether ICTs are actors themselves (instead of tools) is however still debated, since the notion of 'a means to an end' (I9; I17; I17) prevails as opposed to ICTs regarded as actors themselves (I9).

5.3.2 Governance Forms

The question through which governance form digital ecosystems can be governed, this dissertation elaborated a distinction between less formalized and formalized governance forms for coordinating multi-stakeholders interactions in the digital ecosystem. These are determined by a dualism between regulating for sustainable digital development on the one hand and stifling innovation on the other (see Sect. 2.5.4, Sect. 4.4.6).

Firstly, less formalized governance forms identified in this research project are events and workshops as well as self-governed structures. Beginning with the less formalized ones, *events and workshops* are perceived as inclusive and representative of all multi-stakeholders in the Kenyan digital ecosystem according to interviewed experts (Sect. 4.4.6). This governance form emerged from the data analysis and was not identified in the theoretical background. As the ultimate less formalized governance form, *self-governed structures* were discussed in the theoretical background (Sect. 2.5.4) as allowing minimum disruption and intervention in innovation processes and hence are a form of self-coordination without an entity intervening in the digital ecosystem. This is enabled through this participant-driven governance form (e.g., Provan and Kenis 2008). Among interviewed experts, self-governance formats for coordination in the Kenyan digital ecosystem are discussed controversially (see also Sect. 4.4.6): On the one hand, this governance form is argued to allow information sharing, connecting actors and making actors' actions transparent, which is perceived as an intended goal (e.g., I1; I5). This is highlighted as especially important for the digital development context, where a certain amount of space enabled through self-governance is argued to be needed, especially for innovations (e.g., I16; I3; I12). On the other hand, differing interests of multi-stakeholders however are mentioned as impeding sustainable digital development through self-governance (e.g., I11).

Secondly, more formalized governance forms were derived in this dissertation in terms of standards, alliances and associations, committees, working groups in addition to partnership formats of PPPs and MSPs (Sect. 2.5.4), which will be discussed in more detail in the following.

Standards Standards were mainly discussed among expert interviewees (see Sect. 4.4.6): standards for enabling coordination are seen in both a technical but also a regulatory nature, global technical standards for digital topics, standards to regulate the IT industry as well as global and local standards for collaboration and coordination in digital ecosystems (also in the sense of norms). While standards in the theoretical background are mainly discussed in the context of addressing lock-ins (Sect. 2.3.6), the empirical data analysis highlights their importance for sustainable digital development and coordination.

Alliances and Associations Both forms were identified in the reviewed literature and in the interview data analysis as (synonymous) formats predominantly used for pursuing strategic interests. In network governance (Sect. 2.5.4), alliances pursue information sharing and other resource exchange, capacity building, problem solving and service delivery (Agranoff 2007; Raeymaeckers and Kenis 2016) as well as strategic goals, especially from a business view (Whetsell et al. 2020). In the conducted interviews with experts in the Kenyan digital ecosystem, alliances and associations are discussed as being a trend, but also as a best-practice format for pursuing strategic interests (I2) that also for bestow individual actors with a greater voice (I5) due to joint formation (which is discussed as especially relevant for Kenyan tech startups).

Committees In the theoretical background, committees are discussed in terms of ad hoc committees (see e.g., Raeymaeckers and Kenis 2016) in the context of alliances (Sect. 2.5.4). This governance format of committees for coordination of efforts was discussed among expert interviewees as being divided according to thematic areas such as committees for cybersecurity, although similarities to working groups were identified (Sect. 4.4.6).

Working Groups Similarly, working groups as a governance form for coordination were subsumed under alliances in the theoretical background following a definition of Klijn and Koppenjan (2016) in that working groups are formed for top-level consultations (ibid.:165). The experience of expert interviewees with regard to working groups in the Kenyan digital ecosystem however is that they are perceived as less inclusive or relevant to multi-stakeholders as interviewees question their impact (e.g., I7). For instance, donor working groups were argued to pursue the goal of merely informing each other rather than alignment of action (I4, Sect. 4.4.6).

Partnership Formats Partnership formats for coordination multi-stakeholder activities, especially PPPs and MSPs were critically discussed with regard to their impact on sustainable development both in the theoretical and empirical analysis (Sect. 2.5.4, Sect. 4.4.6). Moreover, PPPs and MSP were subsumed under formalized governance forms due to their contractual nature (although more or less binding) between public and private partners. This especially applies to PPPs since they usually take the form of a concession-type contract between public and private actors (e.g., Bayliss and Van Waeyenberge 2018). More importantly, this dissertation contributes to a research gap as stakeholder involvement in PPPs is under-researched (Nederhand and Klijn 2019). Both MSPs and PPPs are also part of SDG 17, emphasizing their relevance for sustainable development.

PPPs In the literature, advantages of PPPs are seen in innovativeness of PPP projects due to (multi-)stakeholder involvement in e.g., improving content and innovativeness or in enhancing democracy (e.g., Klijn and Koppenjan 2016; Sørensen and Torfing 2007; Nederhand and Klijn 2019). Another possible advantage is seen in efficiency and financing in terms of returns to investors on the one hand and service or infrastructure delivery on the other (Bayliss and Van Waeyenberge 2018). This dissertation analyzed PPPs as a format for not only coordinating multi-stakeholders activities for achieving sustainable digital development, but also as a format for addressing asymmetries in the relations between multi-stakeholders in digital ecosystems. Here, there is debate in the literature, and among interviewed experts, as to whether the power asymmetry between actors might be more balanced in ICT projects because of the interdependence of the different actors' resources (e.g., Medaglia et al. 2018, see Sect. 2.5.1). Ismail et al. (2018) for instance voice concerns about private sector domination in PPPs and MSPs, as well as about private sector domination in ICT4D partnerships in many developing countries. Moreover, the effectiveness of partnership formats is a further challenge for sustainable (digital) development (e.g., OECD/UNDP 2019), for instance with regard to poor designs and weak monitoring systems of PPPs and MSPs (OECD 2018b).

The impact of PPPs for coordinating the interaction between public and private actors however is questioned with regard to the Kenyan digital ecosystem (Sect. 4.4.6). On the one hand, benefits of PPPs are highlighted as important coordination mechanism for sustainable digital development in Kenya by I6. Moreover, PPPs and MSPs are seen as important formats for sustainable digital development in leveraging societal impact (I5). Also the learning opportunities, especially for public sector actors, through such formats are highlighted (I20). On the other hand, interviewed expert representatives however assess the potential of PPPs for solving power asymmetries as limited and even as enhancing power asymmetries between public and private actors (e.g. I2; I3). Moreover, a hype around PPPs and the role of private sector actors in the partnership formats are emphasized (e.g., I17; I17) as potential causes for partnership failure and unsustainable development.

MSPs With regard to MSPs as a governance form for multi-stakeholder interactions in digital ecosystems, MSPs can be defined as "manifestations of networked governance" (Pattberg and Widerberg 2016:46). From a development cooperation perspective, MSPs are seen as "potential arenas for coordination and the various actors involved" (Lundsgaarde and Keijzer 2019:221) and as "an expression of a changing approach to addressing development goals" (Lundsgaarde and Keijzer 2019:218). In digital development, MSPs are common formats for instance in infrastructure (Nederhand and Klijn 2019), e-government or national e-IDs (Medaglia et al. 2018).

Among interviewed experts, MSPs as coordination mechanisms for sustainable digital development in the Kenyan digital ecosystem are discussed controversially. On the one hand, experts highlight the importance of MSPs for sustainable digital development in the Kenyan digital ecosystem (e.g., I24; I14; I21). More specifically, the value lies in MSPs as a format for "bringing together" multi-stakeholders across different sectors in the Kenyan digital ecosystem (I6). Further possible potentials of institutionalized inclusivity in multi-stakeholder interactions as well as economic development e.g., through trade opportunities and funding, were counterbalanced by MSP challenges. These included arguments of multi-stakeholder processes being both cost and time consuming and often inefficient as well as complex to manage due to the specific interests, expectations and requirements of multi-stakeholders (Sect. 4.4.6). Moreover, the need for governance of MSPs for sustainable digital development on how to conduct MSPs sustainably is further highlighted (e.g., I15).

As guidelines for these governance formats in addressing these challenges, this dissertation has developed a holistic framework that encompasses preconditions for coordination. These preconditions likewise apply to the discussed governance forms and roles as they highlight the importance of recognizing specific challenges in the digital transformation that need to be considered for each governance form.

5.3.3 Alternative Governance Forms

This dissertation has shown that alternative governance forms (alternative to the ones addressed in the previous question) are needed, especially for the purpose of transparent and inclusive coordination of multi-stakeholder activities in digital ecosystems such as the Kenyan one. These are especially important for coordinating all levels (macro and meso), not only the micro level. Firstly, interviewed experts expressed a need for a coordinative instance (I3), coordinative body (I10) or regulated body (I10). More specifically, the empirical analysis of expert interviews in the Kenyan digital ecosystem expressed a need for 'mapping' and for an overview of multi-stakeholder activities (see Sect. 4.5.1). Similarly, 'mapping' emerged as a precondition for coordination as it enhances visibility and awareness, especially since not all multi-stakeholders have an overview of what others do (again leading to and explaining setting of fragmentation hindering coordination for sustainable dig development). Secondly, from the literature review and the data analysis, this dissertation proposed two possible alternative governance forms for achieving sustainable development of digital ecosystems in developing countries: A *meta organization* and a *collaborative digital platform.*

Meta Organization (Governance) Beginning with the first proposed alternative governance form, meta-governance in the reviewed literature aims at "creating balanced combinations of tools from three different governance styles, notably hierarchical, network and market governance" (Meuleman 2019:5). Especially relevant for this dissertation's analysis of sustainable digital development, meta-governance is discussed as suitable for addressing the numerous 'wicked' problems that the SDGs aim to address (e.g., Monkelbaan 2019). Meta governance literature further argues that traditional (hierarchical) governance mechanisms are to be adapted to networks where governments act as meta-governors of networks exerting network design, framing, management and participation (e.g., Keast et al. 2004; Sørensen and Torfing 2009; Molin and Masella 2016b). However, due to the complexity of these problems, "[..] a central steering role for states very unlikely" (Glasbergen 2011:194). Moreover, the author highlights the question of who governs these networks by referring to the fact that there is no supranational authority setting parameters. Whoever it may be, Provan and Kenis (2008) further claim that it is important for coordinating a network of service providers or stakeholders to have a central or lead organization. Additionally, private regulations are seen as unable "to induce the necessary sustainable improvement" (Glasbergen 2011:191). Besides the discussion on which organizational type should overtake the meta-governing role, meta governance is likewise associated with the performance of partnership formats. However, only little research has been conducted here so far (Beisheim, Ellersiek, Goltermann and Kiamba 2017:107). This dissertation contributes to these (meta) governance research areas through empirical insights from the Kenyan case.

In the empirical analysis, a meta organization was discussed as a central coordinative governance organization (hierarchical form) that is either already present in the ecosystem or a new (type of) organization that functions similar to a 'supranational' entity. While the academic literature review addressed potentials and values of meta organization (understood as part of meta governance, see Sect. 2.5.4), the empirical analysis emphasized a diverse discussion on coordination in Kenyan digital ecosystem through a central coordinative authority (regardless of external or internal). Arguments in favor of such a meta organization are seen in contributing to political stability e.g., through addressing corruption via an external or new governing body (I10, Sect. 4.1.2), in ensuring equal representation of multi-stakeholders (I2) and in deriving economies of scale from a central coordinative body (I2). On the other hand however, the data further suggests arguments against a meta organization. Here, it is argued that a meta organization is not an ideal fit to coordinate, because of too much overhead, friction and cost (I1) highlighting a possible increase instead of a decrease in transaction costs. Moreover, a meta organization is assessed as counterproductive and not compatible with the innovator scene (I5). Instead, if coordination was to be pursued by an organization, the empirical data instead highlights empowerment and reinforcement of existing actors in their governance roles (I5; I4, see Sect. 4.5.2).

Digital Platform The second proposition of a multi-stakeholder digital platform for coordination follows possibilities offered by digital platforms discussed in Sect. 2.2.5. A digital platform would enable the needed mapping efforts of activities and it would enable inclusive multi-stakeholder involvement and participation in a coordinated and transparent manner. Closely linked to the before discussed meta governance format, Ansell and Gash (2018) for instance argue that platforms can overtake the possible role of 'meta-governance' and can be seen as 'meta-organizations' coordinating actors (Ansell and Miura 2020). However, platforms in and for governance still remain "largely experimental" (Ansell and Miura 2020:262). This not only underlines the timely relevance of this dissertation, but also a research gap with regard to ICT4D and digital ecosystems research on platforms. Here it is for instance argued that "there is little attempt to examine the influence of the wider ecosystem members and stakeholders such as regulators, governments and alike on governance regimes" (Bonina et al. 2021:880). Factoring in the development perspective, Lundsgaarde and Keijzer (2019) highlight a shift towards more flexible platforms for promoting collaboration among diverse actors from past approaches for coordination aimed at a top-down, centralized decision-making process. This dissertation contributes to these network formats of coordination for more effective (donor) collaboration (Sect. 2.4.3) while incorporating digital platforms (Sect. 2.5.4).

In the expert interviews, platforms are seen as a possible solution in allowing a mapping for gathering an overview of the Kenyan digital ecosystem and facilitating coordination (e.g., I1; I5) as discussed in more detail in Sect. 4.4.5. Moreover, further values of platforms coordinating activities in the ecosystem are seen in enhancing transparency (I2), bringing multi-stakeholders together, coordinating resources and enabling capacity building. Here however constraints, similar to the derived preconditions and challenges, would need to be addressed in addition to an actual implementation of a digital platform. Especially multi-stakeholder involvement and their willingness are crucial factors to consider for sustainable digital development in the context of coordinative platforms (see further Sect. 4.5.3).

Limitations of Research

6

Synthesizing the previous chapters of the theoretical background (Chap. 2), methodology (Chap. 3), empirical analysis (Chap. 4) and discussion (Chap. 5), limitations of the conducted research are to be further analyzed. In line with the three previous chapters, limitations of research were identified in terms of theoretical, methodological and empirical limitations, each of which will be discussed in the following.

6.1 Theoretical Limitations

Firstly, the research field of ICT4D or digital development is a highly dynamic and recent field. Against this background, and especially given the interdisciplinary nature of both technical (IT) and development perspective poses challenge to the theoretical literature review, where for instance this dissertation's focus on digital ecosystems as a term are not yet established. Adding to complexity, this dissertation's focus on coordination for sustainable digital development further requires both a sustainability and a governance research perspective that are uniquely synthesized with ICT4D. Given this interdisciplinary nature however, and in order to cover all of these research streams, for research pragmatic reasons not all of these streams could be explored in depth. Additionally, given the fact that the digital transformation is a relatively recent and also still ongoing phenomenon and especially since digital ecosystem as a concept is still explored, prior research on digital ecosystem is little. This has been compensated through an extensive ICT4D and network governance literature review as well as through expert interview data.

Secondly, the choice to analyze digital ecosystems from a governance perspective could equally have been conducted through network analysis, ANT or policy theories in this political science research project. ANT would have delivered

C. Stötzel, *The Digital Ecosystem of Kenya*,
https://doi.org/10.1007/978-3-658-45553-8_6

closer insights into network management and enabled a (also statistical and graphical) processing the findings, especially with regard to the preconditions such as relations between stakeholders in terms of closeness or distance (as alluded to in Sect. 2.5.1). This would also have enabled a basis for illustrating a stakeholder activity mapping. In this dissertation however the focus was more on the macro level of governance where these are only parts of the research goal.

Thirdly, limitations of the developed framework that is based on a literature review synthesizing ICT4D and network governance are to be discussed. On a first note with regard to the framework's structure, possible duplications are noted. While the derived 'preconditions' (see Sect. 2.5.1) may remind of 'context' (Sect. 2.1), context refers to general contextual factors such as political or economic environment that influence the digital ecosystem on a macro level whereas 'preconditions' refer to the micro level, where preconditions are relevant for governance and coordination of multi-stakeholder (inter-)actions. A structural duplication however may be perceived with regard to challenges, which are both analyzed in Sect. 2.3 analyzing (general) challenges in terms of ICT4D, whereas challenges in Sect. 2.5.2 of the framework discuss the synthesized challenges of ICT4D and of/for governance (i.e., two theoretical sections on challenges which were synthesized). With regard to this dissertation's research question, a duplication emerged in terms of coordination through governance, especially coordination *of* and *in* the Kenyan digital ecosystem. Here a indistinctness of the research question could be attested, but since this dissertation has argued that both are interrelated and are mutually dependent both perspectives were pursued.

While the theoretical framework of this dissertation identifies interdependence as a precondition, the empirical analysis highlighted that interdependence in sustainable digital development context is rather a challenge to coordination for sustainable digital development, which is why it was analyzed under 'challenge' instead of 'precondition' although it is understood as both (see further below Sect. 6.3).

Fourthly, leading over to a theoretical limitation that emerges from the methodological design and research question, the choice of focussing on developing countries and inclusion of respective development cooperation is accompanied by limitations. On the one hand, as stated beforehand, the digital transformation affects every country and hence every (national) digital ecosystem so that this dissertation's framework could also be applied to developed countries. However, it has been emphasized in this dissertation (Sect. 2.1) that the context in which the digital ecosystem is not only important but also influences the inner workings of the ecosystem. Similarly, differences between different macro factors of countries and especially their development status effect sustainable digital development. Analyzing the unique role of developing countries not only allows an exploration of leapfrogging (Sect. 2.2.2), but also of additional types of stakeholders (donor organizations) that are not present in developed countries.

Moreover, a developing country was chosen, for analyzing sustainable digital development that is still a major challenge for developing countries (arguably to a different degree than for developed countries) and subsequently for analyzing this dissertation's specific research question. Emphasizing the political science research perspective on the research phenomenon of ICT4D, the role of governments is of importance for this dissertation's research question and also a further important reason why a developing country was chosen. It has been emphasized that the political context is crucial, especially in settings with limited state capabilities. The role and more importantly the capacity of governments in the context of fragility in the broadest sense, is hence not comparable to developed countries and inherently pose a specific challenge that helps examining the research question. Moreover, digital technologies and services likewise are an influential factor for governments, their capabilities and capacities that is an ongoing phenomenon that challenges this research project. A special challenge that has arisen during this dissertation's research process is the COVID-19 pandemic. The implications of the COVID-19 pandemic on the Kenyan digital ecosystem and its (inter-)actions however has not been incorporated into this dissertation since it is not part of the research question, but could be pursued in future research. Instead, the context of the pandemic is discussed in terms of its methodological influences and associated limitations for this dissertation in the following.

6.2 Methodological Limitations

As a second limitation, the chosen methods of guideline-based expert interviews and data analysis through qualitative content analysis influenced this research project. Firstly, as mentioned in Sect. 3.1, the qualitative design of this research project had implications on the data and findings. As a common limitation of qualitative research, the method of interviews resulted in subjective data in terms of the interviewed experts' knowledge and experiences, with subjectivity here being a potential limitation for the goal of an objective research. However, the goal of this dissertation was to explore (and explain) exactly these experiences and knowledge of experts, especially for discussing the lack of these factors. Hence, a qualitative instead of quantitative approach was chosen. A quantitative approach could have filled the void of missing graphical illustrations of the ecosystem such as mappings of closeness or distance between multi-stakeholders in the ecosystem, with statistical data being a requirement for illustrations (e.g., using ANT or network analysis tools).

This methodological limitation is closely linked to the before mentioned theoretical limitations of theory choice (Sect. 6.1). Given the active involvement of the researcher in the data gathering phase of qualitative research, a certain subjectivity naturally limits the objectivity of this data.

Secondly, limitations were observed at various stages of the research process. For instance, the research phase of conducting interviews is only a snapshot of the studied reality and thereby poses limits to the context of this specific timeframe. More specifically, limitations prior to the interviews, during interviews and after the interviews (including limitations of data analysis method) were observed and will be discussed in more detail in the following.

6.2.1 Prior to Interviews

As mentioned in Sect. 3.2.1, the sampling group was limited. When contacting organizations or respective representatives, occasionally it was not possible for the researcher either to identify a contact person, the person was not available, the person has already left the job or was new to the job. This limitation can be traced back to the interview participants' status as experts (among other factors). In comparison to other interview forms, the potential time constraints and respective time pressure for the interview (Flick 2014:231–232), posed challenges for the interview setting and guideline questionnaire design. This was anticipated by the researcher and addressed through a timeframe that was offered when requesting for an interview (request of interview meeting was mostly sent one or two weeks prior and offered adjustment to the experts' schedule). Nonetheless, in many instances scheduled interview meetings were postponed, in some occasions even up to three times. Furthermore, due to the expert and often diplomatic status, some requested interview partners declined to participate. This had implications for the empirical data quality (further discussed in Sect. 6.3). Due to the extraordinary circumstances of doing remote digital interviews during the COVID-19 pandemic, this limitation is only minor since otherwise no access to the Kenyan field would have been possible.

Moreover, besides challenges in initially contacting possible interviewees, another challenge in the process of securing interviews with requested experts was the 'declaration of consent' (see Chap. 3). This document was sent to each requested expert as an attachment of the requesting email. The first requested interviewees did not know how to fill this in, even though there were clear instructions made in the document. Here a limitation to 'offline' interviews was noted, since in 'normal' or personal interview settings, the declaration would have been filled out in the face-to-face meeting and with the researcher present for questions. Additionally, almost every participant needed to be reminded to fill out the document prior to the interview, during the interview and often times even after the interview (some of them needed to be reminded over three times).

Already prior to the actual interviewing phase, there were technical constraints, mostly for the researcher, as basically all of the interviews preferred a different conference software in response to suggested tools in the interview requests and information sheet (see Chap. 3). Hence the researcher had to register for and download a range of digital meeting tools (e.g., Microsoft Teams, Zoom, Skype, Google Meet).

6.2.2 During Interviews

Switching to the interview phase of during the interview conduction, disruptions were observed that pose additional limitations to this research project. In almost every conducted interview, there were disturbances in terms of the digital setting, especially due to lost internet connection (for up to four minutes), no video format chosen but rather telephone setting, and background noises that seemed to have a distractional effect on the interviewee. More specifically, participants themselves created distractions, such as background noises (as noted e.g., by Flick 2014:231–232). Fortunately, the audio recording itself was not a source of error as the integrated audio function of Zoom or Microsoft Teams was used. However, lost or only limited internet connection or background noises resulted in occasionally incomplete documentation of the interviews (Döring and Bortz 2016:360–365).

Moreover, due to the interview duration being limited to 60 minutes including a short personal introduction, the actual interview duration was limited to 45–50 minutes. Similar constraints were observed due to the guideline-based interview setting, where the questionnaire only served as a guideline, since some interviewees went into different directions than others which limits comparability of results, but enables a deeper exploration of experts' knowledge and experiences, which ultimately is the goal of this dissertation.

Another limitation during the interviewing phase was observed with regard to language barriers, since English was neither the first language of the researcher nor of many of the participants. While some interviewees struggled for instance in findings words, issues of understanding were more of a terminological nature than understanding of utterances. These is consistent with the empirical findings of Sect. 4.4.4, where difficulties and differences of expert interviewees were observed with regard to the required knowledge of digital development and digital transformation context. During the interviews, the researcher observed that actor groups not involved in development or development cooperation, especially the private sector, had difficulties in answering development questions, and vice versa (i.e., development or public sector actors struggle to answer technical or corporate questions). With the exception of instances that required rephrasing questions (see e.g., "You reframe that question I'm not getting it" I14) and given the research goal of exploring the expert experiences and knowledge, the researcher only adapted questions to a minor degree in order to not interfere or distort findings. The effect of interviewees' difficulties in answering the questions of this research project will be discussed in more detail in Sect. 6.3. Only in instances where statements of the interviewees were unclear, the researcher interfered for instance through asking for clarification or more details. Despite these measures, possible misunderstandings in the communicative interaction between interviewee and researcher cannot be ruled out completely.

Most specifically, during the interviewing phase, a lack of experiences was stated by some interviewees, e.g., I11, in stating limited experience due to having just begun fulfilling the respective position within the organization. This was not communicated to the researcher in the sampling/requesting phase and poses an unavoidable limitation to the meaningfulness of their expert role. A lack of experience in this dissertation is both an empirical finding as well as a limitation, which will be discussed in more detail in Sect. 6.3. Another observation that likewise has an implication for the expert status of the interviewees was made. Some participants were also only able to provide information based on their project scope or their own institution, so that transfer questions, such as on the role of other institutions, were not answered sufficiently. This certainly had an influence on the interview quality. More importantly for this dissertation, a limitation to comment on behalf of expert interviewees was noted (similarly to the previously discussed rejection of interview request due to political and diplomatic research setting). On the one hand, limitations to comment included references to specific names, especially of institutions or companies. On the other hand, limitations to comment on certain interview topics were noted even to a diplomatically sensitive scale (Sect. 4.4.4). This poses another challenge that arose from the chosen method of expert interviewees in the digital development field. Similar instances will be further discussed in Sect. 6.3.

6.2.3 After Interviews

As a last methodological limitation, exiting the field and subsequent analysis of the gathered interview data had an influence on this research project.

Limitations in Phase of Exiting the Field

Firstly, 'the field' here refers to conducting interviews with expert representatives operating in Kenya. Since the interview data could not be gathered in the field in the classical sense (fieldwork physically present in Kenya), exiting the field in the digital interview setting refers to the end of each interview, so that with every new interview the field was entered and exited again (over the period of approximately 12 months, see Chap. 3). More specifically, the applied sampling method of snowballing (Sect. 3.2.1) lead to limitations in forwarding of other contacts. Although almost all interview participants had offered to assist in referring the researcher to additional interview participants in their network, only a few of them actually did. And if so, from this referral mostly only one additional interview participant could be gathered, except for one interview (I17) from which three new participants could be gathered. In some instances, the researcher was introduced to new participants, but the participants never answered. Additionally, after the data collection phase of conducting expert interviews was completed (i.e., after the last interview), the interview list was complied and specific data protection requests of participants as per indication in the declarations of consent were noted (Chap. 3). From these data protection and anonymization degrees, the names and in most instances also the organization's name were anonymized which poses a limit to this research project in so far as the credibility of experts in the respective fields and their expert positions are thereby minimized. Moreover, one interviewee (I20) requested to not directly quote the respective interview, which further limits pointed argumentation.

Methodological Limitations in Data Analysis via Qualitative Content Analysis

Secondly, the interviews were analyzed using qualitative content analysis, where further limitations of this dissertation were identified. As a first methodological limitation, the case study design (in addition to method of qualitative expert interviews) naturally poses a limitation (see Sect. 3.1.4), as it only allows the analysis of one case, here the digital ecosystem of Kenya. Hence, generalizations from the empirical findings are tailored to the Kenyan context and thereby limit their transfer to other countries or contexts (e.g., in contrast to cross-country studies). This dissertation however has developed a (theoretical) framework that can arguably be transferred to the study of further country contexts (Sect. 2.5).

On the one hand, from the methodological approaches of qualitative expert-interviews in the context of digital transformation and development as well as considering digital settings of interviews, the actual sample size posed methodological limitations for the empirical findings. As previously discussed in the beginning of this section and in Chap. 3, an equal amount of expert representatives per actor group was intended, but not achieved. More specifically, given the methodological circumstances around the sampling, a limitation is demonstrated in some interviewees being representatives of not only one, but two organizations. From this limited sample size, empirical limitations emerged in so far as that in the data analysis it was not always clear from which of their organizational roles the interviewee was talking. Similarly, the data of this dissertation is limited due to the fact that none of the (only) two interviewed startups were in contact with donors as they likewise did not participate in donor funded acceleration programs. This for instance is demonstrated in their limited insights on ties between donor organizations and startups (e.g., degree of closeness, see Sect. 4.4.3). Hence, the sample size of two startup representatives here is rather limited for gathering deeper empirical insights and limit generalizations for this actor group.

Furthermore, another methodological limitation of the sample and as discussed in the beginning of this section, the empirical insights on the actor group of the Kenyan government are limited due to a lack of interview recruitment. However, through the methodological approach of this dissertation, this possible limitation is balanced in that every interviewee has been asked about the role of the Kenyan government so that every actor group has commented on this actor group. Thereby at least some expert knowledge could be provided even though in an indirect way. Moreover, it has been noted that not only is every interview subjective and thereby poses limitations even for generalization possibility within an actor group, possible biases stemming from differing natures of the chosen actor groups cannot be ruled out completely.

On the other hand, limitations in the data analysis phase emerged. For instance, all interviews were coded in chronological order after completion of the data gathering phase. As a second step, all coded interviews were scanned through and checked. In this process, the limitation occurred that not all interviews have the exact same codes due to individual and slightly different course of conversation. While this might limit comparability and generalization, it enabled deeper insights into the experts' knowledge and experiences which are the essence of why the method of qualitative expert interviews was chosen. Moreover, in the interview coding review phase, the observation was made that some interviews expressed rather marketing-orientated statements such as selling success stories and praising their own institution thereby posing a limitation to quality and reliability. Additionally, the observation while reviewing the material is made that in some instances, it is not always clear whether the expert refers to digital transformation in general or specifically with regard to Kenya.

6.3 Empirical Limitations

Lastly, both the theoretical and methodological limitations contribute to empirical limitations. With regard to relations between theoretical and empirical limitations, some aspects were identified in the empirical analysis that were not identified in the theoretical analysis (or identified to a lesser degree). This for instance applies to the importance of the (governance) approach of mapping (see e.g., Sect. 4.5.1), which was highlighted by expert interviewees from a practical experience perspective while it was not identified to the same extent in the theoretical literature review (Chap. 2). Similarly, 'common language and understanding' is derived as a precondition as part of 'perceptions' in the theoretical background (Sect. 2.5.1), while in the empirical analysis however it is discussed as a challenge (Sect. 4.4.4). The theoretical precondition refers to 'perceptions' including common language and understanding as well as awareness (Sect. 2.5.1). In the empirical analysis, perceptions are understood as awareness (Sect. 4.4.3).

Another limitation with regard to the research method of interviews is identified with regard to the striking usage of certain terms in an interchangeable and synonymous way. It could be argued that this limitation is due to unclear or imprecise phrasing of questions. This however has been mitigated by the researcher through the conducted pretests (Sect. 3.2.3) and choice of expert interviewees. Here, knowledge of at least the terms mentioned in the questionnaire could therefore be expected. The data analysis however has identified that a reason for this phenomenon can also be explained by a lack of common language and understanding among multi-stakeholders (as discussed in Sect. 4.4.4). The data review phase of the content analysis for instance highlighted an interchangeable reference to terms 'collaboration' and 'cooperation' (sometimes also 'coordination'). Moreover, 'governance' and 'coordination' as well as 'governance' and 'regulation' are used interchangeably (as discussed in Sect. 6.2). In applying the method of content analysis, it also emerged that the terms 'interest', 'focus' and 'objectives' are often used interchangeably by expert interviewees. Gaps in understanding between the multi-stakeholders but also between the interviewees and the researcher also emerged from the data analysis. Here sometimes the private sector is seen in terms of companies and sometimes also of NGOs. Why this is a challenge to this dissertation's data validity will be discussed in the following.

As a major empirical limitation, similarly to the challenges derived in the empirical analysis, common language and understanding (see Sect. 4.4.4) as well as sensitivities to comment (see Sect. 4.4.4) pose limits to this dissertation's empirical data analysis. With regard to common language and understanding (here especially between interviewees and researcher and the expected role of experts), this is for instance highlighted in the data analysis in statements such as "what do you mean by Global North and Global South?" (I10) or "I don't maybe we have different understandings of of of governance" (I14). In the same vein, and as discussed throughout this chapter and in the empirical analysis (Sect. 4.4.4), the expert status of the interviewees was not always as strong as expected. The challenge of required knowledge in the digital development field of both digital knowledge and knowledge of development (cooperation) is greater than expected and vividly emphasizes the need for interdisciplinary knowledge (exchange) among actors for achieving a sustainable digital development (this is almost a universal finding). This is for instance emphasized by I11 acknowledging the relevance of a question while expressing to be unable to an answer: "[...] that is a very good question [...] I don't know I don't know, I'm tempted to say no, but then I could be wrong". Further limits to the expert role and associated data quality are for instance highlighted by statements such as "your questions are hard" (I4), or "[...] that really that's a difficult question" (I10). Due to the expressed challenge of some expert interviewees answering some of the questions, this can be interpreted as an empirical limitation for the empirical analysis. At the same time, this emphasizes the importance of common language and understanding of multi-stakeholders for a sustainable digital development (Sect. 4.4.4).

Conclusion 7

After the discussion of the empirical research results in connection to the theoretical findings (Chap. 2) and limitations of research (Chap. 6), this final chapter will summarize the answers to this dissertation's research question and guiding questions. Moreover, it will discuss the theoretical and empirical contributions to the research field. The work then closes with an outlook on further research opportunities.

7.1 Answering the Research Questions

In this dissertation, one research question and five guiding questions were pursued with the goal of explaining the role of governance and coordination in digital ecosystems for sustainable development using the case of Kenya. Three problem areas have been established in Chap. 1, spanning from problems emerging from the country's development context (problem 1), implications of the digital transformation (problem 2), to unsustainability emerging from problems 1 and 2 and from impeded coordination of actions (problem 3). From the synthesis of the theoretical background, methodological approach of a case study and qualitative guideline-based expert interviews as well as their empirical analysis through qualitative context analysis, the following insights have been obtained in reference to each question. Through the unique multidisciplinary approach of synthesizing research fields ICT4D, sustainable development and governance, a framework for digital ecosystem governance has been developed with five determinants (Fig. 2.6). This framework has been slightly modified (determinant 2) based on the empirical findings (Fig. 5.1).

C. Stötzel, *The Digital Ecosystem of Kenya*,
https://doi.org/10.1007/978-3-658-45553-8_7

1. *What characterizes digital ecosystems as a network form? Are these ecosystems restricted and if so how or by what circumstances?*
   Digital ecosystems are characterized by the context in which they operate (here: macro level), especially with regard to the political, economic, social and infrastructural context (see Sect. 2.1, Fig. 1.1). Studying countries from the Global South, aid fragmentation and (a lack of) donor coordination likewise build a further context. These factors influence the ecosystem's inner workings from the outside. Not only the macro context influences digital ecosystems, but also the digital transformation that is likewise influenced by its context and surroundings. Development theories such as modernization theory, dependence theory and postcolonial theory, but mostly concepts of ICT4D theory such as leapfrogging have built the theoretical angle. From these insights, determinant 1 for governing digital ecosystems was developed and delivered further insights into the nature or characteristics of digital ecosystems (Sect. 2.5.1): network size, level of activity, visibility, transparency, voice, type of actors, actors' interests, willingness, goal consensus, perceptions, trust, network rationale, duration and frequency, intensity (degree closeness or distance), expectations as well as network effectiveness and efficiency. In addition to these theoretical findings, further characteristics defining the ecosystem have been derived from the empirical data in Sect. 4.4.1. More specifically, it has delivered insights into actors' activity levels in the Kenyan digital ecosystem in correspondence with associated influence using both qualitative content analysis and qualitative network analysis (Table 4.1, Chap. 3). In sum, this dissertation has delivered insights for research goal 1 in that it has not only elaborated on the importance of a country's development context for its performance in the digital transformation, but has also identified contextual factors that influence the digital ecosystem.

2. *What factors from both digital transformation and development context enable and hinder sustainable development of digital ecosystem in a developing country?*
   This question refers to research goal 2 (and problem 2) in which the current digital landscape in developing countries, such as respective 'basic' and 'innovative' digital technologies, were analyzed through a theoretical literature review and expert interviews. Here, potential contributions of digital technologies to the development of countries (ICT4D), but also challenges to sustainable development were discussed. Thereby this research project contributed to the current academic and practical discourse around the digital transformation. Concepts of leapfrogging and digital ecosystems as well as digital economy research were synthesized with ICT4D research while embedding digital technologies and services in a wider conceptual framework (meso level, Fig. 1.1).

Firstly, and similarly to research guiding question one, macro factors (and especially in cases where these are flawed) were identified as important challenging factors—for both coordination and sustainable digital development. Such macro factors for instance include political stability and legislation around digital topics, but also how easy (or not) it is to start a business. More importantly, it further discussed the social context, such as digital literacy skills, but also access to digital infrastructure. The role of the Kenyan government here has for instance been highlighted in creating a supportive environment that enables and facilitates participation in the digital transformation and the ecosystem. The political context influences the capacities and capabilities of the Kenyan government and hence can potentially limit its capacities that are especially required in the digital context. This can contribute to unsustainable development if the (only) elected actor is restricted in its ability to act.

Secondly, specific challenges emerged from the ICT4D/digital development context (see Sect. 2.5.2, Sect. 4.4.4). More specifically, these include bureaucracies and traditional structures as well as a lack of flexibility. Moreover, the complexity associated with digital technologies and services, required digital literacy and skills as well as resources, are discussed as lacking. This especially applies to digital capacities of public sector actors (both Kenyan but also donor organizations). Besides these capabilities, financial resources and funding as well as costs (for digital technologies and services) were further identified as challenges. From the perspective of multi-stakeholder interaction in the ecosystem, common language and understanding is emphasized as crucial for both coordination efforts and sustainable digital development (Sect. 2.5.1). However, this is a factor that has been identified as lacking, both in ICT4D discourse and especially in the Kenyan digital ecosystem based on the analyzed interview data. More specifically, a lack of common understanding and language exists with regard to *both* digital transformation and development cooperation. This for instance includes terminologies such as sustainability, ecosystem or governance, but also a lack of a common understanding with regard to certain types of technology such as AI (Sect. 4.4.4, Sect. 4.2.2).

This is further mirrored in an illustrated hype. This refers to overestimating technologies (e.g., AI) and the impact of ICTs on development as well as actors' motivations to proliferate themselves through engagements in the 'shiny topic' of digital transformation (see e.g., Sect. 4.3.1, Sect. 2.2.3). With regard to digital technologies in the Kenyan digital ecosystem, a dualism between 'basic' and 'innovative' digital technologies has been identified (see Sect. 4.2.2). If these listed challenges are not reflected and addressed by multi-stakeholders, these can lead to further unsustainable and damaging development as they amplify the problem.

Closely related to these challenges, the fragmentation and failure of ICT(4)D projects due to failed suitability (misled focus of action) and failed ensuring of sustainability was identified as a further challenge to sustainable digital development. Specifically, a misregard in terms of multi-stakeholders' neglecting the importance of the infrastructure context in the Kenyan digital ecosystem was emphasized (Sect. 4.3.2). This is also linked with capabilities of actor groups in the context of digital complexity and know-how (Sect. 4.4.4) and also to diverging interests of actors in the ecosystem. These two are examples of major challenges for both the coordination of action in the Kenyan digital ecosystem as well as for sustainable action of multi-stakeholders. This is further mirrored in associated dependencies, i.e., dependencies on (donor) protocols, on private sector as well as interdependencies and lock-ins. Lock-ins are for instance created through expensive software. Open source was a discussed but not agreed upon solution to decrease dependencies. A further challenge was highlighted with regard to the usage of hardware and also in terms of interoperability. All these discussed factors culminate in power and knowledge asymmetries on different levels (see Sect. 2.3.6, Sect. 4.3.4): On the private sector level, competition between startups and multinationals and, more importantly for sustainable development, power asymmetries between tech companies and public sector actors are highlighted. More specifically, the analyzed interview data indicated that the tech sector is more dominant in terms of resources (financial and know-how) in the digital transformation, which is experienced in their influence in policy making, even to the extent of overtaking governmental functions. Power of tech actors here is expressed from both the outside but also from the inside of the government (role advisor), leading to dependencies and power imbalances that are also discussed as 'digital colonialism'. Here it is important to highlight that this finding is based on only limited empirical data (Sect. 4.3.7). All in all, it is highlighted in the analysis that the digital ecosystem follows the metaphor of ecological ecosystems, where species (here actors) are interdependent. In the context of this research project, sustainable digital development therefore requires balanced approaches.

3. *Who coordinates in digital ecosystem, hence which governance roles are to be identified and ascribed to which actor?*
Questions three to five contribute to analyzing the identified research problem 3 and research goal 3—unsustainability due to development context, digital transformation and impeded coordination. In light of the discussed problems

(problems 1 and 2), especially failures of ICT4D projects, the research goal was to contribute to (network) governance research in offering coordination 'solutions'. This theoretical lens offered both governance 'solutions' to these problems, but also enabled explaining digital ecosystems as a form of network. Thereby, this dissertation contributed to network governance in expanding the focus on networks as ecosystems where networks also serve as mechanisms of coordination (e.g., Provan and Kenis 2008). The goal here was to identify possible governance roles that multi-stakeholders can overtake in the ecosystem.

From the literature review in the theoretical background, the (network) governance roles of a coordinator, facilitator, broker, network manager and driver of digital transformation were derived (Sect. 2.5.3). With regard to the empirical expert insights into the governance roles in the Kenyan digital ecosystem (see Sect. 4.4.5), these theoretical roles were added by roles of an advisor, balancer, connector, enabler and supporter. These roles were distributed among multi-stakeholder groups as follows: role of advisor (multi-stakeholders), role of balancer (academia, NGOs and Kenyan government), role of broker (multi-stakeholders), role of connector (multi-stakeholders), role coordinator (multi-stakeholders), role of driver of digitalization (private sector and Kenyan government), role of an enabler (ICTs and multi-stakeholders), role of facilitator (Kenyan government), role of a supporter (donors and NGOs) and role of a manger (Kenyan government).

4. *How can ecosystems be governed, i.e., through which governance form?*
For answering the question through which governance form digital ecosystems can be governed, this dissertation followed the distinction between institutionalized ('brokered') and less-institutionalized ('non-brokered') governance forms for coordinating multi-stakeholders' interactions in the digital ecosystem (see Sect. 2.5.4, Sect. 4.4.6). Firstly, less formalized governance forms identified in this research project were events and workshops as well as self-governed structures. Formalized governance forms include standards, alliances and associations, working groups as well as PPPs and MSPs. *Events and workshops* in the Kenyan digital ecosystem were perceived as inclusive and representative of all multi-stakeholders by interviewed experts. As the ultimate less formalized governance form, *self-governed structures* were discussed as a form allowing minimum disruption and intervention in innovation processes and hence a form of self-coordination without an entity intervening in the digital ecosystem (but with notions of unsustainable digital development).

Secondly, more formalized governance forms derived in this dissertation were standards, alliances and associations, committees, working groups in addition to partnership formats of PPPs and MSPs. *Standards* enabling coordination were both of technical but also of regulatory nature. They include global technical standards for digital topics, standards to regulate the IT industry as well as global and local standards for collaboration and coordination in digital ecosystems. *Alliances and associations* were identified as (synonymous) formats predominantly used for pursuing strategic interests. Formations of actors under an alliance or association umbrella were argued as bestowing actors with more power through speaking with a joint voice (which is especially relevant for Kenyan tech startups). *Committees* as a governance format for coordination of efforts was discussed as being divided according to thematic areas such as committees for cybersecurity, although similarities to working groups were identified. *Working groups* as governance forms for coordination of action in the Kenyan digital ecosystem were perceived by expert interviewees as less inclusive or relevant to multi-stakeholders and their impact was questioned (Sect. 4.4.6).

Partnership formats for coordination of multi-stakeholder activities, especially PPPs and MSPs were discussed critically with regard to their impact on sustainable development both in the theoretical and empirical analysis (Sect. 2.5.4, Sect. 2.5.4, Sect. 4.4.6, Sect. 4.4.6). Moreover, PPPs and MSPs were subsumed under formalized governance forms due to their contractual nature between public and private partners (although more or less binding). This dissertation analyzed PPPs as a format for not only coordinating multi-stakeholders activities for achieving sustainable digital development, but also as a format for addressing asymmetries in the relations between multi-stakeholders in digital ecosystems. The impact of PPPs for coordinating the interaction between public and private actors however was questioned with regard to the Kenyan digital ecosystem. On the one hand, this referred to a perceived hype around PPPs and on the other hand, they were experienced as enforcing instead of balancing asymmetries in that these partnerships rather strengthen the private actor side. With regard to MSPs as a governance form for multi-stakeholder interactions in digital ecosystems, possible partnership potentials for institutionalized inclusivity in multi-stakeholder interactions and economic opportunities were counterbalanced by MSP challenges. These included arguments of multi-stakeholder processes being both cost and time consuming, often inefficient and complex to manage due to the specific interests, expectations and requirements of multi-stakeholders.

5. *Are new or alternative forms of governance needed in order to achieve an equitable and sustainable development of digital ecosystems in developing countries?*
This dissertation has shown that alternative governance forms (alternative to

the ones addressed in the previous question) are needed, especially for the purpose of transparent and inclusive coordination of multi-stakeholder activities in digital ecosystems such as the Kenyan one. More specifically, the empirical analysis of expert interviews in the Kenyan digital ecosystem expressed a need for 'mapping' and for an overview of multi-stakeholder activities (see Sect. 4.5.1). Similarly, 'mapping' emerged as a precondition for coordination as it enhances visibility and awareness, especially since not all multi-stakeholders have an overview of what others do (again leading to settings of fragmentation hindering coordination for sustainable digital development). From the literature review and the data analysis, this dissertation proposed two possible alternative governance forms for achieving sustainable development of digital ecosystems in developing countries: A *meta organization* and *a collaborative digital platform*.

A meta organization was discussed as a central coorindative governance organization (hierarchy) that is either already present in the ecosystem or a new organization that functions similar to a 'supranational' entity. While the academic literature review addressed potentials and values of a meta organization (understood as part of meta governance, see Sect. 2.5.4), the majority of the empirical analysis suggested a rejection of coordination in Kenyan digital ecosystem through a central coorindative authority (regardless of external or internal). Instead, if coordination was to be pursued by an organization, the empirical data highlights an empowering and enforcing of existing actors in their governance roles (see Sect. 4.5.2).

The second proposition of a multi-stakeholder digital platform for coordination follows possibilities offered by digital platforms as discussed in Sect. 2.2.5. It is discussed as enabling the needed mapping efforts of activities, an inclusive multi-stakeholder involvement and participation in a coordinated and transparent manner. Here however constraints similar to the derived preconditions and challenges would need to be addressed.

Ultimately, these guiding research questions lead to answering the main research question: "How can digital ecosystems in developing countries be coordinated in such a way that they contribute to a sustainable development?". Coordination of digital ecosystems in developing countries can be achieved through understanding the context in which the ecosystem operates, recognizing challenges to sustainable digital development, multi-stakeholders finding and understanding (or reconsidering) their roles as well as using governance formats, especially digital ones. Furthermore, if preconditions to coordination are not fulfilled, efforts of multi-stakeholders in the digital ecosystem do not counteract unsustainable developments, but rather enforce them.

7.2 Theoretical and Empirical Contributions to Research Fields

The present multidisciplinary research project positions itself at the intersection between ICT4D and governance research. Due to the interwoven structural and argumentative synthesis of the theoretical and empirical background, Chap. 4 and Chap. 5 have demonstrated how this dissertation contributes to an advancement of digital development research. Moreover, this dissertation has emphasized a theoretical research desideratum with regard to *ICT4D* (especially digital ecosystems), *governance* and *sustainable development* research as well as an *empirical* desideratum with regard to developing countries represented by the case of Kenya. The main theoretical research contribution lies in the unique synthesis of these research fields into one framework for coordination of digital ecosystems that addresses various complex interrelations.

7.2.1 ICT4D Research

Beginning with the *ICT4D desideratum*, this dissertation has delivered theoretical and empirical insights into sustainable digital development of developing countries using the case of the Kenyan digital ecosystem. From a theoretical perspective, the challenge of ICT4D as a multi-disciplinary field drawing upon various research fields has been addressed in using theory perspectives ranging from development theories such as post-colonial theory, dependence theory or aid fragmentation to interdisciplinary concepts of leapfrogging or digital colonialism.

More specifically, this dissertation has contributed to *(power) dependence theory* where further research was identified as needed, especially concerning dimensions of power and dependence relationships between public and private actors (e.g., Medaglia et al. 2018). With regard to ICT4D research, it has been established that projects are dependent on bilateral and multilateral donors as well as on private sector organizations for financial resources and technology expertise (e.g., Delponte et al. 2015; Chipidza and Leidner 2019). In the reviewed literature, causes for dependencies are for instance alignment with people, processes and systems on the local level or new technologies leading to new dependencies (e.g., Chawurura et al. 2019). Furthermore, this dissertation has delivered additional five dependency categories based on the data analysis. These were identified as barriers to sustainable digital development in the Kenyan digital ecosystem: ‘new’ dependencies, dependencies on protocols, dependencies on the private sector, interdependencies and lock-ins (Sect. 4.4.4).

Similarly, this dissertation has contributed to *postcolonial theory* through delivering insights into power imbalances between diverse international multi-stakeholders in digital development. It has elaborated the importance of contextual factors, such as historic relations between nations and their power asymmetries (e.g., De' et al. 2018). Empirically with regard to Kenya, post-colonial effects were highlighted especially in terms of inherited political and legislative structures (Sect. 4.1.3), relations with donor organizations (Sect. 4.4.4), and resource dependency (e.g., Chipidza and Leidner 2019; Sect. 4.4.4). These were identified as components for explaining barriers to sustainable digital development (e.g., Sect. 2.3; Sect. 4.3). Further insights into post-colonialism in the context of digital ecosystems have been identified with regard to the concept of digital colonialism.

Both theoretical angles of post-colonial and (power) dependence theory serve as a further basis for the emerging concept of *digital colonialism* (Sect. 2.3.6). The concept combines colonialism with digital development emphasizing a colonization of digital technologies by specific (foreign) actors with inherent power asymmetries (e.g., Kwet 2019). This dissertation has contributed to the research in identifying areas where interviewed experts shared experiencing digital colonialism in the Kenyan digital ecosystem. For instance, the activity of multinational private tech companies is compared to notions of 'scramble' for business in Kenya (Sect. 4.3.4), especially with regard to the value and use of data ('data exploitation', see Sect. 4.2.2).

These theory angles were complemented through the ICT4D related concept of *leapfrogging*. Although the concept is widely discussed in multidisciplinary academic literature, there is only limited research on the nature of innovation activities in developing countries (e.g., Cirera and Maloney 2017). This dissertation contributes to a further understanding of leapfrogging potentials and barriers in developing countries using empirical evidence from Kenya. More specifically, contributions to a further development of the leapfrogging concept are legal framework conditions and structures (I6) and, underlining the significance of this dissertation, coordination of multi-stakeholders (I5). Hence, the leapfrogging potential of the digital transformation as a societal change and a structural leapfrogging can be classified as research contributions (Sect. 4.2.1).

A further contribution was made to the *digital divide* concept. Here, the role of access has been highlighted as crucial for a sustainable digital development, especially in terms of the affordability of digital technologies and services, both on an individual but also governmental level (Sect. 2.5.2). This dissertation has further emphasized that digital infrastructure (Sect. 4.1.3) is a factor that is often underestimated and disregarded in ICT4D projects leading to unsustainable outcomes (Sect. 4.3.2). Moreover, this dissertation has shown that power and knowledge asymmetries further contribute to the digital divide (Sect. 2.3.6). Measures to address these implications of the digital divide have for instance been derived with regard to more investments in infrastructure and (digital) education, which are currently not sufficient (e.g., Chipidza and Leidner 2019; Sect. 4.3.2). This is similarly reflected in an methodological argument: Due to the COVID-19 pandemic, the expert interviews were conducted digitally instead of locally in Kenya. In terms of the importance of digital infrastructure, it is to be noted that several virtual interviews were impacted by disturbances of lost internet connections (Sect. 6.2).

Further synthesizing the development and ICT theory angle, it has contributed to academic literature concerning *aid and donor coordination* by further incorporating governance research. This has offered contributions to further understanding the impact of aid fragmentation on sustainable digital development in developing countries such as Kenya. Firstly, aid fragmentation in the Kenyan digital ecosystem has been emphasized in the data analysis as a barrier to sustainable digital development (Sect. 4.1.5). This specifically includes the different speed of donor activities, favoritism ('aid volatility') towards already known hubs and startups that have an competitive advantage due to knowledge of donor protocols. These examples were extracted as further examples of how aid fragmentation in the Kenyan digital ecosystem amplifies asymmetries and unsustainable development. Moreover, this dissertation identified digital development and ICT4D projects as one area of donor saturation (Sect. 4.3.1). It has also demonstrated that the digital transformation has an impact on the (already existing) aid fragmentation setting due to an increase in more diverse actors. Thereby it has also contributed to governance research in analyzing digital ecosystem as a network, e.g., in terms of the number and type of actors, the effect of overlaps and duplication due to aid fragmentation in the ICT4D context, or network effectiveness and efficiency. For counteracting these challenges to sustainable digital development, this dissertation has argued that coordination is crucial. Theoretically, it has offered coordination mechanisms for sustainable development in digital ecosystems synthesizing the multidisciplinary research fields in the developed framework, especially determinant 3 (governance roles) and determinant 4 (governance forms).

Most importantly, this dissertation has contributed to further understanding and conceptualizing *digital ecosystems*. Considering that the ecosystem metaphor for understanding and analyzing implications of the digital transformation has only recently been adopted in academic literature (Nguyen and Mahundi 2019), this dissertation has made timely research contributions. Firstly, it has offered a new approach to understanding and analyzing digital ecosystems with its three-level-approach: the digital transformation as a force (meso level) between the context or environment of the digital ecosystem (macro level) and its inner workings (micro level). Specifically, it has emphasized that all levels influence each other. Secondly, it has delivered theoretical insights in proposing a holistic ecosystem framework using a network governance approach. The developed digital ecosystem governance framework is specifically tailored to developing countries and their development in delivering 'guidelines' through its four determinants. Empirically, insights into the Kenyan digital ecosystem have been derived, especially with regard to understanding in which context it operates (macro level), how the digital transformation shapes it (meso level) and how both influence the inner workings of the digital ecosystem. Thereby this dissertation contributes to claims in the reviewed literature for a systematic framework for the delivery and management of ICT projects (e.g., Ndegwa et al. 2017; Ominde et al. 2020; Tarus et al. 2015) and of stakeholder engagements (coordination, see Sect. 2.1.6; Sect. 4.1.6). Similarly, it has offered empirical insights into the importance of a framework for multi-stakeholders collaborating in the (Kenyan) digital ecosystem through the conducted expert interviews.

7.2.2 Governance Research

This is closely linked to the theoretical desideratum with regard to *governance* research. This dissertation has contributed especially to network governance in delivering further insights into governance roles in a network (here ecosystem). It has delivered empirical insights into the performance of governance roles overtaken by multi-stakeholders. For each role, respective actors were ascribed to a role based on the conducted expert interviews. Besides derived governance roles from the network governance literature (coordinator, facilitator, broker, network manager and driver of digital transformation; Sect. 2.5.3), this dissertation has identified further governance roles: advisor, balancer, connector, enabler and supporter (Sect. 4.4.5). Moreover, it has contributed to network governance research in transferring this perspective to emerging ecosystem logic. Thereby it has contributed to theory building in demonstrating the mutual exchange between the three levels (macro, meso, micro) and the actors within the ecosystem. Moreover, valuable for governance research are new insights into the role of governmental institutions in shaping the digital transformation, as well as into the role of actors beyond the government (non-governmental and private actors). It has shown that network governance can be used in order to explain both the role of coordination in terms of alignment of action with surroundings, but also coordination of action. Moreover, this work has offered coordination mechanisms of digital ecosystems, especially with regard to the role of the government (similar to network governance research of Provan and Kenis 2008, Kenis and Provan 2009 or Klijn and Koppenjan 2016). Additionally, it has contributed to network governance research in offering connections to digital transformation research and ICTs given their network logic functioning.

Similarly, this dissertation has contributed to network governance research in delivering empirical insights from both the digital transformation context and Kenyan setting into 'brokered' and 'non-brokered' governance forms (Sect. 2.5.4). It has provided empirical insights into the performance of PPP, MSP in ICT4D and in pursuing sustainable digital development. Moreover, this work has contributed to meta governance in discussing the role of a possible 'meta organization' for coordinating in digital ecosystems. As a further alternative governance approach, this dissertation has included a digital or IT research perspective by analyzing governance outcomes through using digital platforms. With regard to this dissertation's focus on ICT4D, it needs to be noted that research into digital platforms is in its early stages (e.g., de Reuver et al. 2018). This research project has contributed to digital platform research in delivering empirical use cases and an expressed benefit of digital platforms for coordination purposes in the digital development context (Sect. 2.5.4).

Ultimately, this dissertation has derived preconditions for a sustainable governance of digital ecosystems. The empirical analysis has provided detailed implications of the digital transformation on each of these factors, such as transparency, voice and ease of entry (Sect. 2.5.1, Sect. 4.4.3). It emphasized that these preconditions are important for a sustainable digital development outcome while highlighting that these can easily become challenges. This dissertation has further provided governance challenges as well as ICT4D challenges. If these are not addressed, they not only contribute to unsustainable digital development, but also can amplify asymmetries. Hence, these two determinants not only contribute to a further understanding of network governance mechanisms for coordination, but also for achieving sustainable digital development. Thereby, it has further contributed to *sustainability governance* research in providing connections to digital transformation.

7.2.3 Sustainability

This research project has contributed to *sustainability governance* research in delivering insights into sustainable digital development, i.e., offering connections to the digital transformation. Here it is worth noting that research contributions include contributions to sustainability in ICT4D research as well as to sustainability in governance research. Thereby it has offered not only to theoretical and empirical insights into the combination of governance research and sustainability (similar to Meuleman 2019 and Monkelbaan 2019), but also into the synthesis of digital sustainability with ecosystem research (similar to concepts such as 'Digital Sustainability' of Linkov et al. 2018). More specifically, it has contributed to the interrelation of how governance influences the SDGs and also how SDGs influence governance. Special insights provided here especially refer to SDG 16, specifically into intergovernmental coordination (Sect. 2.3.5, Sect. 2.5.2, Sect. 4.4.3, Sect. 4.1.6) and SDG 17 (importance and challenges of partnership formats for sustainable digital development; see Sect. 2.5.4 and Sect. 4.4.6). As sustainable digital development is the thread that spins through every action and is the goal to be reached, this dissertation has contributed to a further understanding of why ICT projects fail and how this impacts sustainable digital development.

Secondly, this dissertation has demonstrated the importance of multi-stakeholders' common understanding of the term 'sustainability' in underlining its features of inclusivity and transparency as proposed by the SDGs (Sect. 4.4.4). This is accompanied by power and knowledge asymmetries that contribute to unsustainable digital development. As power imbalances in ICT4D have been identified as not widely covered in the literature (e.g., Chipidza and Leidner 2019), this dissertation has made valuable contributions to both academic literature as well as to decision-makers in handling ICT4D projects. This likewise speaks to a methodological argument as the expert knowledge of the sampled expert interviewees was on the one hand not as strong as expected when it came to required knowledge of both the digital and the development landscape. On the other hand it also emphasizes the importance of a common understanding and language of experts in the field (Sect. 6.2).

Overall, this research project makes both empirical and theoretical contributions that underline the increasingly important understanding of implications of the digital transformation for societies, especially vulnerable ones. The here proposed framework with its determinants delivers valuable insights while highlighting preconditions, possible challenges that need to be considered in the design of governance solutions for securing a sustainable digital development.

7.3 Future Research

This dissertation provided insights into the digital transformation from a multidisciplinary approach synthesizing sustainable ICT4D and governance perspectives. Combining the present research findings and limitations, further research potentials can be pursued, e.g., through other theory angles or by using different cases.

Such different theory angles can for instance apply to comparative studies of two or more digital ecosystems (e.g., countries with similar or opposing macro factors as well as different meso and micro contexts). Moreover, this dissertation and its framework has chosen a macro approach to analyzing digital ecosystems while further incorporating a meso and micro level. The micro level however does not include an analysis of governance for inner-organizational perspectives. This could be a potential for future research in terms of analyzing how the digital ecosystem affects inner-organizational workings (potentially from a public administration perspective). Further research is also needed for conceptualizing digital ecosystems, also in comparison to related concepts such as ICT ecosystems and also into how sub-ecosystems (e.g., entrepreneurial ecosystems) relate to the overall ecosystem.

Moreover, this dissertation follows a development theory angle in choosing the case of a developing country. Since digital ecosystems likewise exist in developed countries (but in different macro, meso and micro contexts), the study of digital ecosystems can be adapted to developed countries ('Global North') and explored in further research with respective adapted theory angles. Similarly, further research potentials lie in analyzing the effects of the political and legislative state on the innovation ability and progress. As one example, future research can deliver further insights into the hypothesis that digital development is 'easier', if a) fixed structures are not already preexisting, and b) policies or regulations are flawed or non-existent while also considering its downside effects (e.g., data exploitation). Similarly, corruption in the context of sustainable digital development requires further research. One approach could be a good governance perspective into how ICTs can counteract these and how they can be promoted in contrast to a preference of 'traditional' or paper-based processes. This refers to both assessing the benefits of digital technologies in bureaucratic settings as well as factors such as motivation and skills. Here, a variety of further theory angles are possible, such as cultural studies that could deliver further valuable insights, for instance considering the discussed role of openness towards innovations for sustainable digital development.

Although digital technologies enable a possibility to transcend boundaries such as geographical or national boundaries, this dissertation has approached digital ecosystems as constructs that operate within national borders and are tangible by a physical presence of multi-stakeholders. Here further research is needed to understand the physicality of digital ecosystems, especially in the context of digital technologies exceeding tangible boundaries. Further expanding on digital ecosystems in a national state context, a specific centralization of activity on the capital Nairobi in the Kenyan digital ecosystem (Sect. 4.4.3) is also identified as an exclusive (rather than inclusive) digital development. Hence the effects of decentralization efforts or decentralized governance of the ecosystem activity on sustainable digital development provides room for further research as the data of this dissertation is limited.

With further regard to governance in the (Kenyan) digital ecosystem, the size of the ecosystem is one example where future research can be valuable. More precisely, this dissertation's findings suggest that a small amount of stakeholders is easier to coordinate, due to the potential of increased efficiency and impact that can possibly lead to more sustainable digital development. On the other hand however, a small amount of stakeholders can also lead to exclusiveness such as silos, less inclusivity and less holistic development. Generally, a quantitative approach to analyzing digital ecosystems could enable more detailed graphical illustrations of the ecosystem such as mappings of closeness or distance between multi-stakeholders in the ecosystem (e.g., using ANT or network analysis tools). Although this dissertation provides some insight into how this balance can be governed, further research is needed here. Further research is also needed with regard to the implications of actors overtaking these roles, but also regarding additional governance roles that specifically apply to digital ecosystems (e.g., driver of the digital transformation in the ecosystem). In the context of governing the digital transformation, the (Kenyan) government has been established by this dissertation as an important, if not the most important actor. However, considering the political stability of developing countries and the severe challenges in the digital transformation context, further research is to be conducted into how to strengthen governmental institutions and promote their democratic legitimacy. Moreover, following a methodological argument, further research could add on this dissertation's findings with interview data from the Kenyan (or any) government, as this dissertation was not able to acquire governmental representatives. The transfer of knowledge (digital capacities) between multi-stakeholders for a sustainable digital transformation is a further area for future research. Among governance roles, governance formats and platforms, the role of standards for sustainable digital development, e.g., for regulating the IT industry (in Kenya and elsewhere) needs to be further explored in future research. Here also the responsibility of implementing and monitoring standards is not clear as it can involve the international and/or national level, with the question remaining of who is responsible. Similarly, more research is needed into the design, implementation, monitoring and value of collaborative digital platforms for coordinating and aligning inner workings, meso factors and macro factors in digital ecosystems.

Given this dissertation's key finding of the importance of common language and understanding in digital development, further research is needed here, especially how this can be achieved and ensured between involved multi-stakeholders (in developing countries, but also globally). This includes an active awareness and understanding of involved multi-stakeholders of *both* sides—the 'ICT' (technical) and the 'D' (development) side. Mirroring the ecosystem metaphor where everyone needs each other for achieving goals and synergies, it is to be further explored how interdisciplinary exchange, transfer of knowledge and understanding, especially in partnership programs such as PPPs or MSPs, can be fostered for sustainable digital development. This further includes more research into how partnerships can be designed more sustainably, i.e., in reaching SDG 17 and its further progress. Another area this dissertation has identified as crucial for a sustainable development and simultaneously as an area that is underestimated, is digital infrastructure. The role of access and quality of digital infrastructure for a variety of sustainable development issues needs further research, both from governance but also development perspectives. Moreover, this dissertation has contributed on the role of AI in sustainable digital development, where further research is needed, especially into the capabilities and capacities as well as suitability of AI in ICT projects.

This is similarly reflected in research on ICTs' potential for decreasing power asymmetries, especially with regard to resources. Open source software as one such technical means is discussed both in academic literature and also among interviewed experts of this dissertation. Since this research project however gathered only limited data, especially in terms of interviews with the Kenyan government (Sect. 6.2), further research can explore the suitability of technical alternatives to costly multinational software licenses that pose the challenge of lock-ins. More importantly, more research needs to be conducted into the (inter-)relation of the international private sector and the (Kenyan) government, especially into how power and knowledge imbalances can be addressed sustainably. This includes further research into how an equal and inclusive representation, involvement and recognition of multi-stakeholders in digital policy processes and projects, especially of NGOs, can be ensured. From a further theory angle, the emerging concept of digital colonialism (Sect. 2.3.6, Sect. 4.3.7) needs further research. This dissertation has elaborated that the perspective of digital colonialism is rather attributed to 'the West', while elements found for instance in SSC are less explored and hence require further research (not only from geopolitical point of view).

Lastly, the impact of the COVID-19 pandemic on the Kenyan digital ecosystem and its interactions on the ground however has not been incorporated into this dissertation as it is not part of its research question. However, this could be pursued in future research, especially considering the entry and exit of (previous) multi-stakeholders. Since the data gathering phase of this dissertation was likewise impacted by the pandemic due to travel restraints, further research can explore the impact of the pandemic on research methodologies and findings.

In conclusion, the digital transformation is a crucial determinant for sustainable development of countries, especially in the Global South—indicating both contributions to development but also hampering development. In order to fully achieve and foster a sustainable digital development, it needs to be clear for everyone involved what 'sustainable' really entails in this context and what development is intended to be achieved, especially by which digital means (not the other way round). This dissertation has demonstrated that sustainable digital development (and digital transformation as such) is about finding a common understanding and a common language of and for everyone—not through 'hyping' digital technologies and services, but through interdisciplinary exchange of diverse multi-stakeholders.

References

Adler, P. S. (2001), 'Market, hierarchy, and trust: The knowledge economy and the future of capitalism', *Organization Science* **12**(2), 215–234.

African Union (2019), 'African leaders call for a common digitalization agenda'. **URL:** https://au.int/en/pressreleases/20190515/african-leaders-call-common-digitalization-agenda, last accessed on 2023-01-29

Agranoff, R. (2007), *Managing within Networks. Adding Value to Public Organizations*, Georgetown University, Washington D.C.

Akamanzi, C., Deutscher, P. and Guerich, B. (2016), 'Silicon Savannah: The Kenya ICT services cluster', *Microeconomics of Competitiveness* (Spring 2016), 1–7.

Aker, J. C. and Mbiti, I. M. (2010), 'Mobile phones and economic development in Africa', *Journal of Economic Perspectives* **24**, 207–232.

Aldasoro, I., Nunnenkamp, P. and Thiele, R. (2010), 'Less aid proliferation and more donor coorindation? The wide gap between words and deeds', *Journal of International Development* **22**, 920–940.

Allen, J. P. (2017), *Technology and Inequality. Concentrated Wealth in a Digital World*, Palgrave Macmillan, Cham.

Alrich, H. (1979), *Organizations and Environments*, Prentice-Hall, Englewood Cliffs.

Alsop, R. and Heinsohn, N. (2005), *Measuring Empowerment in Practice—Structuring Analysis and Framing Indicators*, World Bank, Washington DC.

Anderson, E. (2012), 'Aid fragmentation and donor transaction costs', *Economics Letters* **117**(3), 799–802.

Annen, K. and Kosempel, S. (2007), 'Foreign aid, donor fragmentation, and economic growth', *The B.E. Journal of Macroeconomics* **9**(1), 1–30.

Ansell, C. and Gash, A. (2018), 'Collaborative platforms as a governance strategy', *Journal of Public Administration Research and Theory* **28**(1), 16–32.

Ansell, C. and Miura, S. (2020), 'Can the power of platforms be harnessed for governance?', *Public Administration* **98**(1), 261–276.

Anthopoulos, L., Reddick, C. G., Giannakidou, I. and Mavridis, N. (2016), 'Why e-government projects fail? an analysis of the healthcare.gov website', *Government Information Quarterly* **33**(1), 161–173.

Aoyama, Y. and Parthasarathy, B. (2017), Collaborative social innovation in the hybrid domain organization and rationality, *in* J. Choudri, M. S. Islam, F. Wahid, J. M. Bass and J. E. Priyatma, eds, 'Information and Communication Technologies for Development. 14th IFIP

C. Stötzel, *The Digital Ecosystem of Kenya*,
https://doi.org/10.1007/978-3-658-45553-8

WG 9.4 International Conference on Social Implications of Computers in Developing Countries, ICT4D 2017 Yogyakarta, Indonesia, May 22-24, 2017 Proceedings', Springer Nature, Cham, pp. 257–268.

Atkinson, M. M. and Coleman, W. D. (1992), 'Policy networks, policy communities and the problems of governance', *Governance. An International Journal of Policy and Administration* **5**(2), 154–180.

Aulkemeier, F., Iacob, M. E. and van Hillegersberg, J. (2019), 'Platform-based collaboration in digital ecosystems', *Electronic Markets* **29**(4), 597–608.

Avelino, F. and Rotmans, J. (2009), 'Power in transition: An interdisciplinary framework to study power in relation to structural change', *European Journal of Social Theory* **12**, 543–569.

Avgerou, C. (2010), 'Discourses on ICT and development', *Information Technologies and International Development* **3**(6), 1–18.

Avgerou, C. (2017), Theoretical framing of ICT4D research, *in* J. Choudri, M. S. Islam, F. Wahid, J. M. Bass and J. E. Priyatma, eds, 'Information and Communication Technologies for Development. 14th IFIP WG 9.4 International Conference on Social Implications of Computers in Developing Countries, ICT4D 2017 Yogyakarta, Indonesia, May 22-24, 2017 Proceedings', Springer Nature, Cham, pp. 10–23.

Ayres, R. (1991), 'Evolutionary economics and environmental imperatives', *Structural Change and Economic Dynamics* **2**(2), 255–273.

Bäckstrand, K. (2006), 'Multi-stakeholder partnerships for sustainable development', *European Environment* **16**, 290–306.

Bailur, S. (2007), 'Using stakeholder theory to analyze telecenter projects', *Information Technologies and International Development* **3**(3), 61–80.

Baker, S. E., Edwards, R. and Doidge, M. (2012), 'How many qualitative interviews is enough? Expert voices and early career reflections on sampling and cases in qualitative research', *National Centre for Research Methods Review Paper*.

Baldwin, C. Y. and Clark, K. B. (2006), 'The architecture of participation: Does code architecture mitigate free riding in the open source development model?', *Management Science* **52**(7), 1116–1127.

Bass, J. M., Nicholson, B. and Subrahmanian, E. (2013), 'A framework using institutional analysis and the capability approach in ICT4D', *Information Technologies & International Development* **9**(1), 19–35.

Bass, J. M. and Wall, P. (2020), *Information and Communication Technologies for Development. 16th IFIP WG 9.4 International Conference on Social Implications of Computers in Developing Countries, ICT4D 2020, Manchester, UK, June 10–11, 2020 Proceedings*, Springer Nature Switzerland AG, Cham.

Bayliss, K. and Van Waeyenberge, E. (2018), 'Unpacking the public private partnership revival', *Journal of Development Studies* **54**(4), 577–593.

Beisheim, M., Ellersiek, A., Goltermann, L. and Kiamba, P. (2017), 'Meta-governance of partnerships for sustainable development: Actors' perspectives from Kenya', *Public Administration and Development* **119**(2017), 105–119.

Beisheim, M. and Simon, N. (2016), 'Multi-stakeholder partnerships for implementing the 2030 agenda: Improving accountability and transparency', *COSOC Partnership Forum—March 11, 2016* pp. 1–33.

Bella, J. D., Grant, A., Kindornay, S. and Tissot, S. (2013), 'Mapping private sector engagements in development cooperation'. **URL:** http://www.nsi-ins.ca/wp-content/uploads/2013/09/Mapping-PS-Engagment-in-Development-Cooperation-Final.pdf, last accessed 2019-05-11

Benz, A., Lütz, S., Schimank, U. and Simonis, G. (2007), *Handbuch Governance. Theoretische Grundlagen und empirische Anwendungsfelder*, VS Verlag für Sozialwissenschaften, Wiesbaden.

Berbés-Blázquez, M., González, J.A., Pascual, U. (2016), 'Towards an ecosystem services approach that addresses social power relations. Current opinion in environmental sustainability', *Sustainability Science* **19**, 134–143.

Bevir, M. (2011), *The SAGE Handbook of Governance*, SAGE Publications, London.

Bigsten, A. (2006), *Donor coordination and the uses of aid*, Götheborg University. **URL:** https://gupea.ub.gu.se/bitstream/2077/2723/1/gunwpe0196.pdf%3B, last accessed on 2020-03-09

Bodin, Ö., Alexander, S., Baggio, J., Barnes, M., Berardo, R., Cumming, G., Dee, L., Fischer, A., Fischer, M., Garcia, M., Guerrero, A., Hileman, J., Ingold, K., Matous, P., Morrison, T., Nohrstedt, D., Pittman, J., Robins, G. and Sayles, J. (2019), 'Improving network approaches to the study of complex social-ecological interdependencies', *Nature Sustainability* **2**, 551–559.

Bogner, A. and Menz, W. (2009), The theory-generating expert interview. Epistemological interest, forms of knowledge, interaction, *in* A. Bogner, B. Littig and W. Menz, eds, 'Interviewing Experts', Palgrave Macmillan, Basingstoke, pp. 43–80.

Boley, H. and Chang, E. (2007), Digital ecosystems: Principles and semantics, *in* 'Inaugural IEEE International Conference on Digital Ecosystems and Technologies. Cairns, Australia', pp. 398–403.

Bon, A. and Akkermans, H. (2019), Digital development: Elements of a critical ICT4D theory and praxis, *in* P. Nielsen and H. C. Kimaro, eds, 'Information and Communication Technologies for Development. Strengthening Southern-Driven Cooperation as a Catalyst for ICT4D. 15th IFIP WG 9.4 International Conference on Social Implications of Computers in Developing Countries, ICT4D 2019 Dar es Salaam, Tanzania, May 1–3, 2019 Proceedings, Part II', Springer Nature Switzerland AG, Cham, pp. 26–38.

Bonina, C., Koskinen, K., Eaton, B. and Gawer, A. (2021), 'Digital platforms for development: Foundations and research agenda', *Information Systems Journal* **31**(6), 869–902.

Bracho, G. and Grimm, S. (2016), South-south cooperation and fragmentation: A non-issue?, *in* S. Klingebiel, T. Mahn and M. Negre, eds, 'The Fragmentation of Aid, Concepts, Measurements and Implications for Development Cooperation', Palgrave Macmillan, London, pp. 121–134.

Brass, D. J., Galaskiewicz, J., Greve, H. R. and Tsai, W. P. (2004), 'Taking stock of networks and organizations: A multilevel perspective', *Academy of Management Journal* **47**, 795–817.

Brazys, S. (2016), 'Aid and governance: Negative returns?', *European Journal of Development Research* **28**(2), 294–313.

Briterbridges (n. d.), 'Innovation Maps'. **URL:** https://briterbridges.com/innovation-maps, last accessed on 2021-12-14

Bukht, R. and Heeks, R. (2017), 'Defining, conceptualising and measuring the digital economy', Manchester Centre for Development Informatics, Working Paper.

URL: https://hummedia.manchester.ac.uk/institutes/gdi/publications/workingpapers/di/di_wp68.pdf, last accessed on 2023-01-29

Byrne, B. (2012), Qualitative interviewing, *in* C. Seal, ed., 'Researching Society and Culture', 3rd edn, SAGE Publications, London, pp. 206–226.

Börzel, T. A. (1998), 'Organising babylon: On the different conceptions of policy networks', *Public Administration* **2**(76), 253–273.

Börzel, T. A. and Risse, T. (2010), 'Governance without a state: Can it work?', *Regulation and Governance* **4**(2), 113–134.

Büthe, T. and Cheng, C. (2013), Private transnational governance of economic development: International development aid, *in* M. Moschella and C. Weaver, eds, 'Handbook of Global Economic Governance: Players, Power and Paradigms', Routledge, Abingdon, UK and New York, NY.

Camarinha-Matos, L. M., Afsarmanesh, H. and Antonelli, D. (2019), *Collaborative Networks and Digital Transformation. 20th IFIP WG 5.5 Working Conference on Virtual Enterprises, PRO-VE 2019 Turin, Italy, September 23–25, 2019 Proceedings*, Springer Nature Switzerland AG, Cham.

Cammaerts, B. and Mansell, R. (2020), 'Digital platform policy and regulation: Toward a radical democratic turn', *International Journal of Communication* **14**, 135–154.

Campanelli, P. (2008), Testing survey questions, *in* E. De Leeuw, J. J. Hox and D. A. Dillman, eds, 'International Handbook of Survey Methodology', Taylor & Francis, New York, pp. 176–200.

Carr, M. (2015), 'Power plays in global internet governance', *Millennium: Journal of International Studies* **43**(2), 640–659.

Carrington, P. J. and Scott, J. (2012), Introduction, *in* J. Scott and P. J. Carrington, eds, 'The SAGE Handbook of Social Network Analysis', SAGE Publications, London, pp. 1–8.

Castells, M. (2000), 'Materials for an exploratory theory of the network society', *British Journal of Sociology* **51**(1), 5–24.

Castells, M. (2004), Informationalism, networks, and the network society: A theoretical blueprint, *in* M. Castells, ed., 'The Network Society: A Cross-Cultural Perspective', Edward Elgar Publishing, Cheltenham, pp. 3–48.

Castells, M. (2011), 'A network theory of power', *International Journal of Communication* **5**(1), 773–787.

Cecere, G., Corrocher, N., Gossart, C. and Ozman, M. (2014), 'Lock-in and path dependence: An evolutionary approach to eco-innovations', *Journal of Evolutionary Economics* **24**(5), 1037–1065.

Ceelen, A., Wood, J. and Huesken, S. (2019), *Guidance and Good Practice on CSO Development Effectiveness and Enabling Environment*, Task Team on CSO Development Effectiveness. **URL:** https://www.shareweb.ch/site/Development-Effectiveness/Documents/SmartPages/TSKTM-01C-Guidance.pdf, last accessed on 20323-01-29

Chataway, J., Hanlin, R. and Kaplinsky, R. (2014), 'Inclusive innovation: An architecture for policy development', *Journal of Innovation Development* **4**, 33–54.

Chaudhary, S., McGregor, A., Houston, D. and Chettri, N. (2018), 'Environmental justice and ecosystem services: A disaggregated analysis of community access to forest benefits in Nepal', *Ecosystem Services* **29**, 99–115.

Chawurura, T., Manhibi, R., van Dijk, J. and van Stam, G. (2019), E-Health in Zimbabwe. A case of techno-social development, *in* P. Nielsen and H. C. Kimaro, eds, '15th IFIP WG 9.4

International Conference on Social Implications of Computers in Developing Countries, ICT4D 2019 Dar es Salaam, Tanzania, May 1-3, 2019 Proceedings, Part I', Springer Nature Switzerland AG, Cham, pp. 15–26.

Cheeseman, N., Collord, M. and Reyntjens, F. (2018), 'War and democracy: The legacy of conflict in East Africa', *The Journal of Modern African Studies* pp. 31–61.

China Africa Tech Initiative (n. d.), 'The new age of China Africa collaboration is in technology'. **URL:** https://www.chinaafrica.tech/, last accessed on 2022-10-08

Chipidza, W. and Leidner, D. (2019), 'A review of the ICT-enabled development literature: Towards a power parity theory of ICT4D', *Journal of Strategic Information Systems* **28**, 145–174.

Choi, T. and Chandler, S. M. (2020), 'Knowledge vacuum: An organizational learning dynamic of how e-government innovations fail', *Government Information Quarterly* **37**(1).

Choudri, J., Islam, M. S., Wahid, F., Bass, J. M. and Priyatma, J. E. (2017), *Information and Communication Technologies for Development. 14th IFIP WG 9.4 International Conference on Social Implications of Computers in Developing Countries, ICT4D 2017 Yogyakarta, Indonesia, May 22–24, 2017 Proceedings*, Springer Nature, Cham.

Ciesielska, M. and Janowski, T. (2019), Inter-governmental collaborative networks for digital government innovation transfer—Structure, membership, operations, *in* L. M. Camarinha-Matos, H. Afsarmanesh and D. Antonelli, eds, 'Collaborative Networks and Digital Transformation. 20th IFIP WG 5.5 Working Conference on Virtual Enterprises, PRO-VE 2019 Turin, Italy, September 23–25, 2019 Proceedings', Springer Nature Switzerland AG, Cham, pp. 295–307.

Cirera, X. and Maloney, W. F. (2017), *The Innovation Paradox: Developing-Country Capabilities and the Unrealized Promise of Technological Catch-Up*, The World Bank, Washington, DC.

Clarke, A. (2018), *The Evolving Role of Non-State Actors in Digital Era Government*, Carleton University. **URL:** http://dx.doi.org/10.2139/ssrn.3268084, last accessed on 2022-08-10

Clegg, S., Josserand, E., Mehra, A. and Pitsis, T. S. (2016), 'The transformative power of network dynamics: A research agenda', *Organization Studies* **37**(3), 277–291.

Coase, R. (1991), The nature of the firm (1937), *in* O. E. Williamson and S. G. Winter, eds, 'The Nature of the Firm. Origins, Evolution, and Development', Oxford University Press, New York, pp. 18–33.

Cochrane, L. and Cundill, G. (2018), 'Enabling collaborative synthesis in multi-partner programmes', *Development in Practice* **28**(7), 922–931.

Cohen, B. (2006), 'Sustainable valley entrepreneurial ecosystems', *Business Strategy and the Environment* **15**(1), 1–14.

Cohen, J. E. (2019), *Between Truth and Power: The Legal Construction of Informational Capitalism*, Oxford University Press, Oxford.

Cohen, J. E. (2020), Networks, standards, and network-and-standard-based governance, *in* K. Werbach, ed., 'After the Digital Tornado. Networks, Algorithms, Humanity', Cambridge University Press, Cambridge, pp. 58–80.

Communications Authority of Kenya (2021*a*), '59 million mobile phone devices connected to mobile networks as at September 2021, CA report shows'. **URL:** https://www.ca.go.ke/59-million-mobile-phone-devices-connected-to-mobile-networks-as-at-september-2021-ca-report-shows/, last accessed on 2022-02-01

Communications Authority of Kenya (2021*b*), 'Fourth quarter sector statistics report for the financial year 2020/21. April-June 2021'. **URL:** https://www.ca.go.ke/wp-content/uploads/2021/09/Sector-Statistics-Report-Q4-2020-2021.pdf, last accessed on 2023-01-29

Constantinides, P., Henfridsson, O. and Parker, G. G. (2018), 'Introduction—Platforms and infrastructures in the digital age', *Information Systems Research* **29**(2), 381–400.

Couldry, N. and Mejias, U. A. (2019*a*), *The Cost of Connection. How Data Is Colonizing Human Life and Appropriating It for Capitalism*, Stanford University Press, Stanford.

Couldry, N. and Mejias, U. A. (2019*b*), 'Data colonialism: Rethinking big data's relation to the contemporary subject', *Television and New Media* **20**(4), 336–349.

Creswell, J. W. (2013), *Qualitative inquiry and research design. Choosing among five approaches*, SAGE Publications, Thousand Oaks.

Creswell, J. W. and Creswell, J. D. (2018), *Research Design. Qualitative, Quantitative and Mixed Methods Approaches*, SAGE Publications, Los Angeles.

Cunningham, J. A. and Whalley, J. (2020), *The Internet of Things Entrepreneurial Ecosystems. Challenges and Opportunities*, Palgrave Macmillan, Cham.

Cusumano, M. A., Gawer, A. and Yoffie, D. B. (2019), *The business of platforms: Strategy in the age of digital competition, innovation, and power*, HarperBusiness, New York.

Daniel, J. (2012), *Sampling essentials. Practical guidelines for making sampling choices*, SAGE Publications, Thousand Oaks.

Daugbjerg, C. and Fawcett, P. (2017), 'Metagovernance, network structure, and legitimacy: Developing a heuristic for comparative governance analysis', *Administration and Society* **49**(9), 1223–1245.

Davies, R. B. and Klasen, S. (2013), *Of Donor Coordination, Free-riding, Darlings, and Orphans: The dependence of bilateral aid on other bilateral giving. Discussion Papers 137*, Georg August Universität Göttingen Research Centre: Poverty, Equity and Growth.

Davies, R. B. and Klasen, S. (2019), 'Darlings and orphans: Interactions across donors in international aid', *Scandinavian Journal of Economics* **121**(1), 243–277.

Daw, T., Brown, K., Rosendo, S. and Pomeroy, R. (2011), 'Applying the ecosystem services concept to poverty alleviation: The need to disaggregate human well-being', *Environmental Conservation* **38**, 370–379.

de Reuver, M., Sørensen, C. and Basole, R. C. (2018), 'The digital platform: A research agenda', *Journal of Information Technology* **33**(2), 124–135.

Degnbol-Martinussen, J. and Engberg-Pedersen, P. (2003), *Aid. Understanding International Development Cooperation*, Zed Books Ltd., London.

Delponte, L., Grigolini, M., Moroni, A., Vignetti, S., Giguashvili, N. and Claps, M. (2015), 'ICT in the developing world', European Parliamentary Research Services, Brussels.

DeNardis, L. (2014), *The Global War for Internet Governance*, Yale University Press, New Haven.

Desai, R. M. and Kharas, H. (2018), 'What motivates private foreign aid? Evidence from internet-based microlending', *International Studies Quarterly* **62**(3), 505–519.

Desmarchelier, B., Djellal, F. and Gallouj, F. (2019), Public service innovation networks (PSINs): An instrument for collaborative innovation and value co-creation in public service(s), *in* 'R-D Management conference "The innovation challenge: bridging research, industry and society", Ecole Polytechnique, HEC, Paris, 17–21 June 2019', HAL Open Science, pp. 1–27.

Deutsch, K. W. (1973), *Politische Kybernetik. Modelle und Perspektiven*, 3 edn, Rombach, Freibug im Breisgau.

De', R., Pal, A., Sethi, R., Reddy, S. K. and Chitre, C. (2018), 'ICT4D research: A call for a strong critical approach', *Information Technology for Development* **24**(1), 63–94.

Dhungana, D., Groher, I., Schludermann, E. and Biffl, S. (2013), Guiding principles of natural ecosystems and their applicability to software ecosystems, *in* S. Jansen, M. Cusumano and S. Brinkkemper, eds, 'Software Ecosystems: Analyzing and Managing Business Networks in the Software Industry', Edward Elgar Publisher Inc., Northampton, pp. 43–58.

Diaz-Bone, R. (2007), 'Gibt es eine qualitative Netzwerkanalyse?', *Forum Qualitative Sozialforschung* **8**(1), 2086–2094.

Dietrich, S. (2016), 'Donor political economies and the pursuit of aid effectiveness', *International Organization* **70**(1), 65–102.

Diga, K. and May, J. (2016), 'The ICT ecosystem: The application, usefulness, and future of an evolving concept', *Information Technology for Development* **22**, 1–6.

Dijck, J. v., de Waal, M. and Poell, T. (2018), *The Platform Society: Public Values in a Connective World*, Oxford University Press, Oxford.

Djankov, S., Montalvo, J. and Reynal-Querol, M. (2009), 'Aid with multiple personalities', *Journal of Comparative Economics* **37**(2), 217–229.

DLA Piper (n. d.), 'Data protection laws of the world'. **URL:** https://www.dlapiperdataprotection.com/, last accessed on 2022-02-16

Döring, N. and Bortz, J. (2016), *Forschungsmethoden und Evaluation in den Sozial- und Humanwissenschaften*, Springer Verlag, Berlin.

Dunleavy, P. and Carrera, L. (2013), *Growing the Productivity of Government Services*, Edward Elgar, Cheltenham.

Dunleavy, P., Margetts, H., Bastow, S. and Tinkler, J. (2006), 'New public management is dead—Long live digital-era governance', *Journal of Public Administration Research and Theory* **16**(3), 467–494.

Easterly, W. (2002), 'The cartel of good intentions: The problem of bureaucracy in foreign aid', *Journal of Policy Reform* **5**, 223–250.

Edelenbos, J., Van Buuren, A. and Klijn, E. H. (2013), 'Connective capacities of network managers', *Public Management Review* **15**(1), 131–159.

Effah, J. and Nuhu, H. (2017), 'Institutional barriers to digitalization of government budgeting in developing countries: A case study of Ghana', *Electronic Journal of Information Systems in Developing Countries* **82**(1), 1–17.

Ellersiek, A. and Beisheim, M. (2017), *Partnerships for the 2030 Agenda for Sustainable Development: Transformative, inclusive and accountable?*, SWP Research Paper, Stiftung Wissenschaft und Politik -SWP- Deutsches Institut für Internationale Politik und Sicherheit, Berlin.

Emerson, R. (1962), 'Power-dependence relations', *American Sociological Review* **27**(1), 31–41.

Escobar, A. (1995), *Encountering Development: The Making and Unmaking of the Third World*, Princeton University Press, Princeton.

Eurofound (2018), 'Digital Economy'. **URL:** https://www.eurofound.europa.eu/observatories/eurwork/industrial-relations-dictionary/digital-economy, last accessed on 2021-08-22

Ezeomah, B. and Duncombe, R. (2019), The role of digital platforms in disrupting agricultural value chains in developing countries, *in* P. Nielsen and H. C. Kimaro, eds, 'Information and Communication Technologies for Development. Strengthening Southern-Driven Cooperation as a Catalyst for ICT4D. 15th IFIP WG 9.4 International Conference on Social Implications of Computers in Developing Countries, ICT4D 2019 Dar es Salaam, Tanzania, May 1–3, 2019 Proceedings, Part I', Springer Nature Switzerland AG, Cham, pp. 231–247.

Faulbaum, F., Prüfer, P. and Rexroth, M. (2009), *Was ist eine gute Frage? Die systematische Evaluation der Fragenqualität*, VS Verlag, Wiesbaden.

Fawcett, P. and Daugbjerg, C. (2012), 'Explaining governance outcomes: Epistemology, network governance and policy network analysis', *Political Studies Review* **10**(2), 195–207.

Federation of Kenyan Employer (2021), *The Informal Economy in Kenya*, Federation of Kenyan Employer in partnership with ILO, Nairobi. **URL:** https://www.ilo.org/wcmsp5/groups/public/---ed_emp/---emp_ent/documents/publication/wcms_820312.pdf, last accessed on 2023-01-29

Felipe-Lucia, M., Martín-López, B., Lavorel, S., Berraquero-Díaz, L., Escalera-Reyes, J. and Comín, F. (2015), 'Ecosystem services flows: Why stakeholders' power relationships matter', *PLoS ONE* **10**(7), e0132232. **URL:** https://doi.org/10.1371/journal.pone.01322323

Flick, U. (2009), *An Introduction to Qualitative Research*, SAGE Publications, London.

Flick, U. (2014), *An Introduction to Qualitative Research*, 5th edn, SAGE Publications, London.

Foster, C. and Heeks, R. (2013), 'Conceptualising inclusive innovation: Modifying systems of innovation frameworks to understand diffusion of new technology to low-income consumers', *European Journal of Development Research* **25**(3), 333–355.

Foster, C. and Heeks, R. (2014), 'Nurturing user-producer interaction: Inclusive innovation flows in a low-income mobile phone market', *Journal of Innovation Development* **4**, 221–237.

Foster-Fishman, P. G., Salem, D. A., Allen, N. A. and Fahrbach, K. (2001), 'Facilitating interorganizational collaboration: The contributions of interorganizational alliances', *American Journal of Community Psychology* **29**(6), 875–905.

Franklin, M. (2013), *Digital Dilemmas: Power, Resistance and the Internet*, Oxford University Press, Oxford.

Fransman, M. (2010), *The new ICT ecosystem: Implications for policy and regulation*, Cambridge University Press, Cambridge.

Freedom House (2021*a*), 'Freedom in the World 2021. Kenya'. **URL:** https://freedomhouse.org/country/kenya/freedom-world/2021#PR, last accessed on 2022-02-16

Freedom House (2021*b*), 'Freedom on the Net 2021. Kenya'. **URL:** https://freedomhouse.org/country/kenya/freedom-net/2021, last accessed on 2022-02-16

Freeman, R., Harrison, J., Wicks, A., Parmar, B. and De Colle, S. (2010), *Stakeholder theory: The state of the art*, Cambridge University Press, Cambridge.

Frey, B., Lohmeier, J. H., Lee, S. W. and Tollefson, N. (2006), 'Measuring collaboration among grant partners', *American Journal of Evaluation* **27**(3), 383–392.

Friederici, N. (2019), Hope and hype in Africa's digital economy: The rise of innovation hubs, *in* M. Graham, ed., 'Digital Economies at Global Margins', MIT Press, Cambridge, pp. 193–222.

Friederici, N., Ojanperä, S. and Graham, M. (2017), 'The impact of connectivity in Africa: Grand visions and the mirage of inclusive digital development', *Electronic Journal of Information Systems in Developing Countries* **79**(2), 1–20.

Fuchs, A., Nunnenkamp, P. and Öhler, H. (2015), 'Why donors of foreign aid do not coordinate: The role of competition for export markets and political support', *The World Economy* pp. 255–285.

Fuchs, A. and Öhler, H. (2020), 'Does private aid follow the flag? An empirical analysis of humanitarian assistance', *The World Economy* **44**, 601–854.

Fuhse, J. A. (2018), *Soziale Netzwerke. Konzepte und Forschungsmethoden*, UVK Verlagsgesellschaft, Konstanz.

Fukuyama, F. (2016), 'Governance: What do we know, and how do we know it?', *Annual Review of Political Science* **19**(1), 89–105.

Gandomi, A. and Haider, M. (2015), 'Beyond the hype: Big data concepts, methods, and analytics', *International Journal of Information Management* **35**(2), 137–144.

Gartner (2015), 'Hype cycle for emerging technologies 2015'. **URL:** https://www.gartner.com/doc/3100227, last accessed on 2020-03-06

Gartner (2021), 'Hype cycle for emerging technologies 2021'. **URL:** https://www.gartner.com/smarterwithgartner/3-themes-surface-in-the-2021-hype-cycle-for-emerging-technologies, last accessed on 2022-05-24

Gauld, R. and Goldfinch, S. (2006), *Dangerous Enthusiasms: E-Government, Computer Failure, and Information System Development*, Otago, Dunedin.

Gehring, K., Michaelowa, K., Dreher, A. and Spörri, F. (2017), 'Aid fragmentation and effectiveness: What do we really know?', *World Development* **99**, 320–334.

Gerring, J. (2017), *Case Study Research. Principles and Practices*, Cambridge University Press, Cambridge.

Ghazawneh, A. and Henfridsson, O. (2013), 'Balancing platform control and external contribution in third-party development: The boundary resources model', *Information Systems Journal* **23**(2), 173–192.

Gitogo, W. (2020), 'Census 2019 data shows Kenya has a youthful rural population', *The Kenyan Wall Street* . **URL:** https://kenyanwallstreet.com/census-2019-datashows-kenya-has-a-youthful-rural-population/, last accessed on 2021-12-14

Glasbergen, P. (2011), 'Mechanisms of private meta-governance: An analysis of global private governance for sustainable development', *International Journal of Strategic Business Alliances* **2**(3), 189–206.

Global Partnership for Effective Development Co-operation (2019), *Kampala Principles. On effective private sector engagement in development co-operation key terms and definitions*, Global Partnership for Effective Development Co-operation. **URL:** https://www.oecd.org/dac/effectiveness/Kampala-Principles-on-effective-private-sector-engagement-development-cooperation.pdf, last accessed on 2022-08-22

Goldsmith, S. and Eggers, W. D. (2004), *Governing by network: The new shape of the public sector*, Brookings Institution Press, Washington D.C.

Graham, M. (2019), *Digital Economies at Global Margins*, MIT Press, Cambridge.

Graham, M. and Mann, L. (2013), 'Imagining a Silicon Savannah? Technological and conceptual connectivity in Kenya's BPO and software development sectors', *Electronic Journal of Information Systems in Developing Countries* **56**(1), 1–19.

Grävingholt, J., Ziaja, S. and Kreibaum, M. (2015), 'Disaggregating state fragility: A method to establish a multidimensional empirical typology', *Third World Quarterly* **36**(7), 1281–1298.

GSMA (2019), *The mobile gender gap report*, GSMA.

GSMA (2020), *State of mobile internet connectivity report 2020*, GSMA.

GSMA (2021*a*), *The mobile economy Sub-Saharan Africa 2021*, GSMA.

GSMA (2021*b*), *The state of mobile internet connectivity 2021*, GSMA.

Hall, R. H. (1999), *Organizations: Structures, Processes, and Outcomes*, 7th edn, Prentice Hall, Upper Saddle River.

Hanneman, R. A. and Riddle, M. (2012), Concepts and measures for basic network analysis, *in* J. Scott and P. J. Carrington, eds, 'The SAGE Handbook of Social Network Analysis', SAGE Publications, London, pp. 341–369.

Hatakka, M., Øystein, S. and Thapa, D. (2019), A framework to explain the relation between ict and development: Combining affordances and the capability approach, *in* P. Nielsen and H. C. Kimaro, eds, '15th IFIP WG 9.4 International Conference on Social Implications of Computers in Developing Countries, ICT4D 2019 Dar es Salaam, Tanzania, May 1–3, 2019 Proceedings, Part II', Springer Nature Switzerland AG, Cham, pp. 60–71.

Heeks, R. (2002), 'i-development not e-development: Special issue on ICTs and development', *Journal of International Development* **14**(1), 1–11.

Heeks, R. (2007), 'Theorizing ICT4D research', *Information Technologies and International Development* **3**(3), 1–4.

Heeks, R. (2008), 'Success and failure rates of e-government in developing/transitional countries'. **URL:** http://www.egov4dev.org/success/sfrates.shtml, last accessed on 2019-10-02

Heeks, R. (2010), 'Do information and communication technologies (ICTs) contribute to development?', *Journal of International Development* **22**(5), 625–640.

Heeks, R. (2016), *Examining "Digital Development": The Shape of Things to Come?*, Centre for Development Informatics. **URL:** https://www.gdi.manchester.ac.uk/research/publications/di/di-wp64/, last accessed on 2022-08-10

Heeks, R. (2017), *Decent Work and the Digital Gig Economy: A Developing Country Perspective on Employment Impacts and Standards in Online Outsourcing, Crowdwork, Etc.*, Centre for Development Informatics. Development Informatics Working Paper Series. **URL:** https://papers.ssrn.com/sol3/papers.cfm?abstract_id=3431033, last accessed on 2022-08-10

Heeks, R. (2020*a*), 'ICT4D 3.0? Part 1—The components of an emerging "digital-for-development" paradigm', *Electronic Journal of Information Systems in Developing Countries* **86**(3), 1–15.

Heeks, R. (2020*b*), 'ICT4D 3.0? Part 2—The patterns of an emerging "digital-for-development" paradigm', *Electronic Journal of Information Systems in Developing Countries* pp. 1–13.

Heeks, R., Amalia, M., Kintu, R. and Shah, N. (2013), *Inclusive Innovation: Definition, Conceptualisation and Future Research Priorities*, Centre for Development Informatics. Development Informatics Working Paper Series. **URL:** http://dx.doi.org/10.2139/ssrn.3438439, last accessed on 2022-08-10

Heeks, R., Foster, C. and Nugroho, Y. (2014), 'New models of inclusive innovation for development', *Innovation and Development* **4**(2), 175–185.

Heeks, R. and Stanforth, C. (2007), 'Understanding e-government project trajectories from an actor-network perspective', *European Journal of Information Systems* **16**(2), 1–23.

Heeks, R. and Stanforth, C. (2015), 'Technological change in developing countries: Opening the black box of process using actor-network theory', *Development Studies Research* **2**(1), 33–50.

Heffernan, C., Lin, Y. and Thomson, K. (2012), 'Drawing from development: Towards unifying theory and practice of ICT4D', *Journal of International Development* **28**, 902–918.

Hein, A., Schreieck, M., Riasanow, T., Setzke, D. S., Wiesche, M., Böhm, M. and Krcmar, H. (2020), 'Digital platform ecosystems', *Electronic Markets* **30**(1), 87–98.

Helfferich, C. (2014), Leitfaden- und Experteninterviews, *in* N. Baur and J. Blasius, eds, 'Handbuch Methoden der empirischen Sozialforschung', Springer Fachmedien, Wiesbaden, pp. 559–574.

Hertting, N. and Vedung, E. (2012), 'Purposes and criteria in network governance evaluation: How far does standard evaluation vocabulary takes us?', *Evaluation* **18**(1), 27–46.

Herz, A., Peters, L. and Truschkat, I. (2015), 'How to do qualitative structural analysis. The qualitative interpretation of network maps and narrative interviews', *Forum Qualitative Sozialforschung* **16**(1).

Hill, R. (2014), Internet governance: The last gasp of colonialism, or imperialism by other means?, *in* R. Radu, J.-M. Chenou and R. H. Weber, eds, 'The Evolution of Global Internet Governance', Springer Verlag GmbH Berlin Heidelberg, Heidelberg, pp. 79–94.

Hofmann, J. (2016), 'Multi-stakeholderism in internet governance: Putting a fiction into practice', *Journal of Cyber Policy* **1**(1), 29–49.

Hollstein, B. (2010), Qualitative Methoden und Mixed-Method-Designs, *in* C. Stegbauer and R. Häußling, eds, 'Handbuch Netzwerkforschung', Band 4 edn, VS Verlag für Sozialwissenschaften, Wiesbaden, pp. 458–470.

Hollstein, B. (2012), Qualitative approaches, *in* J. Scott and P. J. Carrington, eds, 'The SAGE Handbook of Social Network Analysis', SAGE Publications, London, pp. 404–416.

Hollstein, B., Matiaske, W. and Schnapp, K.-U. (2017), *Networked Governance*, Springer International Publishing AG, Cham.

Hussain, F. and Brown, S. (2018), 'Developing a comparative framework of ICT4D initiatives in the Global South: A critical realist approach', *The Electronic Journal of Information Systems in Developing Countries* p. e12055.

Iansiti, M. and Levien, R. (2004), *The keystone advantage: What the new dynamics of business ecosystems mean for strategy, innovation, and sustainability*, Harvard Business Press, Boston.

IGF (2019), *The Internet of Trust. Thirteenth Internet Governance Forum (IGF) 12–14 2018 Paris, France. Chair's Summary & IGF Messages*, United Nations Publication.

IMF (2018), 'Measuring the digital economy. IMF staff report'. **URL:** https://www.imf.org/~/media/Files/Publications/PP/2018/022818MeasuringDigitalEconomy.%20ashx, last accessed on 2021-08-22

International Trade Administration (2021), 'Kenya—Country Commercial Guide'. **URL:** https://www.trade.gov/country-commercial-guides/kenya-information-communications-and-technology-ict, last accessed on 2022-02-16

International Trade Centre (2019), *Tech hubs in Africa: How can they support tech start-ups across the continent*, ITC, Geneva.

Isett, K., Mergek, I., LeRoux, K., Mischen, P. and Rethemeyer, R. (2011), 'Networks in public administration scholarship: Understanding where we are and where we need to go', *Journal of Public Administration Research and Theory* **21**(Suppl. 1), i157–i173.

Ismail, S. A., Heeks, R., Nicholson, B. and Aman, A. (2018), 'Analyzing conflict and its management within ICT4D partnerships: An institutional logics perspective', *Information Technology for Development* **24**(1), 165–187.

ITU (2015), *ICT indicators for the SDG monitoring framework. Technical information sheets prepared by ITU*, ITU, New York.

ITU (2018*a*), *Measuring the Information Society Report 2018. Volume 1*, ITU Publishing.

ITU (2018*b*), *Measuring the Information Society Report 2018. Volume 2*, ITU Publishing.

ITU (2019*a*), *ICT Centric Innovation Ecosystem Kenya: Country Review*, ITU Publishing.

ITU (2019*b*), 'Measuring digital development. Facts and figures 2019', *ITU Publishing* pp. 1–15.

ITU (2020), 'Measuring digital development. Facts and figures 2020', *ITU Publishing* pp. 1–19.

ITU (n. d.), 'Digital Development Dashboard'. **URL:** https://www.itu.int/en/ITU-D/Statistics/Dashboards/Pages/Digital-Development.aspx, last accessed on 2023-01-10

Jackson, T. (2021), *African tech startup funding passes $700m in record-breaking 2020*, Disrupt Africa. **URL:** https://disrupt-africa.com/2021/01/21/african-tech-startup-funding-passes-700m-in-record-breaking-2020/, last accessed on 2021-12-14

Jacobides, M. G., Cennamo, C. and Gawer, A. (2018), 'Towards a theory of ecosystems', *Strategic Management Journal* **39**(8), 2255–2276.

Jansen, D. (2006), *Einführung in die Netzwerkanalyse. Grundlagen, Methoden, Forschungsbeispiele*, VS Verlag für Sozialwissenschaften, Wiesbaden.

Jansen, D. (2017), Networks, social capital, and knowledge production, *in* B. Hollstein, W. Matiaske and K.-U. Schnapp, eds, 'Networked Governance', Springer International Publishing AG, Cham, pp. 15–42.

Jansen, D. and Wald, A. (2007), Netzwerktheorien, *in* A. Benz, S. Lütz, U. Schimank and G. Simonis, eds, 'Handbuch Governance. Theoretische Grundlagen und empirische Anwendungsfelder', VS Verlag für Sozialwissenschaften, Wiesbaden, pp. 188–199.

Jansen, S., Cusumano, M. and Brinkkemper, S. (2013), *Software Ecosystems: Analyzing and Managing Business Networks in the Software Industry*, Edward Elgar Publisher Inc., Northampton.

Janus, H., Klingebiel, S. and Paulo, S. (2014), 'Beyond aid: A conceptual perspective on the transformation of development cooperation', *The Journal of International Development* **27**(2), 155–169.

Jessop, B. (2002), *The Future of the Capitalist State*, Polity Press, Cambridge.

Jiménez, A. and Zheng, Y. (2017), A spatial perspective of innovation and development: Innovation hubs in Zambia and the UK, *in* J. Choudri, M. S. Islam, F. Wahid, J. M. Bass and J. E. Priyatma, eds, 'Information and Communication Technologies for Development. 14th IFIP WG 9.4 International Conference on Social Implications of Computers in Developing Countries, ICT4D 2017 Yogyakarta, Indonesia, May 22–24, 2017 Proceedings', Springer Nature, Cham, pp. 171–181.

Kallinikos, J., Aaltonen, A. and Marton, A. (2013), 'The ambivalent ontology of digital artifacts', *MIS Q* **37**, 357–370.

Katz, M. and Shapiro, C. (1986), 'Technology adoption in the presence of network externalities', *Journal of Political Economy* **94**, 822–841.

Keast, R. and Brown, K. (2010), 'The government service delivery project: A case study of the push and pull of central government coordination', *Public Management Review* **4**, 1–36.

Keast, R., Brown, K. and Mandell, M. (2007), 'Getting the right mix: Unpacking integration meanings and strategies', *International Public Management Journal* **10**(1), 9–33.

Keast, R., Mandell, M. P. and Brown, K. (2006), 'Mixing state, market and network governance modes: The role of government in crowded policy domains', *International Journal of Organization Theory and Behavior* **9**(1), 27–50.

Keast, R., Mandell, M. P., Brown, K. and Woolcock, G. (2004), 'Network structures: Working differently and changing expectation', *Public Administration Review* **64**(3), 363–371.

Keijzer, N., Klingebiel, S., Örnemark, C. and Scholtes, F. (2018), *Seeking Balanced Ownership in Changing Development Cooperation Relationships*, EBA Report 2018, Stockholm.

Kenis, P. and Provan, K. G. (2009), 'Towards an exogenous theory of public network performance', *Public Administration* **87**(3), 440–456.

Kenya National Bureau of Statistics (2016), *Micro, Small and Medium Establishment (MSME) Survey. Basic Report*, Kenya National Bureau of Statistics, Nairobi. **URL:** https://www.knbs.or.ke/?wpdmpro=2016-msme-basic-report#, last accessed on 2023-01-29

Kern, T. and Willcocks, L. (2000), 'Exploring information technology outsourcing relationships: Theory and practice', *The Journal of Strategic Information Systems* **9**(4), 321–350.

Kimura, H., Mori, Y. and Sawada, Y. (2012), 'Aid proliferation and economic growth: A cross-country analysis', *World Development* **40**(1), 1–10.

King'ori Ndegwa, A., Kiriri, P. and Achoki, G. (2016), 'Factors affecting adoption of donor funded ICT Projects in the public sector in Kenya', *International Journal of Project Management* **1**(1), 1–18.

Kleine, D. (2010), 'ICT4WHAT?—Using the choice framework to operationalise the capability approach to development', *Journal of International Development* **22**(1), 674–692.

Klievink, B., Bharosa, N. and Tan, Y. H. (2016), 'The collaborative realization of public values and business goals: Governance and infrastructure of public-private information platforms', *Government Information Quarterly* **33**(1), 67–79.

Klievink, B. and Janssen, M. (2014), 'Developing multi-layer information infrastructures: Advancing social innovation through public-private governance', *Information Systems Management* **31**(3), 240–249.

Klijn, E. H. (2008), 'Governance and governance networks in europe: An assessment of ten years of research on the theme', *Public Management Review* **10**(4), 505–525.

Klijn, E. H. and Edelenbos, J. (2007), Meta-governance as network management, *in* E. Sørensen and J. Torfing, eds, 'Theories of democratic network governance', Edward Elgar, Cheltenham, chapter 11, pp. 199–214.

Klijn, E. H. and Koppenjan, J. (2012), 'Governance network theory: Past, present and future', *Policy and Politics* **40**(4), 187–206.

Klijn, E. H. and Koppenjan, J. (2016), *Governance Networks in the Public Sector*, Routledge, New York.

Klijn, E. H., Koppenjan, J. and Termeer, K. (1995), 'Managing networks in the public sector: A theoretical study of management strategies in policy networks', *Public Administration* **73**(3), 437–454.

Klijn, E. H. and Teisman, G. R. (2003), 'Institutional and strategic barriers to public-private partnership: An analysis of Dutch cases', *Public Money & Management* **23**, 137–146.

Klingebiel, S. (2013), *Entwicklungszusammenarbeit—eine Einführung*, Deutsches Institut für Entwicklungspolitik, Bonn.

Klingebiel, S., Mahn, T. and Negre, M. (2016), *The Fragmentation of Aid, Concepts, Measurements and Implications for Development Cooperation*, Palgrave Macmillan, London.

Klingebiel, S., Negre, M. and Morazan, P. (2016), 'Costs, benefits and the political economy of aid coordination: The case of the European Union', *European Journal of Development Research* **29**, 144–159.

Knack, S. (2001), 'Aid dependence and the quality of governance: Cross-country empirical tests', *Southern Economic Journal* **68**(2), 310–329.

Knack, S. and Rahman, A. (2007), 'Donor fragmentation and bureaucratic quality in aid recipients', *Journal of Development Economics* **83**(1), 176–197.

Konrad Adenauer Stiftung (2019), *The Kenyan Digital Economy. Which framework does it need?*, Konrad Adenauer Stiftung.

Koppenjan, J. and Klijn, E. H. (2004), *Managing Uncertainties in Networks: A Network Approach to Problem Solving and Decision Making*, Routledge, London.

Koskinen, K., Bonina, C. and Eaton, B. (2019), Digital platforms in the Global South: Foundations and research agenda, *in* P. Nielsen and H. C. Kimaro, eds, 'Information and Communication Technologies for Development. Strengthening Southern-Driven Cooperation as a Catalyst for ICT4D15th IFIP WG 9.4 International Conference on Social Implications of Computers in Developing Countries, ICT4D 2019 Dar es Salaam, Tanzania, May 1–3, 2019 Proceedings, Part I', Springer Nature Switzerland AG, Cham, pp. 319–330.

Kowal, S. and O'Connell, D. C. (2004), The transcription of conversations, *in* U. Flick, E. von Kardorff and I. Steinke, eds, 'A Companion to Qualitative Research', SAGE Publications, London, pp. 248–252.

Krueger, E. H. and Levin, S. A. (2022), 'Governing sustainable transformations of urban social-ecological-technological systems', *Urban Sustainability* **2**(10), 1–12.

Kruse, J. (2014), *Qualitative Interviewforschung. Ein integrativer Ansatz*, Beltz Juventa, Weinheim, Basel.

Kuckartz, U. and Rädiker, S. (2019), *Analyzing Qualitative Data with MAXQDA. Text, Audio, and Video*, Springer Nature Switzerland, Cham.

Kwet, M. (2019), 'Digital colonialism: US empire and the new imperialism in the Global South', *Race and Class* **60**(4), 3–26.

Lamnek, S. (2010), *Qualitative Sozialforschung*, Beltz Verlag, Weinheim.

Lang, D. J., Wiek, A., Bergmann, M., Stauffacher, M., Martens, P., Moll, P., Swilling, M. and Thomas, C. J. (2012), 'Transdisciplinary research in sustainability science: Practice, principles, and challenges', *Sustainability Science* **7**(Suppl 1), 25–43.

Larkin, P. J., Dierckx de Casterlé, B. and Schotmans, P. (2007), 'Multilingual translation issues in qualitative research. reflections on a metaphorical process', *Qualitative Health Research* **4**(17), 468–476.

Larson, J. and Munger, M. (2017), Reimagine what you already know: Toward new solutions to longstanding problems, *in* B. Ndemo and T. Weiss, eds, 'Digital Kenya: An Entreprenurial Revolution in the Making', Palgrave Macmillian, pp. 133–153.

Lau, R. K.-S. (2020), Africa-China relations in the context of Belt and Road Initiative: Realizing African-Chinese dreams for common development?, *in* J. A. Berlie, ed., 'China's

Globalization and the Belt and Road Initiative', Springer International Publishing, Cham, pp. 157–178.

Lee, K. (2019), *The Art of Economic Catch-Up. Barriers, Detours, and Leapfrogging In Innovation Systems*, Cambridge University Press, Cambridge.

Lewis, D. and Madon, S. (2004), 'Information systems and nongovernmental development organizations: Advocacy, organizational learning, and accountability', *Information Society* **20**(2), 117–126.

Li, W., Badr, Y. and Biennier, F. (2012), 'Digital ecosystems: Challenges and prospects', *Proceedings of the International Conference on Management of Emergent Digital EcoSystems, MEDES 2012* pp. 117–122.

Li, Y.-R. (2009), 'The technological roadmap of Cisco's business ecosystem', *Technovation* **29**(5), 379–386.

Liamputtong, P. (2010), *Performing Qualitative Cross-Cultural Research*, Cambridge University Press, New York.

Lincoln, Y. S. and Guba, E. G. (1985), *Naturalistic Inquiry*, SAGE Publications, Newbury Park.

Lindgren, I. and Van Veenstra, A. F. (2018), Digital government transformation. A case illustrating public e-service development as part of public sector transformation, *in* 'Proceedings of the 19th Annual International Conference on Digital Government Research', Association for Computing Machinery, New York.

Linkov, I., Trump, B. D., Poinsatte-Jones, K. and Florin, M. V. (2018), 'Governance strategies for a sustainable digital world', *Sustainability (Switzerland)* **10**(2), 1–8.

Lithman, E. (2014), 'Transaction costs and development aid through SIDA: A pre-study for a strategic evaluation', *SIDA Studies in Evaluation* .

Löblich, M. and Pfaff-Rüdiger, S. (2011), 'Network analysis. A qualitative approach to empirical studies on communication policy', *International Communication Gazette* **73**(7), 630–647.

Lock, I. and Seele, P. (2017), 'Theorizing stakeholders of sustainability in the digital age', *Sustainability Science* **12**(2), 235–245.

Lundsgaarde, E. and Keijzer, N. (2019), 'Development cooperation in a multilevel and multistakeholder setting: From planning towards enabling coordinated action?', *European Journal of Development Research* **31**(2), 215–234.

Lyon, D. (2018), *The Culture of Surveillance: Watching as a Way of Life*, Polity Press, Cambridge.

Maail, A., Kurnia, S. and Chang, S. (2017), Enhancing social inclusion through optimal community participation levels in ICT4D projects, *in* A. Maail, S. Kurnia and S. Chang, eds, 'Social Inclusion and Usability of ICT-enabled Services', Routledge, London and New York, pp. 81–106.

Macharia, D. (2016), *Are Consortia a Good Way Forward?*, Oxfam GB, Oxford.

Madianou, M. (2019), 'Technocolonialism: Digital innovation and data practices in the humanitarian response to refugee crises', *Social Media + Society* **5**(3), 1–13.

Mansell, R. (2010), Technology, innovation, power, and social consequence, *in* P. Kalantzis-Cope and K. Gherab-Martín, eds, 'Emerging Digital Spaces in Contemporary Society. Properties of Technology', Palgrave Macmillan, New York, pp. 305–323.

Mansell, R. (2011), 'Power and interests in information and communication an development: Exogenous and endogenous discourses in contention', *Journal of International Development* **26**, 109–127.

Mansell, R. (2015*a*), 'Futures of knowledge societies—destabilization in whose interest?', *Information Communication and Society* **18**(6), 627–643.

Mansell, R. (2015*b*), 'Platforms of power', *Intermedia* **43**(1), 20–24.

Mansell, R. (2015*c*), 'The public's interest in intermediaries', *Info* **17**(6), 8–18.

Mansell, R. (2016), 'Power, hierarchy and the internet: Why the internet empowers and disempowers', *The Global Studies Journal* **9**(2), 19–25.

Mansell, R. (2017), 'Bits of power: Struggling for control of information and communication networks', *The Political Economy of Communication* **5**(1), 2–29.

Mansell, R. and Tremblay, G. (2013), *Renewing the Knowledge Societies Vision for Peace and Sustainable Development*, UNESCO.

Marchant, E. R. (2017), Organizational cultural hybrids: Nonprofit and for-profit cultural influences in the Kenyan technology sector, *in* B. Ndemo and T. Weiss, eds, 'Digital Kenya: An Entreprenurial Revolution in the Making', Palgrave Macmillian, pp. 303–329.

Margetts, H. and Dunleavy, P. (2013), 'The second wave of digital-era governance: A quasi-paradigm for government on the web', *Philosophical Transactions of The Royal Society* pp. 1–17.

Markus, M. L. and Bui, Q. (2012), 'Going Concerns: The Governance of Interorganizational Coordination Hubs', *Journal of Management Information Systems* **28**(4), 163–198.

Matias, K., Kidokoro, T., Menoni, S., Mejri, O. and Aminoltaheri, N. (2017), Open data reuse, recycling and sharing as potential solution to data and information resource inadequacies, *in* J. Choudri, M. S. Islam, F. Wahid, J. M. Bass and J. E. Priyatma, eds, 'Information and Communication Technologies for Development. 14th IFIP WG 9.4 International Conference on Social Implications of Computers in Developing Countries, ICT4D 2017 Yogyakarta, Indonesia, May 22–24, 2017 Proceedings', Springer Nature, Cham, pp. 221–231.

Mayntz, R. (1987), Politische Steuerung und gesellschaftliche Steuerungsprobleme—Anmerkungen zu einem theoretischen Paradigma, *in* T. Ellwein, ed., 'Jahrbuch zur Staats- und Verwaltungswissenschaft', Nomos, Baden-Baden, pp. 89–111.

Mayring, P. (2014), *Qualitative Content Analysis: Theoretical Foundation, Basic Procedures and Software Solution*, Klagenfurt. **URL:** https://nbn-resolving.org/urn:nbn:de:0168-ssoar-395173, last accessed on 2023-01-29

Mayring, P. (2015), Qualitative content analysis. Theoretical background and procedures, *in* A. Bikner-Ahsbahs, C. Knipping and N. Presmeg, eds, 'Approaches to Qualitative Research in Mathematics Education. Examples of Methodology and Methods', Springer Science + Business Media, Dordrecht, pp. 365–380.

Mayring, P. and Fenzl, T. (2014), Qualitative Inhaltsanalyse, *in* N. Baur and J. Blasius, eds, 'Handbuch Methoden der empirischen Sozialforschung', Springer Fachmedien, Wiesbaden, pp. 543–556.

McGuire, M. and Agranoff, R. (2011), 'The limitation of public management networks', *Public Administration* **89**(2), 265–284.

Meadowcroft, J. (2007), Democracy and accountability: The challenge for cross-sectoral partnerships, *in* P. Glasbergen, F. Biermann and A. Mol, eds, 'Partnerships, Governance and Sustainable Development: Reflections on Theory and Practice', Edward Elgar, Cheltenham, pp. 194–213.

Meadowcroft, J. (2011), Sustainable development, *in* M. Bevir, ed., 'The SAGE Handbook of Governance', SAGE Publications, London, pp. 535–551.

Medaglia, R., Hedman, J. and Eaton, B. (2018), It takes two to tango: Power dependence in the governance of public-private e-government infrastructures, *in* 'ICIS 2017 Seoul: Transforming Society with Digital Innovation', pp. 1–12.

Meuleman, L. (2008), *Public Management and the Metagoverance of Hierarchies, Networks, and Markets: The Feasibility of Designing and Managing Governance Style Combinations*, Springer, Heidelberg.

Meuleman, L. (2019), *Metagovernance for Sustainability. A Framework for Implementing the Sustainable Development Goals*, Routledge, New York.

Meuser, M. and Nagel, U. (2009), The expert interview and changes in knowledge production, *in* A. Bogner, B. Littig and W. Menz, eds, 'Interviewing Experts', Palgrave Macmillan, Basingstoke, pp. 17–42.

Meyerhoff Nielsen, M. and Jordanoski, Z. (2020), Digital Transformation, governance and coordination models: A comparative study of Australia, Denmark and the Republic of Korea, *in* 'The 21st Annual International Conference on Digital Government Research', dg.o '20, Association for Computing Machinery, New York, pp. 285—293.

Mimbi, L. and Kyobe, M. (2017), 'Public perceptions of the role of IT in resolving governance challenges in a transition state: The case of Tanzania', *The Electronic Journal of Information Systems in Developing Countries* **78**(2), 1–25.

Ministry of Foreign Affairs of the People's Republic of China (2018), 'Beijing declaration—Toward an even stronger China-Africa community with a shared future'. **URL:** https://www.fmprc.gov.cn/mfa_eng/wjdt_665385/2649_665393/201809/t20180910_679540.html, last accessed on 2020-03-05

Mirza, I. (2013), 'The provision of information services in a new political economy: A case study from Hyderabad, India', *Journal of Contemporary Asia* **43**(1), 148–172.

Molin, M. D. and Masella, C. (2016*a*), 'From fragmentation to comprehensiveness in network governance', *Public Organization Review* **16**(4), 493–508.

Molin, M. D. and Masella, C. (2016*b*), 'Networks in policy, management and governance: A comparative literature review to stimulate future research avenues', *Journal of Management and Governance* **20**(4), 823–849.

Monkelbaan, J. (2019), *Governance for the Sustainable Development Goals: Exploring an integrative framework of theories, tools, and competencies*, Springer Nature Singapore Pte Ltd., Singapore.

Moore, M. (2016), *Tech giants and civic power*, Centre for the Study of Media, Communication and Power. The Policy Institute at Kings College London, London.

Moore, M. and Tambini, D. (2018), *Digital Dominance: The Power of Google, Amazon, Facebook, and Apple*, Oxford University Press, Oxford.

Morriss, P. (1987), *Power: A Philosophical Analysis*, 1st edn, Manchester University Press, Manchester.

Mukhopadhyay, S. and Bouwman, H. (2019), 'Orchestration and governance in digital platform ecosystems: A literature review and trends', *Digital Policy, Regulation and Governance* **21**(4), 329–351.

Mulder, F. (2020), 'Governing the humanitarian knowledge commons', *Social Inclusion* **8**(4), 407–420.

Nagano, A. (2018), Economic growth and automation risks in developing countries due to the transition toward digital modernity, *in* A. Kankanhalli, A. Ojo and D. Soares, eds, 'Proceedings of the 11th International Conference on Theory and Practice of Electronic Governance, Galway, Irleand, April 2018', ACM, New York, pp. 42–50.

Nan, W. V. and Markus, M. L. (2019), Is inclusive digital innovation inclusive? An investigation of M-Shwari in Kenya, *in* P. Nielsen and H. C. Kimaro, eds, '15th IFIP WG 9.4 International Conference on Social Implications of Computers in Developing Countries, ICT4D 2019 Dar es Salaam, Tanzania, May 1–3, 2019 Proceedings, Part I', Springer Nature Switzerland AG, Cham, pp. 460–471.

Nancy, G. and Yontcheva, B. (2006), 'Does NGO aid go to the poor? Empirical evidence from Europe', *International Monetary Fund Working Paper* **39**(6), 19–35.

Navarrete, C., Gil-Garcia, J. R., Mellouli, S., Pardo, T. A. and Scholl, H. J. (2010), 'Multinational e-government collaboration, information sharing, and interoperability: An integrative model', *Proceedings of the 43rd Hawaii International Conference on System Sciences (HICSS-43)* pp. 1–10.

Ndegwa, A., Kiriri, P. and Achoki, G. (2017), 'Factors affecting adoption of donor funded ICT projects in the public sector in Kenya', *International Journal of Project Management* **1**(1), 1–18.

Ndemo, B. and Weiss, T. (2017), *Digital Kenya: An Entreprenurial Revolution in the Making*, Palgrave Macmillian, London.

Nederhand, J. and Klijn, E. H. (2019), 'Stakeholder involvement in public-private partnerships: Its influence on the innovative character of projects and on project performance', *Administration and Society* **51**(8), 1200–1226.

Nguyen, S. P. and Mahundi, M. H. (2019), 'The dynamics of national ICT ecosystems', *Electronic Journal of Information Systems in Developing Countries* **85**(1), 1–10.

Nguyen, S. P., Nielsen, P. and Sæbø, J. I. (2017), The role of global standardization communities in shaping national health information architectures, *in* J. Choudri, M. S. Islam, F. Wahid, J. M. Bass and J. E. Priyatma, eds, 'Information and Communication Technologies for Development. 14th IFIP WG 9.4 International Conference on Social Implications of Computers in Developing Countries, ICT4D 2017 Yogyakarta, Indonesia, May 22–24, 2017 Proceedings', Springer Nature, Cham, pp. 93–103.

Nielsen, P. and Kimaro, H. C. (2019*a*), *Information and Communication Technologies for Development. Strengthening Southern-Driven Cooperation as a Catalyst for ICT4D. 15th IFIP WG 9.4 International Conference on Social Implications of Computers in Developing Countries, ICT4D 2019 Dar es Salaam, Tanzania, May 1–3, 2019 Proceedings, Part II*, Springer Nature Switzerland AG, Cham.

Nielsen, P. and Kimaro, H. C. (2019*b*), *Information and Communication Technologies for Development. Strengthening Southern-Driven Cooperation as a Catalyst for ICT4D. 15th IFIP WG9.4 International Conference on Social Implications of Computers in Developing Countries, ICTD4D 2019, Dar es Salaam Tanzania, May 1–3, 2019 Proceedings, Part I*, Springer Nature Switzerland AG, Cham.

Nunnenkamp, P., Rank, M. and Thiele, R. (2016), Aid fragmentation and donor coordination in Uganda: A district-level analysis, *in* S. Klingebiel, T. Mahn and M. Negre, eds, 'The Fragmentation of Aid, Concepts, Measurements and Implications for Development Cooperation', Palgrave Macmillan, London, pp. 153–164.

Nunnenkamp, P., Sotirova, A. and Thiele, R. (2016), 'Do aid donors specialize and coordinate within recipient countries? The case of Malawi', *Development Policy Review* **34**(6), 831–849.

OECD (2014), 'Aid fragmentation and aid orphans'. **URL:** http://www.oecd.org/dac/financing-sustainable-development/fragmentation-orphans.htm, last accessed on 2022-08-10

OECD (2018*a*), 'Digital Economy Framework'. **URL:** http://www.oecd.org/going-digital/framework/, last accessed on 2022-02-17

OECD (2018*b*), 'Measuring the results of private-sector engagement through development co-operation', pp. 1–15. **URL:** https://www.oecd.org/dac/results-development/docs/results-pse-results-workshop-apr-18.pdf, last accessed on 2023-01-29

OECD (2019*a*), *Development Co-operation Report 2019. A Fairer, Greener, Safer Tomorrow*, OECD Publishing, Paris.

OECD (2019*b*), 'ODA Press Release 2019. Development aid drops in 2018, especially to neediest countries'. **URL:** https://www.oecd.org/dac/financing-sustainable-development/development-finance-data/ODA-2018-detailed-summary.pdf, last accessed on 2022-12-12

OECD (2019*c*), 'ODA to Africa—Summary', pp. 1–16. **URL:** https://www.oecd.org/dac/financing-sustainable-development/development-finance-topics/Africa-Development-Aid-at-a-Glance.pdf, last accessed on 2022-10-08

OECD (2020), *Multilateral Development Finance 2020*, OECD, Paris.

OECD (n. d.), 'Aid at a glance by recipient'. **URL:** https://public.tableau.com/views/OECDDACAidataglancebyrecipient_new/Recipients?:embed=y&:display_count=yes&:showTabs=y&:toolbar=no?&:showVizHome=no, last accessed on 2022-10-08

OECD/UNDP (2019), *Making Development Co-operation More Effective. 2019 Progress Report*, OECD Publishing, Paris.

Oh, D. S., Phillips, F., Park, S. and Lee, E. (2016), 'Innovation ecosystems: A critical examination', *Technovation* **54**, 1–6.

Öhler, H. (2017), A micro-level analysis of the effects of aid fragmentation and aid alignment, Technical report, Deutsches Institut für Entwicklungspolitik, Bonn.

Ojanperä, S., Graham, M. and Zook, M. (2016), 'Measuring the hidden contours of the global knowledge economy with a digital index', pp. 1–19.

Ojo, A. and Mellouli, S. (2016), 'Deploying governance networks for societal challenges', *Government Information Quarterly* **35**(4, Supplement), S106–S112.

Okeyo, W. and Kioko, S. (2016), 'Integration and use of information and communication technologies in the management of county governments in Kenya: A case of Machakos county', *International Journal of Management and Leadership Studies* **6**(12), 67–72.

Ominde, D., Ochieng, E. and Omwenga, V. (2020), 'Enhancing delivery of information technology projects through stakeholder sense-making', *International Journal of Business Process Integration and Management* **10**(1), 29–41.

Ominde, D., Ochieng, E. and Owenga, V. (2021), 'Refitting stakeholder integration strategies: Case ICT Projects in Kenya', *SSRN Electronic Journal* pp. 1–41.

Onyango, G. and Ondiek, J. O. (2021), 'Digitalization and integration of sustainable development Goals (SGDs) in public organizations in Kenya', *Public Organization Review* .

Osiakwan, E. M. (2017), The KINGS of Africa's digital economy, *in* B. Ndemo and T. Weiss, eds, 'Digital Kenya: An Entreprenurial Revolution in the Making', Palgrave Macmillian, pp. 55–83.

Otoo, S., Agapitova, N. and Behrens, J. (2009), The capacity development results framework: A strategic and results-oriented approach to learning for capacity development, Technical report, The World Bank. **URL:** https://openknowledge.worldbank.org/handle/10986/23037, last accessed on 2022-05-23

Overseas Development Institute (2018), 'FOCAC 2018: Top takeaways from the China-Africa summit'. **URL:** https://www.odi.org/comment/10679-focac-2018-top-takeaways-chinaafrica-summit, last accessed on 2020-03-06

Owen-Smith, J. and Powell, W. W. (2004), 'Knowledge networks as channels and conduits: The effects ofspillovers in the Boston biotechnology community', *Organization Science* **15**(1), 5–21.

Palagashvili, L. (2019), Evaluating aid agencies: Challenges, comparisons, and causes of best aid practices, *in* N. Dutta and C. R. Williamson, eds, 'Lessons on Foreign Aid and Economic Development. Micro and Macro Perspectives', Palgrave Macmillan, Cham, pp. 85–108.

Pattberg, P. and Widerberg, O. (2016), 'Transnational multistakeholder partnerships for sustainable development: Conditions for success', *Ambio* **45**(1), 42–51.

Pfaff-Rüdiger, S. and Löblich, M. (2018), Blickpunkt Netzwerk. Die Auswertung qualitativer Netzwerkanalysen, *in* A. M. Scheu, ed., 'Auswertung qualitativer Daten. Strategien, Verfahren und Methoden der Interpretation nicht-standardisierter Daten in der Kommunikationswissenschaft', Springer Fachmedien, Wiesbaden, pp. 81–96.

Pieterse, J. N. (2010), Digital capitalism and development: The unbearable lightness of ICT4D, *in* P. Kalantzis-Cope and K. Gherab-Martín, eds, 'Emerging Digital Spaces in Contemporary Society. Properties of Technology', Palgrave Macmillan, New York, pp. 305–323.

Pistor, K. (2020), 'Statehood in the digital age', *Wiley Constellations* **27**(3), 3–18.

Pomerantz, P. R. (2011), Chapter 11: Development theory, *in* M. Bevir, ed., 'The SAGE Handbook of Governance', SAGE Publications Ltd, London, pp. 160–178.

Prieto-Martin, P., Faith, B., Hernandez, K. and Ramalingam, B. (2017), *Doing digital development differently: Lessons in adaptive management from technology for governance initiatives in Kenya*, Institute of Development Studies, Brighton.

Principles for Digital Development (n. d.) . **URL:** https://digitalprinciples.org/, last accessed on 2020-03-06

Provan, K. G. and Kenis, P. (2008), 'Modes of network governance: Structure, management, and effectiveness', *Journal of Public Administration Research and Theory* **18**, 229–252.

Provan, K. G. and Milward, H. B. (2001), 'Do networks really work? A framework for evaluating public sector organizational networks', *Public Administration Review* **61**, 414–423.

Przyborski, A. and Wohlrab-Sahr, M. (2014*a*), Forschungsdesigns für die qualitative Sozialforschung, *in* N. Baur and J. Blasius, eds, 'Handbuch Methoden der empirischen Sozialforschung', Springer Fachmedien, Wiesbaden, pp. 117–134.

Przyborski, A. and Wohlrab-Sahr, M. (2014*b*), *Qualitative Sozialforschung. Ein Arbeitsbuch*, Oldenbourg Verlag, München.

Qureshi, S. (2015), 'Are we making a better world with information and communication technology for development (ICT4D) research? Findings from the field and theory building', *Information Technology for Development* pp. 511–522.

Raeymaeckers, P. and Kenis, P. (2016), 'The influence of shared participant governance on the integration of service networks: A comparative social network analysis', *International Public Management Journal* **19**(3), 397–426.

Raeymaeckers, P., Vermeiren, C., Noël, C., Van Puyvelde, S. and Willems, J. (2020), 'The governance of public-nonprofit service networks: A comparison between three types of governance roles', *Voluntas* **31**(5), 1037–1048.

Ramadani, L., Kurnia, S. and Breidbach, C. F. (2018), 'In search for holistic ICT4D research: A systematic literature review', *Hawaii International Conference on System Sciences* pp. 2424–2433.

Renken, J. (2019), What motivates ICT4D champions?, *in* P. Nielsen and H. C. Kimaro, eds, 'Information and Communication Technologies for Development. Strengthening Southern-Driven Cooperation as a Catalyst for ICT4D. 15th IFIP WG 9.4 International Conference on Social Implications of Computers in Developing Countries, ICT4D 2019 Dar es Salaam, Tanzania, May 1–3, 2019 Proceedings, Part I', Springer Nature Switzerland AG, Cham, pp. 307–318.

Republic of Kenya (2015), 'Speech by His Excellency Hon. Uhuru Kenyatta, C.G.H., President and Commander In Chief of the defence forces of the Republic of Kenya during the official opening of the 6th global entrepreneurship summit, U.N. Headquarters, Gigiri, 25th July, 2015'. **URL:** http://www.president.go.ke/2015/07/25/speech-by-his-excellency-hon-uhuru-kenyatta-c-g-h-president-and-commander-in-chief-of-the-defence-forces-of-the-republic-of-kenya-during-the-official-opening-of-the-6th-global-entrepreneurship-sum/, last accessed on 2021-12-14

Republic of Kenya (2019), *Digital Economy Blueprint. Powering Kenya's Transformation*, Republic of Kenya. **URL:** https://www.ict.go.ke/wp-content/uploads/2019/05/Kenya-Digital-Economy-2019.pdf, last accessed on 2023-01-29

Republic of Kenya (2020*a*), *Economic Survey 2020*, Kenya National Bureau of Statistics, Nairobi.

Republic of Kenya (2020*b*), 'Kenya gazette supplement no. 163 (senate bills no. 16)'. **URL:** http://parliament.go.ke/sites/default/files/2020-10/STARTUP%20BILL%2C%202020.pdf, last acessed on 2022-02-16

Rhodes, R. (1997), *Understanding governance: Policy networks, governance, reflexivity and accountability*, Open University Press, Buckingham.

Riddell, R. C. and Robinson, M. (1995), *Non-Governmental Organizations and Rural Poverty Alleviation*, Clarendon Press, Oxford.

Rieth, L. (2009), Humanitarian assistance and corporate social responsibility, *in* J. Steets and D. S. Hamilton, eds, 'Humanitarian Assistance: Improving US-European Cooperation', Center for Transatlantic Relations, The Johns Hopkins University/Global Public Policy Institute, pp. 293–317.

Rogers, E. M. (2003), *Diffusion of innovations*, Free Press, New York.

Rostow, W. (1960), *The Stages of Economic Growth: A Non-Communistic Manifesto*, Cambridge University Press, Cambridge.

Ryen, A. (2002), Cross-cultural interviewing, *in* J. F. Gubrium and J. A. Holstein, eds, 'The SAGE Handbook of Interview Research. Context and Method', SAGE Publications, Thousand Oaks, pp. 335–354.

Sands, R. G., Bourjolly, J. and Roer-Strier, D. (2007), 'Crossing cultural barriers in research interviewing', *Qualitative Social Work* **3**(6), 353–372.

Sanner, T. A. and Nielsen, P. (2019), Software platforms for inclusive innovation, *in* P. Nielsen and H. C. Kimaro, eds, 'Information and Communication Technologies for Development. Strengthening Southern-Driven Cooperation as a Catalyst for ICT4D. 15th IFIP WG 9.4 International Conference on Social Implications of Computers in Developing Countries, ICT4D 2019 Dar es Salaam, Tanzania, May 1–3, 2019 Proceedings, Part I', Springer Nature Switzerland AG, Cham, pp. 218–230.

Scharpf, F. W. (1975), *Politische Durchsetzbarkeit innerer Reformen*, Schwarze, Göttingen.

Scharpf, F. W. (1997), *Games Real Actors Play. Actor-centered institutionalism in policy research*, Westview Press, Boulder.

Scheule, R. M. (2005), 'Das "Digitale Gefälle" als Gerechtigkeitsproblem', *Informatik-Spektrum* **28**(6), 474–488.

Schou, L., Høstrup, H., Lyngsø, E., Larsen, S. and Poulsen, I. (2012), 'Validation of a new assessment tool for qualitative research articles', *Journal of Advanced Nursing* **68**(9), 2086–2094.

Schreier, M. (2013), Qualitative Forschungsmethoden, *in* W. Hussy, M. Schreier and G. Echterhoff, eds, 'Forschungsmethoden in Psychologie und Sozialwissenschaften für Bachelor', Springer Verlag, Berlin/Heidelberg, pp. 189–221.

Schreier, M. (2014), Qualitative content analysis, *in* U. Flick, ed., 'The SAGE Handbook of Data Collection', SAGE Publications, London, pp. 171–183.

Schwab, K. (n. d.), *The Fourth Industrial Revolution*, World Economic Forum, Geneva.

Sein, M. K., Thapa, D., Hatakka, M. and Øystein, S. (2019), 'A holistic perspective on the theoretical foundations for ICT4D research', *Information Technology for Development* **25**(1), 7–25.

Seyle, C. (2018), The role of the private sector in support of reporting under SDG 16, Technical report, SDG16 Hub. **URL:** https://www.sdg16hub.org/system/files/2018-08/Private_Sector_Role_SDG16_Report%20R2.pdf, last accessed on 2022-08-10

Shapiro, C. and Varian, H. R. (1999), *Information Rules: A Strategic Guide to the Network Economy*, Harvard Business School Press, Cambridge.

Silverman, D. (2006), *Interpreting qualitative data. Methods for Analyzing Talk, Text and Interaction*, SAGE Publications, London.

Simons, H. (2009), *Case Study Research in Practice*, SAGE Publications, London.

Simons, H. (2014), Case study research. in-depth understanding in context, *in* P. Leavy, ed., 'The Oxford Handbook of Qualitative Research', Oxford University Press, New York, pp. 455–470.

Smith, S. (2018), Living differences. Experiences from Botswana, *in* G. Griffin, ed., 'Cross-cultural Interviewing. Feminist experiences and reflections', Routledge, New York, pp. 79–93.

Song, A. K. (2019), 'The digital entrepreneurial ecosystem—A critique and reconfiguration', *Small Business Economics* **53**(3), 569–590.

Sørensen, E. (2014), The metagovernance of public innovation in governance networks, *in* 'Policy & Politics Conference, Bristol, 16th–17th of September 2014'.

Sørensen, E. and Torfing, J. (2005), 'The democratic anchorage of governance networks', *Scandinavian Political Studies* **3**(28), 195–218.

Sørensen, E. and Torfing, J. (2007), Theoretical approaches to metagovernance, *in* E. Sørensen and J. Torfing, eds, 'Theories of democratic network governance', Palgrave MacMillan, Houndsmills, pp. 169–182.

Sørensen, E. and Torfing, J. (2009), 'Making governance networks effective and democratic through metagovernance', *Public Administration* **87**(2), 234–258.

Sørensen, E. and Torfing, J. (2011), 'Enhancing collaborative innovation in the public sector', *Administration and Society* **43**(8), 842–868.

Sørensen, E. and Torfing, J. (2017), 'Metagoverning collaborative innovation in governance networks', *American Review of Public Administration* **47**(7), 826–839.

Span, K. C., Luijkx, K. G., Schols, J. M. and Schalk, R. (2012), 'The relationship between governance roles and performance in local public interorganizational networks: A conceptual analysis', *The American Review of Public Administration* **42**(2), 186–201.

Spivak, G. C. (1988), Can the subaltern speak?, *in* C. Nelson and L. Grossberg, eds, 'Marxism and the interpretation of culture', Columbia University Press, New York, pp. 271–313.

Srnicek, N. (2017), *Platform Capitalism*, Polity Press, Cambridge.

Stakes, R. E. (2005), Qualitative case studies, *in* N. K. Denzin and Y. S. Lincoln, eds, 'The SAGE Handbook of Qualitative Research', SAGE Publications, London, pp. 443–466.

Standish Group (2014), 'Chaos report'. **URL:** http://www.standishgroup.com/, last accessed on 2021-08-22

Startup Genome (n. d.), 'Ecosystem Kenya. Nairobi'. **URL:** https://startupgenome.com/ecosystems/nairobi, last accessed on 2021-12-14

Stegbauer, C. and Häußling, R. (2010), *Handbuch Netzwerkforschung*, Band 4 edn, VS Verlag für Sozialwissenschaften, Wiesbaden.

Steinke, I. (2004), Quality criteria in qualitative research, *in* U. Flick, E. von Kardorff and I. Steinke, eds, 'A Companion to Qualitative Research', SAGE Publications, London, pp. 184–190.

Stockmann, R., Menzel, U. and Nuscheler, F. (2016), *Entwicklungspolitik. Theorien—Probleme—Strategien*, De Gruyter Oldenbourg, Oldenbourg.

Straus, F. (2010), Netzwerkkarten. Netzwerke sichtbar machen, *in* C. Stegbauer and R. Häußling, eds, 'Handbuch Netzwerkforschung', Band 4 edn, VS Verlag für Sozialwissenschaften, Wiesbaden, pp. 527–538.

Stuermer, M., Abu-Tayeh, G. and Myrach, T. (2017), 'Digital sustainability: Basic conditions for sustainable digital artifacts and their ecosystems', *Sustainability Science* **12**(2), 247–262.

Sumner, A. and Tribe, A. (2008), *International Development Studies: Theories and Methods in Research and Practice*, SAGE Publications, Los Angeles.

Sutherland, E. (2020), Internet of things: Governance and metagovernance of networking everything, *in* J. A. Cunningham and J. Whalles, eds, 'The Internet of Things Entrepreneurial Ecosystems. Challenges and Opportunities', Palgrave Macmillan, Cham, pp. 95–120.

Swart, D. (2011), *Africa's technology futures. Three scenarios*, Boston University Press, Boston.

Sydow, J. (2017), Managing inter-organizational networks: Governance and practices between path dependence and uncertainty, *in* B. Hollstein, W. Matiaske and K.-U. Schnapp, eds, 'Networked Governance', Springer International Publishing AG, Cham, pp. 43–54.

Tan, C. and Sia, S. K. (2006), 'Managing flexibility in outsourcing', *Journal of the Association for Information Systems* **7**(4), 179–205.

Tarus, J., Gichoya, D. and Muumbo, A. (2015), 'Challenges of implementing e-learning in Kenya: A case of Kenyan public universities', *The International Review of Research in Open and Distributed Learning* **16**(1), 21–23.

Thatcher, J., O'Sullivan, D. and Mahmoudi, D. (2016), 'Data colonialism through accumulation by dispossession: New metaphors for daily data', *Environment and Planning D: Society and Space* **34**(6), 990–1006.

The Economist (2021), 'Global democracy has a very bad year'. **URL:** https://www.economist.com/graphic-detail/2021/02/02/global-democracy-has-a-very-bad-year, last accessed on 2022-02-16

The World Bank (2009), *Information and Communications for Development. Extending Reach and Increasing Impact*, The World Bank, Washington. **URL:** https://elibrary.worldbank.org/doi/abs/10.1596/978-0-8213-7605-8, last accessed 2023-01-29

The World Bank (2014), *Platform ecosystems: Aligning architecture, governance, and strategy*, Morgan Kaufmann, Waltham.

The World Bank (2015), 'Digital governance projects (database)'. **URL:** http://data.worldbank.org/data-catalog/dg-projects-database, last accessed on 2021-08-22

The World Bank (2016*a*), *Digital Dividends*, The World Bank, Washington DC.

The World Bank (2016*b*), *Public-Private Dialogue (PPD) Stakeholder Mapping Toolkit*, World Bank Group, Washington D.C.

The World Bank (2017), *World Development Report 2017: Governance and the Law*, The World Bank, Washington D.C.

The World Bank (2019*a*), *Kenya Digital Economy Assessment. Summary Report*, The World Bank Group, Washington, DC.

The World Bank (2019*b*), 'Kenya economic update: Securing future growth, policies to support Kenya's digital transformation'. **URL:** https://dc.sourceafrica.net/documents/119707-Kenya-Economic-Update-Securing-Future-Growth.html, last accessed on 2022-07-14

The World Bank (n. d. *a*), 'Access to electricity (% of population)—Kenya'. **URL:** https://data.worldbank.org/indicator/EG.ELC.ACCS.ZS?locations=KE, last accessed on 2021-12-14

The World Bank (n. d. *b*), 'Access to electricity (% of population)—Sub-Saharan Africa'. **URL:** https://data.worldbank.org/indicator/EG.ELC.ACCS.ZS?locations=ZG, last accessed on 2022-08-30

The World Bank (n. d. *c*), 'Agriculture, forestry, and fishing, value added (% of GDP)—Kenya'. **URL:** https://data.worldbank.org/indicator/NV.AGR.TOTL.ZS?locations=KE, last accessed on 2021-12-14

The World Bank (n. d. *d*), 'Doing Business 2020. Comparing Business Regulation in 190 Economies'. **URL:** https://openknowledge.worldbank.org/bitstream/handle/10986/32436/9781464814402.pdf, last accessed on 2022-07-14

The World Bank (n. d. *e*), 'Employment in agriculture (% of total employment) (modeled ILO estimate)—Kenya'. **URL:** https://data.worldbank.org/indicator/SL.AGR.EMPL.ZS?locations=KE,last accessed on 2021-12-14

The World Bank (n. d. *f*), 'Foreign direct investment, net inflows (% of GDP)'. **URL:** https://data.worldbank.org/indicator/BX.KLT.DINV.WD.GD.ZS?locations=KE, last accessed on 2021-12-14

The World Bank (n. d. *g*), 'GDP per capita, PPP (current international $)—Kenya'. **URL:** https://data.worldbank.org/indicator/NY.GDP.PCAP.PP.CD?locations=KE, last accessed on 2021-12-14

The World Bank (n. d. *h*), 'Individuals using the Internet (% of population)—Kenya'. **URL:** https://data.worldbank.org/indicator/IT.NET.USER.ZS?locations=KE, last accessed on 2022-08-10

The World Bank (n. d. *i*), 'Individuals using the Internet (% of population)—Sub-Saharan Africa'. **URL:** https://data.worldbank.org/indicator/IT.NET.USER.ZS?locations=ZG, last accessed on 2022-08-30

The World Bank (n. d. *j*), 'Labor force participation rate, total (% of total population ages 15–64), modeled ILO estimate—Kenya'. **URL:** https://data.worldbank.org/indicator/SL.TLF.ACTI.ZS?locations=KE, last accessed on 2021-12-14

The World Bank (n. d. *k*), 'Literacy rate, adult total (% of people ages 15 and above)—Kenya'. **URL:** https://data.worldbank.org/indicator/SE.ADT.LITR.ZS?locations=KE, last accessed on 2021-12-14

The World Bank (n. d. *l*), 'Mobile cellular subscriptions (per 100 people)—Kenya'. **URL:** https://data.worldbank.org/indicator/IT.CEL.SETS.P2?locations=KE, last accessed on 2021-12-17

The World Bank (n. d. *m*), 'People using safely managed drinking water services (% of population)—Sub-Saharan Africa'. **URL:** https://data.worldbank.org/indicator/SH.H2O.SMDW.ZS?locations=ZG, last accessed on 2022-08-30

The World Bank (n. d. *n*), 'People using safely managed sanitation services (% of population)—Sub-Saharan Africa'. **URL:** https://data.worldbank.org/indicator/SH.STA.SMSS.ZS?locations=ZG, last accessed on 2022-08-30

The World Bank (n. d. *o*), 'Population in total—Kenya'. **URL:** https://data.worldbank.org/indicator/SP.POP.TOTL?locations=KE, last accessed on 2022-08-10

The World Bank (n. d. *p*), 'Rural population (% of total population)—Kenya'. **URL:** https://www.worldbank.org/en/country/kenya/overview#1, last accessed on 2021-12-14

The World Bank (n. d. *q*), 'Services, value added (% of GDP)—Kenya'. **URL:** https://data.worldbank.org/indicator/NV.SRV.TOTL.ZS?locations=KE, last accessed on 2021-12-14

The World Bank (n. d. *r*), 'The World Bank in Kenya. Overview'. **URL:** https://data.worldbank.org/indicator/SP.RUR.TOTL.ZS?locations=KE&view=chart, last accessed on 2021-12-14

The World Bank (n. d. *s*), 'Unemployment, total (% of total labor force), modeled ILO estimate—Kenya'. **URL:** https://data.worldbank.org/indicator/SL.UEM.TOTL.ZS?locations=KE, last accessed on 2021-12-14

The World Bank (n. d. *t*), 'Worldwide governance indicators'. **URL:** https://info.worldbank.org/governance/wgi/Home/Reports, last accessed on 2021-12-14

Torfing, J. (2005), 'Governance network theory: Toward a second generation', *European Political Science* **4**, 305–315.

Transparency International (n. d.), 'Corruption perceptions index 2020'. **URL:** https://www.transparency.org/en/cpi/2020/index/ke, last accessed on 2021-12-14

Trezzini, B. (2010), Netzwerkanalyse, Emergenz und die Mikro-Makro-Problematik, *in* C. Stegbauer and R. Häußling, eds, 'Handbuch Netzwerkforschung', Band 4 edn, VS Verlag für Sozialwissenschaften, Wiesbaden, pp. 193–204.

Tsebelis, G. (1990), *Nested Games: Rational choice in comparative politics*, University of California Press, Berkeley.

TWI2050 (2018), *Transformations to Achieve the Sustainable Development Goals*, International Institute for Applied Systems Analysis (IIASA), Laxenburg.

Uimonen, P. and Hellström, J. (2015), ICT4D donor agencies and networks, *in* P. H. Ang and R. Mansell, eds, 'The International Encyclopedia of Digital Communication and Society', 1 edn, John Wiley & Sons, pp. 1–9.

UN (n. d.), 'Sustainable development goals'. **URL:** https://sustainabledevelopment.un.org/?menu=1300, last accessed on 2023-01-29

UNCTAD (2019), 'Donor support to the digital economy in developing countries. A 2018 survey of public and private organizations', *UNCTAD Technical Notes on ICT for Development* **No 13**.

UNCTAD (2021*a*), *Digital economy report 2021. Cross-border data flows and development: For whom the data flow*, United Nations Publications, New York.

UNCTAD (2021*b*), *E-Commerce and Digital Economy Programme: Year in Review 2030: Facilitating inclusive digital economies in challenging times*, UNCTAD. **URL:** https://unctad.org/system/files/official-document/_dtlstictinf2021d2_en.pdf, last accessed on 2023-01-29

UNCTAD (n. d.), 'Cyberlaw tracker: The case of Kenya'. **URL:** https://unctad.org/page/cyberlaw-tracker-country-detail?country=ke, last accessed on 2022-02-16

UNDP (1995), *Public Sector Management, Governance, and Sustainable Human Development*, UNDP, New York.

UNDP (2020), *Human Development Report 2020. The next frontier. Human development and the Anthropocene*, UNDP, New York.

UNEP (2011), Decoupling natural resource use and environmental impacts from economic growth, *in* M. Fischer-Kowalski, M. Swilling, E. U. von Weizsäcker, Y. Ren, Y. Moriguchi, W. Crane, F. Krausmann, N. Eisenmenger, S. Giljum, P. Hennicke, P. Romero Lankao and A. Manalang, eds, 'A report of the working group on decoupling to the international resource panel', UNEP.

UNESCO (2018), 'R&D data release'. **URL:** http://uis.unesco.org/en/news/rd-data-release, last accessed on 2023-01-29

USAID (2020), *Digital Ecosystem Country Assessment Kenya*, United States Agency for International Development. **URL:** https://www.usaid.gov/sites/default/files/documents/DECA_Report_KENYA_EXTERNAL_2OCT2-2-_2.pdf, last accessed on 2023-01-30

USAID (2021), 'Economic Growth and Trade. Kenya'. **URL:** https://www.usaid.gov/kenya/economic-growth-and-trade, last accessed on 2021-12-14

Vakil, A. (1997), 'Confronting the classification problem: Towards a taxonomy of NGOs', *World Development* **25**(12), 2057–2071.

Vallet, A., Locatelli, B., Barnaud, C., Makowski, D., Quispe Conde, Y. and Levrel, H. (2020), 'Power asymmetries in social networks of ecosystem services governance', *Environmental Science and Policy* **114**, 329–340.

Vangen, S. and Huxham, C. (2003), 'Nurturing collaborative relations: Building trust in interorganizational collaboration', *Journal of Applied Behavioural Science* **39**, 5–31.

Verhoef, P. C., Broekhuizen, T., Bart, Y., Bhattacharya, A., Qi Dong, J., Fabian, N. and Haenlein, M. (2021), 'Digital transformation: A multidisciplinary reflection and research agenda', *Journal of Business Research* **122**(July 2018), 889–901.

Wakunuma, K. (2019), 'Power as an ethical concern in the Global South's digital transformation', *TATuP Zeitschrift für Technikfolgenabschätzung in Theorie und Praxis* **28**(2), 29–34.

Walsham, G. (2017), 'ICT4D research: Reflections on history and future agenda', *Information Technology for Development* **23**(1), 18–41.

Wamoto, F. O. (2015), 'E-government implementation in Kenya, an evaluation of factors hindering or promoting e-government successful implementation', *International Journal of Computer Applications Technology and Research* **4**(12), 906–915.

Wankuru, P. C. (2019), 'For the first time, the relative economic size of Kenya's counties is clear', *World Bank Blogs* . **URL:** https://blogs.worldbank.org/africacan/for-the-first-time-the-relative-economic-size-of-kenyas-counties-is-clear, last accessed on 2021-12-14

Wareham, J., Fox, P. B. and Cano Giner, J. L. (2014), 'Technology ecosystem governance', *Organization Science* **25**(4), 1195–1215.

WBGU (2018), *Digitalization: What we need to talk about*, WBGU—German Advisory Council on Global Change, Berlin.

Weber, R. H. (2014), Visions of political power: Treaty making and multistakeholder understanding, *in* R. Radu, J.-M. Chenou and R. H. Weber, eds, 'The Evolution of Global Internet Governance', Springer Verlag GmbH Berlin Heidelberg, Heidelberg, pp. 95–114.

Wehrmann, D. (2018), *Incentivising and regulating multi-actor partnerships and private-sector engagement in development cooperation*, Deutsches Institut für Entwicklungspolitik, Bonn.

Weichbold, M. (2014), Pretest, *in* N. Baur and J. Blasius, eds, 'Handbuch Methoden der empirischen Sozialforschung', Springer Fachmedien, Wiesbaden, pp. 299–304.

Wennersten, R. and Qie, S. (2018), United nations sustainable development goals for 2030 and resource use, *in* W. Leal Filho, ed., 'Handbook of Sustainability Science and Research', Springer International Publishing AG, Cham, pp. 317–340.

Weyer, J. (2000), Soziale Netzwerke als Mikro-Makro-Scharnier: Fragen an die soziologische Theorie, *in* J. Weyer, ed., 'Soziale Netzwerke: Konzepte und Methoden der sozialwissenschaftlichen Netzwerkforschung', Oldenbourg Verlag, München, pp. 193–204.

Whetsell, T. A., Leiblein, M. J. and Wagner, C. S. (2020), 'Between promise and performance: Science and technology policy implementation through network governance', *Science and Public Policy* **47**(1), 78–91.

Whitfield, L. and Fraser, A. (2010), 'Negotiating aid: The structural conditions shaping the negotiating strategies of african governments', *International Negotiation* **15**, 341–366.

Williamson, O. E. (1985), *The Economic Institutions of Capitalism. Firms, Markets and Relational Contracts*, Free Press, New York.

Wilson, C. E., Morrison, T. H. and Everingham, J.-A. (2017), 'Multi-scale meta-governance strategies for addressing social inequality in resource dependent regions', *Sociologia Ruralis* **58**(3).

Wolf, C. (2010), Egozentrierte Netzwerke: Datenerhebung und Datenanalyse, *in* C. Stegbauer and R. Häußling, eds, 'Handbuch Netzwerkforschung', Band 4 edn, VS Verlag für Sozialwissenschaften, Wiesbaden, pp. 471–484.

Wolff, S. (2004), Ways into the field and their variants, *in* U. Flick, E. von Kardorff and I. Steinke, eds, 'A Companion to Qualitative Research', SAGE Publications, London, pp. 195–202.

World Economic Forum (2019), 'Platforms and ecosystems: Enabling the digital economy', (February).

World Economic Forum (2021), *Global gender gap report 2021*, World Economic Forum, Geneva.

World Intellectual Property Organization (2021), *Global Innovation Index 2021: Tracking innovation through the COVID-19 crisis*, World Intellectual Property Organization, Geneva.

Yin, R. K. (2014), *Case Study Research. Designs and Methods*, SAGE Publications, Thousand Oaks.

Yin, R. K. (2018), *Case Study Research and Applications. Design and Methods*, SAGE Publications, Thousand Oaks.

Young, J. C. (2019), 'The new knowledge politics of digital colonialism', *Environment and Planning A* **51**(7), 1424–1441.

Zheng, Y., Hatakka, M., Sahay, S. and Andersson, A. (2018), 'Conceptualizing development in information and communication technology for development (ICT4D)', *Information Technology for Development* **24**(1), 1–14.

Zuboff, S. (2019), *The Age of Surveillance Capitalism*, Public Affairs, New York.

SPRINGER NATURE

GPSR Compliance

The European Union's (EU) General Product Safety Regulation (GPSR) is a set of rules that requires consumer products to be safe and our obligations to ensure this.

If you have any concerns about our products, you can contact us on ProductSafety@springernature.com

In case Publisher is established outside the EU, the EU authorized representative is:

Springer Nature Customer Service Center GmbH
Europaplatz 3
69115 Heidelberg, Germany

The manufacturer's authorised representative in the EU is Springer Nature Customer Service Centre GmbH, Europaplatz 3, 69115 Heidelberg, Germany. If you have any concerns regarding our products, please contact ProductSafety@springernature.com

Printed and bound by CPI Group (UK) Ltd, Croydon, CR0 4YY
01/10/2026
02225861-0007